Entrepreneurship in Integrated Farming System

Entrepreneurship in Integrated Farming System

Editors

Dr. Adyant Kumar

Assistant Professor cum Junior Scientist,
Department of Agronomy, Tilka Manjhi Agriculture College, Godda
(Ex-Research Associate, NAHEP-CAAST Project on IFS),
Birsa Agricultural University, Kanke, Ranchi, Jharkhand

Dr. Milind B. Katare

Subject Matter Specialist (Fisheries), Krishi Vigyan Kendra
(Central Agricultural University, Imphal), East Garo Hills, Meghalaya

Dr. Babloo Sharma

Subject Matter Specialist (Soil Science), Krishi Vigyan Kendra,
(Swami Keshwanand Rajasthan Agricultural University, Bikaner),
Sriganganagar, Rajasthan

Dr. Sudhanand Prasad Lal

Assistant Professor, Department of Extension Education,
Post Graduate College of Agriculture,
Dr. Rajendra Prasad Central Agricultural University, Pusa, Samastipur, Bihar

Prof. (Dr.) M.S. Malik

Dean, Faculty of Forestry and Principal Investigator,
NAHEP-CAAST Project on Integrated Farming System,
Birsa Agricultural University, Kanke, Ranchi, Jharkhand

2023

Daya Publishing House®
A Division of

Astral International Pvt. Ltd.

New Delhi – 110 002

ISBN: 9789359191225

Published by : **Daya Publishing House**
A Division of
Astral International Pvt. Ltd.
– ISO 9001:2015 Certified Company
4736/23, Ansari Road, Darya Ganj
New Delhi-110 002
Ph. 011-43549197, 23278134
E-mail: info@astralint.com
Website: www.astralint.com

बिरसा कृषि विश्वविद्यालय
BIRSA AGRICULTURAL UNIVERSITY
काँके, राँची - 834006, झारखण्ड, भारत
Kanke, Ranchi - 834006, Jharkhand (INDIA)

डॉ. ओंकार नाथ सिंह
Dr. Onkar Nath Singh
कुलपति
Vice-Chancellor

Foreword

The opening of the Indian economy to the global market during early nineties due to globalization of trade and agriculture and the policy reforms at the national level, the opportunities for entrepreneurship development in agricultural sector have significantly increased. Agribusiness has offered enormous scope for value addition, packaging, retailing, and exports of agricultural products, agricultural input and marketing with advancements in technology and IT tools. Business opportunities are available in agricultural production, agro-processing and value addition, agro-produce manufacturing, agricultural marketing, agro-inputs manufacturing and marketing, agro-services, agro-tourism etc.

Agri-entrepreneurship is the need of the hour to make agriculture more lucrative and profitable business enterprise for empowering the rural unemployed youth when other sectors are not able to provide employment opportunities to the desired extent. In our country, agriculture and allied sector have good scope for entrepreneurship which can be harnessed by effective management of natural resources, and agro-inputs on the farm to meet the market demand. The constraints such as lack of financial capital, non-remunerative prices, lack of agro-inputs and market linkages must be addressed to make agriculture an avenue for business development.

Agri-preneurship is a sustainable employment strategy that will ensure self-reliance and economic self-sufficiency for the entrepreneur. There are several climatic zones in India having different scope and prospects for integrated farming. The development of agricultural entrepreneurship refers to the

Tel. : 0651-2450500 (O), 2450115, 2450623 (R)
E-mail : vc@bauranchi.org, Website : www.bauranchi.org

promotion of entrepreneurial skills amongst common individuals and building the entrepreneurial approach in the field of agriculture. Therefore, the integrated farming systems have a great scope for agri-preneurship in a sustainable way, which utilize not only the wastes but also contribute in doubling the farmers' income.

Dr. M.S. Malik, Principal Investigator, NAHEP-CAAST in Integrated Farming System, Dr. Adyant Kumar, Research Associate, NAHEP-CAAST Project, B.A.U., Ranchi and his team have come up with this vital publication *"Entrepreneurships in Integrated Farming Systems"*. The book provides important information regarding entrepreneurial development in integrated farming systems throughout the country, where agriculture and allied components are extensively presented under the same umbrella. The contents were received from subject experts from across the country and it has been well edited.

I congratulate the entire team members for their efforts in compiling a large volume of useful information for this publication. I hope the book will be useful for the students, researchers, scientists, subject experts, policymakers, agri-entrepreneurs and extension workers.

05/11/2022

O.N. Singh

Preface

India is a global agricultural powerhouse, which contributes about 18.60 per cent to the country GDP (2021-22). It is the world's largest producer of milk, pulses, spices; and the second largest producer of rice, wheat, cotton, sugarcane, fruits, vegetables, tea, farmed fish, sheep and goat meat. It employs 48.90 per cent of the total work force and is the principal source of livelihood for 58.00 per cent of the Indian population. The population of India has already crossed 1.30 billion (2020) and expected to increase 1.50 billion by 2035. Thus, the demand for food grains would increase from 293 MT (2020) to 334.90 MT (2025). Whereas, the demand for high value commodities such as fruits, vegetables, livestock products, fish, poultry, *etc.* are increasing faster than that of food grains and is expected to increase by more than 100.00 per cent by 2030.

According to the reports of Indian Council of Medical Research (ICMR), the per capita daily requirement for an adult is about 420 grams cereals, 40 grams pulses, 50 grams leafy vegetables, 60 grams other vegetables, 150 ml milk and 40 grams fat and oils, which has been worked out to get 2738 calories of energy and 65 grams of protein to perform voluntary as well as involuntary functions of body. Hence, the country has to produce more food and other agricultural commodities. But, the average size of land holdings in India has declined to 0.68 ha in 2020 from 1.16 ha during 2010-11, to 1.41 hectares in 1995-96 to 1.84 hectares in 1980-81, and 2.28 ha in 1970-71. According to Agriculture Census Report (2010-11), if this trend continues, then the average size of holding in India would be further reduced to 0.32 ha in 2030. This is due to fragmentation, rapid urbanization, creation of infrastructure facilities like roads, railway tracks, dams *etc.* With gradual declining trend in size of land holding, it poses a serious challenge to the sustainability and profitability of farming community because of the execution of unorganized ineffective farming system and unutilized resources. The government had also introduced many schemes to upgrade the farming community and increase the income but due to the fragmented land holding and conventional practices, the farm mechanization and new technologies are not able to implement. This situation in India, calls for an integrated effort to address the emerging livelihood issues. Thus, it is imperative to develop strategies and agricultural technologies that enable adequate income and employment generation for small and marginal farmers', who constitute more than 85.00 per cent of the farming community. Thus, in today's farming context, a system or approach should suitable and considered to be the most powerful tool for the fragment land holding, cutting down the input cost, provision for the reuse of the waste as the input for the another as well as for enhancing the profitability of small and marginal farmers. The integration of enterprises lead to greater dividends than single enterprise based farming, especially for small and marginal farmers.

The Integrated Farming System (IFS), as one of the bio-economy models, operates with a variety of agricultural activities interaction to optimize benefits through the synergistic transfer of resources. Bio-economy is a strategic approach to sustainable agri-business development and plays a vital role as an integrated economic-ecological model, which is beneficial for soil health, weeds and pest control, and increase the efficiency of other natural resources and its quality. From the economic aspect, IFS may improve the economy of small and marginal farmers as the production and cost increase, by optimizing resource utilization and recycling waste materials. Despite its numerous advantages, the implementation of IFS in smallholder levels is still low. The small and marginal farmers are facing obstacles in implementation of IFS through integration of the production system, agri-business management requirements and risk management. The IFS need to be operated by individuals who aspire to generate values through the creation or expansion of economic activity, by identifying and exploring new products, processes or markets. This condition requires the actors in IFS to have a willingness to innovate, which will accommodate risk-taking behaviour for future benefits and sustaining a business, which is related to entrepreneurship behaviour. The entrepreneurial behaviour is a combination of several factors like innovativeness, achievement motivation, decision making ability, risk orientation, co-ordination ability, planning ability, information seeking behaviour, cosmopoliteness, self-confidence and profit orientation. Understanding the role of these factors is essential for creating an environment which can facilitate the development of entrepreneurial behaviour. The IFS needs specific characteristics in strategy, competencies and management that align with the entrepreneurship aspect in agriculture.

The entrepreneurship in agriculture is a crucial concept when someone is crafting strategies in the hopes of doing something new and exploiting opportunities that other organizations cannot. It refers to the processes, practices, and decision-making styles of organizations that act entrepreneurially. Farmer entrepreneurship generally relates to a method of recombining agricultural resources innovatively to create opportunities for value creation that consists of five dimensions – autonomy, competitive aggressiveness, innovativeness, proactiveness and risk-taking. Autonomy describes a condition someone has the freedom to develop an entrepreneurial idea and then tries to realize it. Competitive aggressiveness is the spirit to intensely and directly challenge competitors rather than trying to avoid them. Innovativeness is the desire to pursue creativity and experiments. Proactiveness is the willingness to anticipate and act on future needs rather than reacting to events after they evolve. At the same time, risk-taking describes behaviour to engage in bold rather than cautious actions.

The concept of integrated farming can be adopted in any situational basis as it does not involve any particular pattern and it can be formulated as the facilities available and land type. Integrated farming system adoption can be better business opportunities for youths as by better planning and self-development as entrepreneur as well as providing employment for the rural population. It will also be a better enterprise option for women entrepreneur as it involves low cost and yields high income.

With this background, the book has been designed to discuss the aspects and elements of entrepreneurship that must be present in developing the integrated farming system models and its suitability under different agro-climatic situations. We hope that this book have been presented diversified and suitable IFS models based research findings and policy suggestions which will be useful for policymakers, financial institutions, and researcher, in identifying the challenges and innovation confronting integrated farming system for the livelihood security of Indian farmers and helpful for self-reliant (*Atmanirbhar*).

Ranchi, Jharkhand — ***Adyant Kumar***

East Garo Hills, Meghalaya — ***Milind B. Katare***

Sriganganagar, Rajasthan — ***Babloo Sharma***

Pusa, Bihar — ***Sudhanand Prasad Lal***

Ranchi, Jharkhand — ***M. S. Malik***

Contents

The Editors

Dr. Adyant Kumar is working as Assistant Professor cum Junior Scientist, Department of Agronomy, Tilka Manjhi Agriculture College, Godda at Birsa Agricultural University, Ranchi, Jharkhand. He had completed his Doctorate and Master degree in Agriculture with specialization in Agronomy from BCKV, West Bengal and RAU, Pusa (Bihar), respectively. He is the recipient of ICAR-Fakhruddin Ali Ahmed Award for Outstanding Research in Tribal Farming Systems, 2021 given by ICAR under MoA and FW, GoI. He is also awarded with various awards and accolades *viz.* Best Oral Presentation, Best Thesis, Young Scientist and Young Achiever award as well as Valued Reviewer from Springer Nature *etc.* He performed as the Co-PI in the project "Study of Millets production and consumption pattern in last 20 years in Jharkhand" sanctioned by DRDM Tribal Welfare Research Institute, Ranchi. He had also contributed as a Mentor under the GOAL (Going Online as Leaders) Program for the year 2020-21 which was organized by Ministry of Tribal Affairs, Govt. of India in collaboration with Meta (Facebook). He is the Lifetime Member in Indian Society of Agronomy, and Crop and Weed Science Society as well as in the Board of Advisors of NAVADRISHTI, Ranchi, Jharkhand. He has guided six M.Sc. students during his teaching period. He had written >20 Research articles, 01 Review articles, 01 Book, 01 Souvenir, >04 Book chapters and >10 Popular articles in English and regional language for the academician, researcher, students and farming communities.

Dr. Milind B. Katare, Ph.D. (Aquaculture) is presently working as Subject Matter Specialist (Fisheries) at Krishi Vigyan Kendra, East Garo Hills-794111 Meghalaya under the administrative control of Central Agricultural University, Imphal-795004, Manipur. He pursued his B.F.Sc. (Bachelor of Fishery Science) from College of Fishery Science, Maharashtra Animal and Fishery Sciences University, Nagpur and followed by M.F.Sc. (Master of Fishery Science) in Aquaculture from College of Fisheries, Mangalore under Karnataka Veterinary, Animal and Fisheries Sciences University, Bidar. He has completed the Doctorate (Ph.D.) in Aquaculture from ICAR-Central Institute of Fisheries Education, Mumbai, Maharashtra. His publications reflected in the library of FAO-AGRIS of the United Nations and ICAR, New Delhi, especially in the area of fish endocrinology, physiology, genetics and breeding. He has honoured with several awards *viz.*, Young Scientist, Best KVK Scientist and Excellence in Extension Scientist Award *etc.* He is member of the Board of Studies for Centurion University of Technology and Management, Odisha-761211, INDIA (Established Vide Odisha Act 4 of 2010). He is presently acting as a Valued Reviewer for esteemed Springer and Elsevier journals.

Dr. Babloo Sharma is working as Subject Matter Specialist at Krishi Vigyan Kendra (SKRAU, Bikaner) from last four year. He had done his doctorate and master degree in Agriculture with specialization in Soil and water conservation from BCKV, West Bengal and Banaras Hindu University Varanasi, respectively. He has also good knowledge about application of remote sensing and GIS and development of different thematic maps and generates the quantitative data which are helpful for the planning of watershed development through implementation of different water conservation activities. He is awarded as Best Young Scientist, Best Ph.D. Thesis and *etc.* by the different reputed societies. He is trying to make some contribution to scientific societies by publishing his research works from various peer reviewed international and national journals. He is also published more than 5 books, 10 book chapters and many popular articles in English and regional language for the academician, researcher, students and farming communities. He is presented his work in different international and national conference organized by reputed institutes and Societies.

Dr. Sudhanand Prasad Lal is currently an Assistant Professor cum Scientist, Department of Extension Education, PG College of Agriculture, RPCAU, Pusa, MoA and FW, Government of India. He pursued his B.Sc. Ag. (Hons.) from CSA University, Kanpur in 2012 and received Chancellor's gold medallist. He did his MSc from, ICAR- National Dairy Research Institute, Karnal-132001, (Haryana) in 2014 and was awarded merit certificate in Social Science and Management group. He did his Ph.D. Agricultural Extension Education on 12th March 2018 and received Ph.D. best thesis Director's

Gold medal @ICAR-NDRI, Karnal from Dr. Trilochan Mohapatra, Secretary (DARE) and Director General (ICAR) Krishi Bhavan, New Delhi. He is the recipient of NTS in B.Sc., JRF in M.Sc., and SRF in Ph.D. He is the recipient of Certificate of Excellence Award from Elsevier Researcher Academy, Amsterdam, Netherlands. He has written 40 Research articles, 01 Review articles, 02 Short communication, 03 Books, 06 Book chapters and 20 Popular articles. He was invited as an expert by 'The Times Higher Education Academic Reputation Survey in partnership with Elsevier for the 2019 Times Higher Education World Reputation Rankings and the 2019 World University Rankings, supported by UNICEF. Dr. Lal has received citations from DRDO, IIT-Delhi, CAB Direct and FAO.

Dr. M.S. Malik, an eminent researcher and administrator in the field of Agroforestry, born in Vaishali District of Bihar. He obtained his Master degree in Forestry from Sri Lanka in 1987 and did his Ph.D. in Forestry from Tamil Nadu Agricultural University, Coimbatore, Tamil Nadu in 1995. He is presently serving as Dean of the Faculty of Forestry, Birsa Agricultural University, Ranchi, Jharkhand, being a University Professor and heading the Department of Silviculture and Agroforestry for more than 16 years. His major contribution as a researcher is in the field of Forestry, Agroforestry, Silviculture, and developing economically sustainable models of Integrated Farming systems. He has authored four books, several research papers, scientific articles, laboratory manual, bulletin *etc.* He has handled more than 10 research projects including World Bank-NAHEP-CAAST project on IFS and AICRP on Agroforestry. He has been involved in developing academic curricula in Forestry and revising syllabus for ICAR-ARS exam in Agroforestry. He has also served as University Coordinator-ICAR, member of Board of Studies and Academic council *etc.* He has been honoured with many awards *viz.*, ICAR–Fakhruddin Ali Ahmed Award, Life Time Achievement Award, Professional Excellence Award, Distinguished Scientist Award, Outstanding Achievement Award, and Meritorious Service Award.

The Contributors

Sl. No.	*Name of the Contributor*	*Affiliation of the Contributor*
1	**Sudhanand Prasad Lal**	Assistant Professor cum Scientist, PG Department of Extension Education (PGCA), Dr. Rajendra Prasad Central Agricultural University Pusa, Samastipur – 848125, Bihar, India
2	**Abhishek Mahendra**	Research Scholar, PG Department of Extension Education (PGCA), Dr. Rajendra Prasad Central Agricultural University Pusa, Samastipur – 848125, Bihar, India
3	**Afsana Rozi**	Research Scholar, PG Department of Extension Education (PGCA), Dr. Rajendra Prasad Central Agricultural University Pusa, Samastipur – 848125, Bihar, India
4	**Monika Verma**	Research Scholar, PG Department of Extension Education and Communication Management (C.C.Sc.), Dr. Rajendra Prasad Central Agricultural University Pusa, Samastipur – 848125, Bihar, India
5	**Visha Kumari Venugopalan**	Scientist (Agronomy), Division of Crop Science, ICAR-Central Research Institute for Dryland Agriculture, Santosh Nagar, Hyderabad – 500059, Telangana, India

Sl. No.	Name of the Contributor	Affiliation of the Contributor
6	**Kajal Sengupta**	Professor and Head, Division of Agronomy, Faculty of Agriculture and Rural Development (FARD), Ramakrishna Mission Vivekananda Educational and Research Institute (RKMVERI) [Deemed to-be-University], Narendrapur Campus, Kolkata-700103, West Bengal, India
7	**Sunil Dutt**	Ph.D. Scholar, Livestock Production Management Division, ICAR- National Dairy Research Institute, Karnal – 132001, Haryana, India
8	**Nishant Kumar**	Senior Scientist (Animal Reproduction), Livestock Production Management Division, ICAR-National Dairy Research Institute, Karnal – 132001, Haryana, India
9	**T.K. Mohanty**	Principal Scientist (Animal Reproduction), Livestock Production Management Division, ICAR- National Dairy Research Institute, Karnal – 132001, Haryana, India
10	**Rajeev Kumar Srivastava**	Assistant Professor (Agronomy), Directorate of Seed and Farms, Tirhut College of Agriculture Campus, Dholi (Dr. Rajendra Prasad Central Agricultural University, Pusa, Samastipur), Muzaffarpur – 843121, Bihar, India
11	**Samrat Sikdar**	Assistant Professor at School of Agriculture, Sanskriti University, Mathura – 281401, Uttar Pradesh, India
12	**Akramul Hoque**	Research Scholar, Department of Agronomy, Uttar Banga Krishi Viswavidyalaya, Pundibari, Coochbehar – 736165, West Bengal, India
13	**Narayan Lal**	ICAR-Indian Institute of Soil Science, Bhopal – 462038, Madhya Pradesh, India
14	**Asha Sahu**	ICAR-Indian Institute of Soil Science, Bhopal – 462038, Madhya Pradesh, India
15	**A.K. Vishwakarma**	ICAR-Indian Institute of Soil Science, Bhopal – 462038, Madhya Pradesh, India
16	**B.L. Lakaria**	ICAR-Indian Institute of Soil Science, Bhopal – 462038, Madhya Pradesh, India
17	**Nisha Sahu**	ICAR-Indian Institute of Soil Science, Bhopal – 462038, Madhya Pradesh, India

Sl. No.	Name of the Contributor	Affiliation of the Contributor
18	**D.K. Jayswal**	ICAR-National Agricultural Higher Education Project, KAB-II, Pusa, New Delhi – 110012, India
19	**A.B. Singh**	ICAR-Indian Institute of Soil Science, Bhopal – 462038, Madhya Pradesh, India
20	**A.K. Biswas**	ICAR-Indian Institute of Soil Science, Bhopal – 462038, Madhya Pradesh, India
21	**A.K. Patra**	ICAR-Indian Institute of Soil Science, Bhopal – 462038, Madhya Pradesh, India
22	**Mukesh Kumar**	Senior Scientist (Agronomy), Dr Rajendra Prasad Central Agricultural University (RPCAU), Pusa, Samastipur – 848125, Bihar, India
23	**A.K. Ghorai**	Principal Scientist (Agronomy), ICAR-Central Research Institute for Jute and Allied Fibres, Barrackpore, Kolkata – 700120, West Bengal, India
24	**S. Mitra**	Principal Scientist (Agronomy), ICAR-Central Research Institute for Jute and Allied Fibres, Barrackpore, Kolkata – 700120, West Bengal, India
25	**Arunbeer Singh**	Assistant Professor, Directorate of Extension Education, Guru Angad Dev Veterinary & Animal Sciences University, Ludhiana – 141004, Punjab
26	**Pragya Bhadauria**	Scientist, Agricultural Technology Application Research Institute (ATARI), Zone-I, PAU Campus, Near Gate No. 6, Ludhiana – 141004, Punjab, India
27	**Shabnam Sidhu**	Assistant Professor, Guru Angad Dev Veterinary and Animal Sciences University, PAU Campus, Ludhiana – 141004, Punjab, India
28	**Amandeep Singh**	Assistant Professor, Directorate of Extension Education, Guru Angad Dev Veterinary & Animal Sciences University, Ludhiana – 141004, Punjab, India
29	**Y.S. Jadoun**	Assistant Professor, Directorate of Extension Education, Guru Angad Dev Veterinary & Animal Sciences University, Ludhiana – 141004, Punjab, India
30	**Gurpreet Kour**	Ph.D. Scholar, Department of Animal Genetics & Breeding, Guru Angad Dev Veterinary & Animal Sciences University, Ludhiana – 141004, Punjab, India

Sl. No.	Name of the Contributor	Affiliation of the Contributor
31	**Harpreet Kour**	Ph.D. Scholar, Department of Veterinary Pharmacology & Toxicology, Guru Angad Dev Veterinary & Animal Sciences University, Ludhiana – 141004, Punjab, India
32	**Avinash Talukdar**	Department of Fisheries Resource Management, College of Fisheries, Assam Agricultural University (AAU), Raha – 782103, Assam, India
33	**Dharitri Baruah**	World Bank, Guwahati, Assam, India
34	**Imtiaz Ahmed**	Department of Fisheries Resource Management, College of Fisheries, Assam Agricultural University (AAU), Raha – 782103, Assam, India
35	**Kankana Borah**	Department of Fisheries Resource Management, College of Fisheries, Assam Agricultural University (AAU), Raha – 782103, Assam, India
36	**Chinmoy Kashyap**	Department of Aquatic Environment Management, College of Fisheries, Assam Agricultural University (AAU), Raha-782 103, Assam, India
37	**S. Sangavi**	Assistant Professor, Trichy Centre for Sustainable Aquaculture, Directorate of Sustainable Aquaculture, Tamil Nadu Dr. J Jayalalitha Fisheries University, Trichy – 639101, Tamil Nadu, India
38	**C. Judith Betsy**	Assistant Professor, Fisheries College and Research Institute, Tamil Nadu Dr. J Jayalalitha Fisheries University, Tuticorin – 628008, Tamil Nadu, India
39	**J. Stephen Sampath Kumar**	Professor, Director- Directorate of Sustainable Aquaculture, TamilNadu Dr. J Jayalalitha Fisheries University, Nagapattinam – 611002, Tamil Nadu, India
40	**S. S. Rathore**	College of Fisheries, Karnataka Veterinary, Animal and Fisheries Sciences University, Mangalore – 575002, Karnataka, India
41	**N. K. Suyani**	College of Fisheries, Karnataka Veterinary, Animal and Fisheries Sciences University, Mangalore – 575002, Karnataka, India
42	**M.I.R. Khan**	College of Fisheries, Central Agricultural University, Lembucherra, Agartala – 799210, Tripura, India

Sl. No.	Name of the Contributor	Affiliation of the Contributor
43	**A. Devivaraprasad Reddy**	Scientist, Krishi Vigyan Kendra, Dr. Y.S.R. Horticultural University, Venkataramannagudem, West Godavari District – 534112, Andhra Pradesh, India
44	**T. Vijaya Nirmala**	Scientist, Krishi Vigyan Kendra, Dr. Y.S.R. Horticultural University, Venkataramannagudem, West Godavari District – 534112, Andhra Pradesh, India
45	**CH. Balakrishna**	Scientist, Krishi Vigyan Kendra, Acharya N.G. Ranga Agricultural University, Amadalavalsa, Srikakulam District – 532185, Andhra Pradesh, India
46	**J. Venkat Satish**	Subject Matter Specialist, Krishi Vigyan Kendra, Dr. Y.S.R. Horticultural University, Venkataramannagudem, West Godavari District – 534112, Andhra Pradesh, India
47	**E. KarunaSree**	Director of Extension, Krishi Vigyan Kendra, Dr. Y.S.R. Horticultural University, Venkataramannagudem, West Godavari District – 534112, Andhra Pradesh, India
48	**K. Venkata Subbaiah**	Scientist, Krishi Vigyan Kendra, Dr. Y.S.R. Horticultural University, Venkataramannagudem, West Godavari District – 534112, Andhra Pradesh, India
49	**V. Deepthi**	Scientist, Krishi Vigyan Kendra, Dr. Y.S.R. Horticultural University, Venkataramannagudem, West Godavari District – 534112, Andhra Pradesh, India
50	**B. Srinivasulu**	Director (PME), Administrative Office, Dr. Y.S.R. Horticultural University, Venkataramannagudem, West Godavari District – 534101, Andhra Pradesh, India
51	**J.V. Prasad**	Director, ICAR-Agricultural Technology Application Research Institute (ATARI), Zone-10, Hyderabad – 500059, Telangana, India
52	**Banlam Jingshai Marbaniang**	Research Scholar, ICAR-Central Institute of Fisheries Education, Panch Marg, Off. Yari Road, Versova, Andheri (West), Mumbai – 400061, Maharashtra, India

Sl. No.	*Name of the Contributor*	*Affiliation of the Contributor*
53	**Gouranga Biswas**	Senior Scientist, ICAR-Central Institute of Fisheries Education (Kolkata Centre), 2, GN Block, Sector V, Bidhannagar, Kolkata – 700091, West Bengal, India
54	**Rajinder Kaur**	Krishi Vigyan Kendra, Barnala, Guru Angad Dev Veterinary and Animal Sciences University, Ludhiana – 141004, Punjab, India
55	**P.S. Tanwar**	Krishi Vigyan Kendra, Barnala, Guru Angad Dev Veterinary and Animal Sciences University, Ludhiana – 141004, Punjab, India
56	**Tarang Kumar Shah**	Department of Fisheries, Dr. G.C. Negi College of Veterinary and Animal Sciences, C.S.K. Himachal Pradesh Agriculture University, Palampur – 176061, Himachal Pradesh, India
57	**Prasun Chatterjee**	Post-Graduate Student, Division of Agronomy, Faculty of Agriculture and Rural Development (FARD), Ramakrishna Mission Vivekananda Educational and Research Institute (RKMVERI) [Deemed to-be-University], Narendrapur Campus, Kolkata-700103, West Bengal, India
58	**Adyant Kumar**	Assistant Professor cum Junior Scientist, Department of Agronomy, Tilka Manjhi Agriculture College, (Birsa Agricultural University, Ranchi), Godda - 814133, Jharkhand
59	**S. Dam Roy**	ICAR-Central Island Agricultural Research Institute, Garacharama, Port Blair – 744 105, Andaman and Nicobar Islands, India
60	**T.P. Swarnam**	ICAR-Central Island Agricultural Research Institute, Garacharama, Port Blair – 744 105, Andaman and Nicobar Islands, India
61	**A. Velmurugan**	ICAR-Central Island Agricultural Research Institute, Garacharama, Port Blair – 744 105, Andaman and Nicobar Islands, India
62	**Jham Lal**	Department of Aquaculture, College of Fisheries, Central Agricultural University (Imphal), Lembucherra – 799210, Tripura, India

Sl. No.	Name of the Contributor	Affiliation of the Contributor
63	**Nitesh Kumar Yadav**	Department of Aquaculture, College of Fisheries, Central Agricultural University (Imphal), Lembucherra – 799210, Tripura, India
64	**Soibam Khogen Singh**	Department of Aquaculture, College of Fisheries, Central Agricultural University (Imphal), Lembucherra – 799210, Tripura, India
65	**Pradyut Biswas**	Department of Aquaculture, College of Fisheries, Central Agricultural University (Imphal), Lembucherra – 799210, Tripura, India
66	**R. Somu Sunder Lingam**	Assistant Professor, Krishnagiri-Barur Centre for Sustainable Aquaculture, Tamil Nadu Dr. J. Jayalalithaa Fisheries University, Barur, Krishnagiri – 635201, Tamil Nadu, India
67	**S. Aanand**	Assistant Professor, Erode-Bhavanisagar Centre for Sustainable Aquaculture, Tamil Nadu Dr. J. Jayalalithaa Fisheries University, Bhavanisagar, Erode – 638451, Tamil Nadu, India
68	**P. Velmurugan**	Assistant Professor, Kanyakumari-Parakkai Centre for Sustainable Aquaculture, Tamil Nadu Dr. J. Jayalalithaa Fisheries University, Parakkai, Kanyakumari – 629601, Tamil Nadu, India
69	**Deepak Kumar**	Ph.D. Scholar, Department of Soil and Water Conservation Engineering, College of Agricultural Engineering and Technology, Junagadh Agricultural University, Junagadh – 362001, Gujarat, India
70	**Rashmi**	Research Scholar, Department of Agricultural Economics, Institute of Agricultural Sciences, Banaras Hindu University, Varanasi – 221005, Uttar Pradesh, India
71	**Utkarsh Kumar**	Crop Production Division, ICAR-Vivekananda Parvatiya Krishi Anusandhan Sansthan, Almora - 263601, Uttarakhand
72	**Aditya Pratap Singh**	Research Scholar, Department of Genetics and Plant Breeding, Bidhan Chandra Krishi Viswavidyalaya, Mohanpur, Nadia – 741252, West Bengal, India
73	**Y. Linthoingambi Devi**	Research Scholar, Department of Genetics and Plant Breeding, Indira Gandhi Krishi Viswavidyalaya, Raipur – 492012, Chhattisgarh, India

Sl. No.	Name of the Contributor	Affiliation of the Contributor
74	**Pramod Kumar Pandey**	Assistant Maize Breeder, AICRP on Maize, School of Crop Improvement, College of Post Graduate Studies in Agricultural Sciences, Central Agricultural University (Imphal), Umiam – 793103, Meghalaya, India
75	**Telem M. Chanu**	Assistant Professor, Department of Social Science, College of Horticulture and Forestry, Central Agricultural University (Imphal), Pasighat – 791102, Arunachal Pradesh, India

Chapter 1

Origin and History of Integrated Farming System: Few Grassroots Experiences and its Revival through Agro-tourism

Sudhanand Prasad Lal[1]*, Abhishek Mahendra[2], Afsana Rozi[2] and Monika Verma[3]

[1]Assistant Professor-cum-Scientist, [2]Research Scholar, PG Department of Extension Education (PGCA), [3]Research Scholar, PG Department of Extension Education and Communication Management (C.C.Sc.)
Dr. Rajendra Prasad Central Agricultural University Pusa, Samastipur – 848125, Bihar
Corresponding author: sudhanand.lal@rpcau.ac.in

"A technological revolution on the farm has led to an output explosion, but we have not yet learned to harness that explosion usefully, while protecting our farmers' right to full parity income."

– John F. Kennedy

Abstract

Agriculture is both the oldest and largest of all identified and un-identified industries in the globe and possibly the sole industry that prevails in every country; and IFS has given distinct momentum to it. The origin of Integrated Farming system happened at Lautenbach, F.R.G. (Federal Republic of Germany) in the district Ortenau of Baden-Württemberg, Germany in 1978. An IFS is centrally based on the concept of "there is no waste" and "waste is only a misplaced resource," which can be used as a valued component in a different product. IFS can be defined *as a third way or middle course between organic and conventional farming. The key of IFS is a set of principles and procedures that must be followed, rather than a prescriptive approach. Crop, livestock, and fisheries are the major components of integrated farming system. In this manuscript, some outstanding models of IFS are dealt i.e., IFS practiced by Hara foundation like Agroforestry with crop cultivation; Eucalyptus honey is one of the successful IFS model of Hosiyarpur in Punjab. Some innovative farmers are now linking their integrated farms with agro-tourism, which is giving an economic boost to their business. India's agricultural exports crossed the $50 billion mark for the first time in its history during the financial year 2021-22, which implicit the possibility of enormous agro-tourism in India.*

Keywords: *Agro-tourism, Conventional farming, Environment, Income generation, Integrated farming, Nutrition management.*

Introduction

Agriculture is both the oldest and largest of all identified and un-identified industries in the globe and possibly the sole industry that prevails in every country, and IFS has given momentum to it. There is a scanty manuscript that deals with the origin and history of the Integrated farming system, so an attempt has been made in this manuscript; in addition, remunerative linkage to associate IFS with agrotourism has also been made. Agrotourism nowadays is gaining high momentum due to its better economic feasibility. An Integrated Farming System (IFS) can provide food security, and job opportunities even in susceptible climatic condition as it reduces the risk. In many countries like India, IFS has changed traditional crop farming, cattle farming, aquaculture, horticulture, agro-industry, and allied sectors from its isolation to its integration. IFS provide manure, compost, feed, fodder and

Table 1.1: Definition of Integrated Farming System

Sl.No.	*Definition*	*Source*
1.	"An integrated farming system is proposed as means of reducing the cost of fuel, feed and fertilizer with minimum capital investment."	(Chan, 1985)
2.	"Integrated Farm Management (IFM) is a site-specific farm business approach that uses the best of modern technology and traditional methods."	(LEAF-Linking Environment and Farming, 1991)
3.	"Integrated Farming Systems is a holistic pattern of land use, which integrates natural regulation processes into farming activities to achieve a maximum replacement of off-farm inputs and to sustain farm income."	(Titi, 1993)
4.	"Integrated Farming Systems defined as an output from one subsystem in an integrated farming system, which otherwise may have been wasted, becomes an input to another subsystem resulting in a greater efficiency of output of desired products from the land/water area under a farmer's control".	(FAO, 2008)

a healthy environment along with balance ecology (Gill *et al.*, 2010). Recycling of organic waste from the farm and using it as organic manure decreases the cost of cultivation and increases soil fertility. Edwards *et al.* (1986) explain IFS basic concept as "there is no waste" and "waste is only a misplaced resource," which can be used as a valued component in different products.

History of Integrated Farming System

In 1978, at the estate of Lautenbach, F.R.G. (Federal Republic of Germany) in the district Ortenau of Baden-Württemberg, Germany, a clear-cut distinction between a low-input farming system (Integrated farming) with a "traditional" farming system was started (Figure 1.1). The Lautenbach Project began in 1978 on the 245 ha Lautenbach farm. The Lautenbach project was designed to show a differentiation between the productivity and the ecological effects of the two farming systems, *i.e.*, "integrated" and "conventional" (Table 1.2). The effect of both farming practices on soil fauna components was studied by Titi and Ipach (1989).

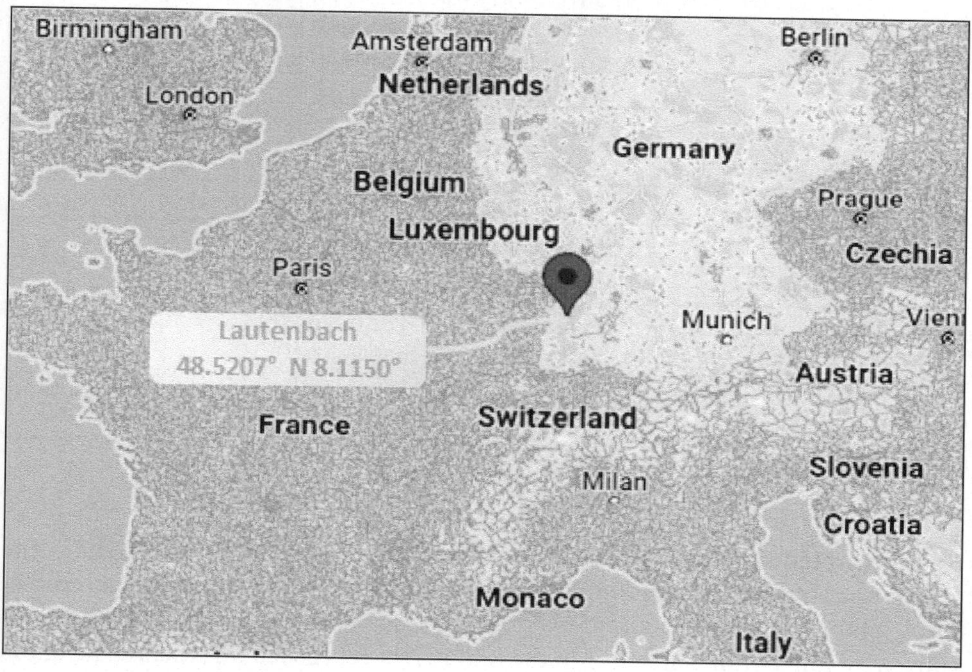

Figure 1.1: Lautenbatch Farm, Germany (Centre of Origin of Integrated Farming System).

The population of common earthworms (*Lumbricus terrestris*) was more in the integrated farm than on the conventional farm. Titi and Ipach (1989) research on Lautenbach farm revealed that the Plant parasite species *Heterodera avenae* and *Ditylenchus dipsaci* have lower population densities in the integrated farming system, while saprophytic and predatory nematodes were present in larger densities in the conventional farming system.

Differentiation between conventional and integrated farming has been executed in the Table 1.2. In integrated farming use of fertilizer is minimum while in conventional one optimum dose is used. Under sowing of Berseem or Clover (*Trifolium alexandrinum*), which is also used as a forage crop, and sowing of catch crops like vegetables between the main crops to get extra income. Plant protection is mainly done by chemicals in the conventional method, but in integrated farming, it is done by both mechanical and chemical method.

Table 1.2: Integrated and Conventional Farming Systems in Lautenbach, F.R.G.

Sl.No.	Particulars	Integrated	Conventional
1.	Tillage	Broadshare cultivator	Reversible Drilling
2.	Sowing	Double rows (6cm within, 24 between)	16cm between
3.	Fertilization	Minimum amount	Optimal amount
4.	N,P,K and Ca	Soil analysis Nitrogen minimum and other methods should be in suboptimal amount/slow acting.	High soluble
5.	Organic manure	20t/ha every third year	Not necessary
6.	Under sowing	Clover in cereal or catch crop	Catch crop every 4 year
7.	Plant protection	Mechanical or chemical	Chemical
a.	Weeds	Mechanical or chemical	Ibid.
b.	Diseases	Mechanical or chemical	Ibid.
c.	Insects	Selective or in reduced dosage, Economic thresholds	Ibid.
8.	Food chain relationship	Consumers are closer to producers.	Consumers are distant from producers.

Source: Titi and Ipach, 1989; Morris and Winter, 1999.

Integrated farming systems were developed in Europe towards the end of the 1970s as a strategy for building sustainable farming systems that were tailored to local conditions. It was done to make people aware of the environmental issues caused by conventional farming in the late 1970s. It is based on research from the 1920s on Integrated Pest Management. This sparked an array of long-term, larger-scale, and coordinated, integrated agricultural trials across Europe (Holland, *et al.*, 1994), particularly in the Netherlands (Vereijken and Royle, 1989). In 1981, inside the UK a project started to show the prospective of integrated farming system; 7 of the projects merged in 1994 to become the Integrated Arable Crop Production Alliance (IACPA), which aimed to coordinate research, share information, remove duplication of effort, and enable quick distribution of research findings to farmers (IACPA, 1996).This does not imply that the IACPA's member organizations have reached an agreement. Although the policies of IFS have yet to be wholly tabulated, it is clear from the research that agricultural scientists who are at the forefront of IFS research use a 'stronger' definition of IFS than those who are meant to demonstrate and promote the concept to farmers *viz.*, agro-chemical companies and LEAF.

IFS can be defined as a third way or middle course between organic and conventional farming, as shown in Figure 1.2 (Morris and Winter, 2000).

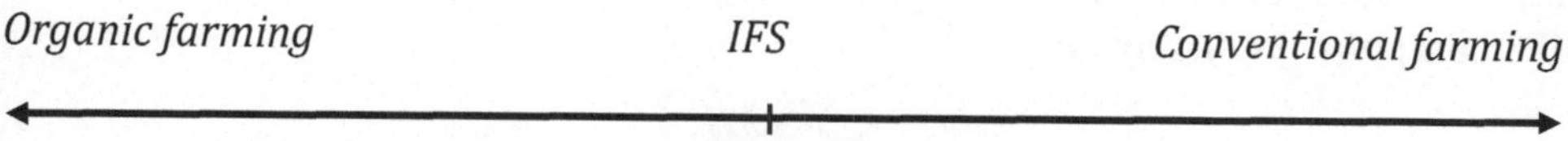

Figure 1.2: Integrated Farming System Continuum.

Principles of Integrated Farming System

The key of IFS is that it is a set of principles and procedures that must be followed rather than a prescriptive approach. In Figure 1.3, different principles of integrated farming have been shown. Multifunctional crop rotation maintains the fertility of the soil. Integrated nutrient management is essential for optimum nutrient supply to the crops. It also balances nutrient uptake by the plant. Overall all the principles of IFS work in a row to balance all the resources on the farm.

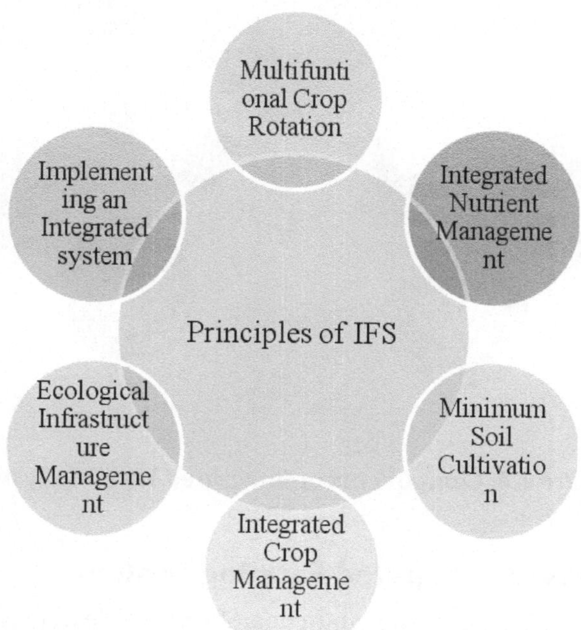

Figure 1.3: Principles of Integrated Farming System (Holland, 2020).

Components of Integrated Farming System

Crop, livestock, and fisheries are the major components of an integrated farming system (Figure 1.4). Various other enterprises like mushroom, agroforestry, bio-gas are also integrated to get a higher returns and also for nutrient and waste management. Components of IFS are selected according to the soil type, rainfall, and its distribution, length of the crop growing season, local market conditions *etc.* The needs and resource base of the farmers are also factors in the selection of IFS components on any farm.

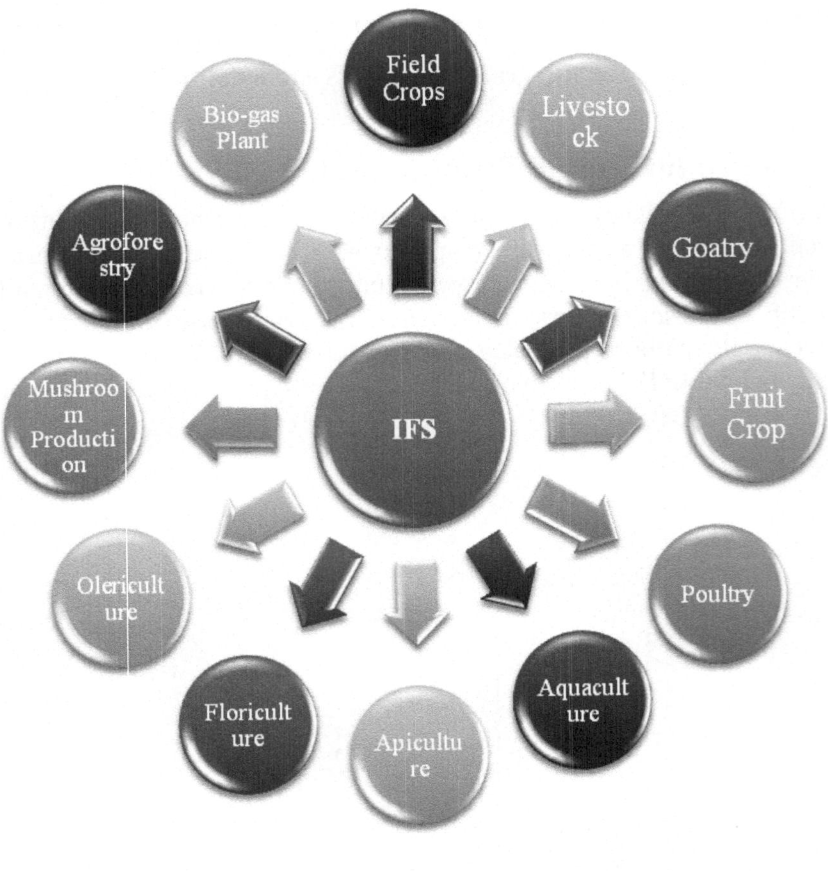

Figure 1.4: Components of Integrated Farming System.

Pros and Cons of Integrated Farming System

In Figure 1.5, different pros and cons of IFS are described. Advantages of integrated farming are that it increases the farm productivity and so profitability also increases, but on the other hand, one needs a high initial investment in IFS. It provides balance food from a single farm. But due to a lack of infrastructure and skilled labor, an integrated farming system is not adopted frequently in India.

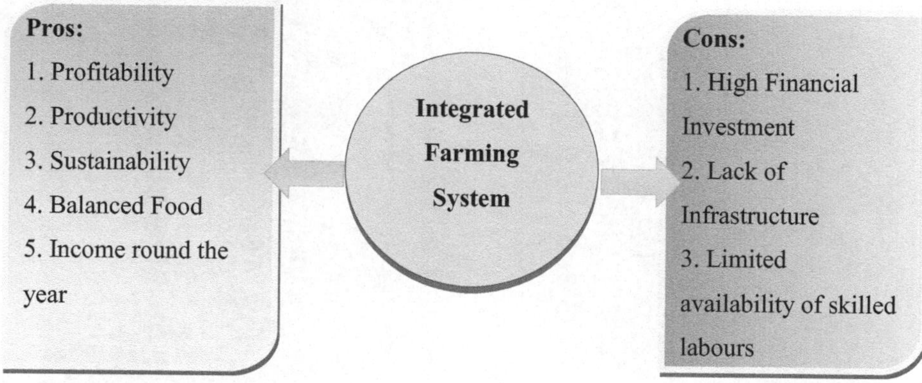

Figure 1.5: Pros and Cons of Integrated Farming System.

Exclusively Selected Integrated Farming System Models from PAN-India

A. Hara Foundation IFS Models of Haryana

Surinder Singh Hara from Haryana's Yamunanagar district is gaining worldwide recognition for his novelties in agro-forestry. Hara has been growing "money on trees" since 1980. They have successfully grown crops of fruits, turmeric, wheat, and paddy, beneath the poplar trees required for the timber industry. The scientific community regards Hara's farm as one of the best in the world for agro-forestry. The yield from the farm is ten times more than what American farmers produce. Dr Norman Borlaug, the first agriculture scientist to get a Nobel Prize, visited Hara's farm in 2005 and said, "Hara has taken farming a step beyond what I had thought". Hara's efforts to grow poplar trees have now spawned into 400 plywood factories and timber worth $3 billion is produced every year in Haryana, Himachal, Uttarakhand, Uttar Pradesh and Punjab. In Yamunanagar market alone, 15,000 tonnes of timber are brought every day. Haryana's forest cover has also increased from 3 per cent to 8 per cent within 20 years because of Hara's efforts (The Sunday Guardian, 2013).

Integrated Farming System Models Practiced in Hara Foundation

1. **Poplar + Wheat + Turmeric:** It is an agri-silviculture model where wheat, turmeric, and poplar trees are grown on the same land and at the same time (Figure 1.6A). Agroforestry is beneficial for the environment; it is ecologically safe and also economic to all; as it gives high output with less inputs. When the wheat doesn't grow, the poplars do. The organic composition of the soil is improved by the leaves which fall from the poplars. However, simply planting trees was insufficient because the agricultural crop rotation ranged from five to ten years. As a result, other

Figure 1.6A-B: Integrated Farming System Models of Hara Foundation.

crops had to be cultivated to provide an alternate source of revenue. Dwarf varieties of wheat and turmeric became the main underwood crops. Turmeric has become the new favourite because it is shade tolerant. When wheat grows between November and March, the poplars trees are on quiescent stage.

2. **Mango + Poplar + Turmeric:** It is a three-storey building farming system where mango, poplar, and turmeric are grown on the same land. In this

system of farming, solar energy and soil layers are utilized efficiently. Turmeric is a shade-loving plant, so it is grown between the interspaces of principal mango and poplar tree. The objective of this system of farming is to utilize the vertical space efficiently.

B. Doubling of Income through Integrated Farming System by a Tribal Woman in Karnataka

Smt. Anita, who owns 4.5 acres in Bettampady village, Puttur Taluk in Dakshin Kannad District of Karnataka, India, cultivated diverse crops such as coconut, arecanut, pepper, vegetables, and livestock (cattle, goats, poultry, and birds) using a traditional agricultural approach. However, low farm production and poor economic returns are an issue. She discovered a sustainable agricultural technique for secure livelihood with the support of ATMA Yojana and ZBNF (Zero Budget Natural Farming). She also enlisted other farmers by creating the Samruddhi Raitham Gampu SHG. Smt. Anita earns Rs. 7.60 lakhs from cattle, vegetables, paddy, vermicompost, and cow dung cake, and among other things, by adopting IFS. She also gives Advisory Services to 400 farmers and has inspired 15 farmers to switch to organic farming and 6 farmers to goat farming. She has received numerous honours on various occasions (ICAR-KVK, 2022).

C. Integrated Farming System Models in Hosiyarpur District of Punjab

1. **Eucalyptus + Apiculture:** Eucalyptus and apiculture make a fine combination of Integrated farming systems (Figure 1.7A). Eucalyptus tree is a fast-growing tree and very useful for India's pulp and paper industry. Honeybees pollinate by blossoms of the eucalyptus tree and thus producing the Eucalyptus honey. Furthermore, honeybees have been observed foraging honeydew off the bark of eucalyptus trees. When a honey bee gathers eucalyptus nectar from a bloom, it is mixed with an enzyme called invertase, also known as "the bee enzyme," which is secreted by honeybee glands in the mouth. The flavour may vary in earthiness, menthol, and peppermint depending on the location where it is being produced. Thus, eucalyptus and apiculture are economically as well as environmentally beneficial (BeesWiki, 2022).
2. **Vermicomposting:** Vermicomposting is a process in which various species of earthworm, *viz., Eisenia fetida, Eudrilus eugeniae, Perionyxex cavatius,* with microorganism converts the organic waste to usable manure as an end product. In recent decades, many individuals have taken vermicomposting as an agri-enterprise, and agro-based companies have started commercial production of vermicompost. The average cost of production is ₹ 3.50 kg^{-1}, and the general selling price is ₹ 5.00 kg^{-1}, so this is a pretty profitable business. High-density polyethylene vermicomposting beds (Figure 1.7B) are recommended for farmers, who can't build permanent structures and invest handsome amounts of money. Seven farming system treatment was experimented at 3 locations

Figure 1.7A-B: Integrated Farming System Models of Hosiyarpur District in Punjab.

of Eastern India *viz.*, Patna, Vaishali and Munger districts of Bihar by ICAR Research Complex for Eastern Region, Patna. In that Inorganic fertilizers combined with organic wastes obtained from various components of IFS-recycling, poultry manure, duck manure, goat manure, and bovine dung as FYM, composted/cereal residues, and vermicompost, each @ 10 t ha^{-1}, were applied to different crops grown in various IFS models to maintain soil fertility and high productivity (Figure 1.8) (Kumar *et al.*, 2012).

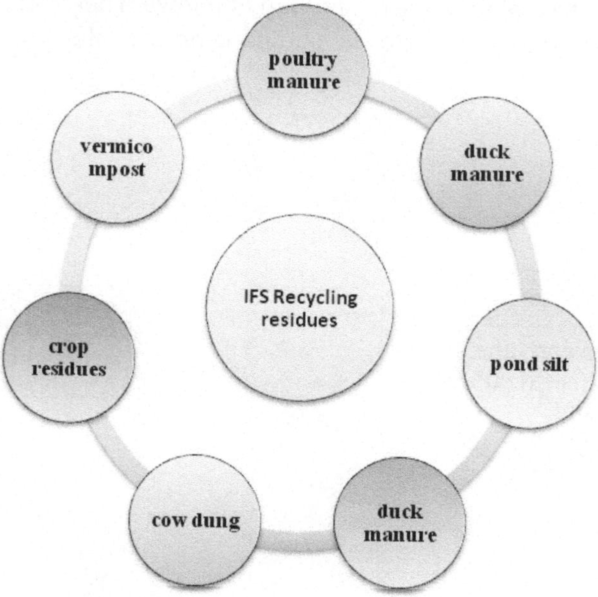

Figure 1.8: Recycling of Residues in Integrated Farming System.

A New Era of Integrated Farming System: Agro-tourism

The origin of the term "Agro-tourism" is credited to Italy, where several "agriturismos," or farm lodging, became a widespread way to see the countryside in the 1970s and 1980s. In Italy, Agro-tourism is called "agritourismo" or referred to as "sleeping in the straw" in Switzerland. Agro-tourism is defined as a business performed by farmers to entertain and educate tourists on their functioning agricultural, horticultural, or agribusiness operations. Entertainment, recreation, involvement, and education are all available. The term Agro-tourism is defined as "travel that combines agricultural or rural settings with products of agricultural operations – all within a tourism experience." As per Agri Tourism Development Corporation (ATDC, Pune) official, Agro-Tourism is the Agri-Business venture in which native farmers of the region offers tours to their agriculture farm to allow the tourist to view and acquaint them through sowing, growth period, harvesting, and processing village grown products, or any agriculture produce that the tourist would not encounter in their cities. In addition, the farmers also provide lodging, fooding opportunity as well as education (Kumbhar, 2009). Thus, agro-tourism is an agricultural-related activity that combines tourism with agriculture. Finally, agro-tourism has the potential to provide farmers with additional sources of revenue and quantum job opportunities.

How to Choose a Location for Agro-tourism?

Agro-tourism is emerging as a new type of tourism; people nowadays are searching for places that are far from the cities, pollution-free places where they can spend some free time in the kosher lap of nature. Therefore, when selecting a site for an agro-tourism centre, we should keep the following points in mind (Figure 1.9).

Figure 1.9: How to Choose a Location for Agro-tourism?

Benefits of Agro-tourism Spots

Agro-tourism has the prospect of transforming traditional agriculture's economic landscape. The advantages of developing agro-tourism are numerous. It would assist farmers and rural residents in various ways, both directly and indirectly. The benefits of Agro-tourism are the following:

a) Farmers, especially farm family members and youth, have employment opportunities.

b) Social moral ideals and cultural transition between urban and rural peoples.

c) Farmers can enhance living standards by interacting with city dwellers.

d) Urban people can better understand rural living and agricultural activities.

e) It aids in the development of rural and agricultural areas.

Figure 1.10: Agro-tourism Centre of ICAR-Indian Veterinary Research Institute, Izatnagar.

Conclusion

From the above content, facts, and figures, it can be concluded that IFS provides an opportunity to integrate all the farming system components. The integration of components ensures optimal utilization of farm resources and family labours. IFS maintain the ecological balance by waste management and nutritional recycling. From grassroots experiences, one can conclude that IFS helps farmers to increase their income and provide wider opportunities than the conventional farming. The concept of '*Vocal for Local*' given by our honourable Prime Minister has made Agro-tourism a lucrative prospect. As agro-tourism is much related with local products, natural attractions and other services in order to develop the local economy. It creates additional scope for employment and income generation in that particular area. The structure of agro-tourism would be more potent if government recommends policies and provide subsidies to the farmers.

REFERENCES

BeesWiki. (2022). Eucalyptus Honey. All Rights Reserved. BeesWiki.com is an encyclopaedic website. Stable URL: https://beeswiki.com/eucalyptus-honey/

Bungenstab, D.J. and Almeida, R.G. (2014). Integrated crop-livestock-forestry systems, a Brazilian experience for sustainable farming, pp. 11-18. Stable URL: https://www.researchgate.net/publication/277247305_Integrated_systems_what_they_are_their_advantages_and_limitations

Helly, G. and Ellen, R. (2011). Agri-tourism and Nature Tourism in California, Richmond: University of California: Agriculture and Natural Resources, pp. 143-144 (ISBN: 9781601077424). Stable URL: https://agris.fao.org/agris-search/search.do?recordID=US201300002561

Holland, J.M. (2020). Integrated Farming Systems. Managing Soils and Terrestrial Systems, pp. 171-175). DOI:10.1201/9780429346255-23.

ICAR-KVK. (2022). Doubling of Income through IFS by a Tribal Woman, Dakshina Kannada, Mangaluru. Stable URL: https://icar.org.in/content/doubling-income-through-ifs-tribal-women

Insam, H., Franke, I. and Goberna, W.M. (2009).Vermicomposting: Earthworms Enhance the Work of Microbes. Microbes at Work, pp. 93-114 (ISBN: 978-3-642-04043-6).

Kumar, S., Subhash, N., Shiwani, Singh, S. and Dey, A. (2012). Evaluation of different components under integrated farming system (IFS) for small and marginal farmers under semi-humid climatic environment. *Experimental Agriculture*. **48**(3): 399-413. doi: 10.101 7/SOO l 44 797 12000087

Kumbhar, V.M. (2009). Agro-tourism: Scope and Opportunities for the farmers in Maharashtra, Y.C. College, Pachwad, Satara, Maharashtra. Stable URL: http://citeseerx.ist.psu.edu/viewdoc/summary?doi=10.1.1.607.585

Moore, D. 2018. An Introduction to Agri-tourism: A Power Tool for Storytelling and Revenue Generation. *AIANTA's Tribal Agri-tourism Consultant*. Stable URL: https://www.aianta.org/an-introduction-to-agritourism-a-power-tool-for-storytelling-and-revenue-generation/#:~:text=Agritourism per cent 2C per cent 20a per cent 20portmanteau per cent 20of per cent 20agriculture,way per cent 20to per cent 20experience per cent 20the per cent 20countryside

Morris, C. and Winter, M. (1999). Integrated farming systems: the third way for European agriculture? *Land Use Policy*. 16(4): 193-205. Stable URL: https://www.sciencedirect.com/science/article/pii/S0264837799000204

Sharma, S. (2013). Innovator in agro-forestry impresses world. *The Sunday Guardian. Chandigarh.* Stable URL: http://www.sunday-guardian.com/news/innovator-in-agro-forestry-impresses-world

Singh, J., Kumar, S., Sarma, K., Kumar, A., Tripathi, M.K, and Kumar, P. (2021). Animal as a Linking Component of the Integrated Farming System (IFS) For Resilient Production. *Food and Scientific Reports.* ISSN 2582-5437. Stable URL: https://www.researchgate.net/profile/Pankaj-Kumar-137/publication/354684300_Animal_as_a_linking_component_of_the_integrated_farming_system_IFS_for_resilient_production/links/61466995a595d06017d78b5f/Animal-as-a-linking-component-of-the-integrated-farming-system-IFS-for-resilient-production.pdf

Soto, D. (2009). Integrated mariculture: A global review. *FAO Fisheries and Aquaculture Technical paper 529.* ISSN 2070-7010.FAO Vialedelle Terme di Caracalla, 00153 Rome, Italy. URL: https://www.fao.org/docrep/pdf/012/i1092e/i1092e01.pdf

Titi, E. and Ipach, U. (1989). Soil fauna in sustainable agriculture: Results of an integrated farming system at Lautenbach, F.R.G. *Agriculture, Ecosystems and Environment.* **27:** 561-572.

Wijnands, F., Malavolta, C., Alaphilippe, A., Gerowitt, B. and Baur, R. (2018). Integrated Production IOBC-WPRS Objectives and principles. Stable URL: https://www.iobc-wprs.org/ip_integrated_production/IOBC-WPRS_IP_objectives_and_principles_4th_edition_2018_EN.pdf

Leong, E. and Jalaludin, S. (1983). The Poultry Industries of South East Asia – The Need for an Integrated Farming System for Small Poultry Producers. *World's Poultry Science Journal,* **38** (3): 213-219. DOI: https://doi.org/10.1079/WPS19820017

Raina, R., Hara, S.S., Hara, V. and Irland, L.C. (2011). Twenty-first century forest plantations: development of agroforestry in Haryana. *Journal of Resources, Energy, and Development.* **8**(2): 67-74.

Chapter 2

Business Opportunities in Integrated Farming System

Visha Kumari Venugopalan[1] and Kajal Sengupta[2]*

[1]Scientist (Agronomy), Division of Crop Science, ICAR-Central Research Institute for Dryland Agriculture, Santosh Nagar, Hyderabad – 500059, Telangana
[2]Professor and Head, Division of Agronomy, Faculty of Agriculture and Rural Development (FARD), Ramakrishna Mission Vivekananda Educational and Research Institute (RKMVERI) [Deemed to-be-University], Narendrapur Campus, Kolkata – 700103, West Bengal
Corresponding author: visha.venugopal@gmail.com

Abstract

Integrated farming comprises crop production and other agricultural production techniques which maintain both ecology and economic sustainability. Integrated farming system (IFS) with field crops, horticultural crops and livestock not only enhanced the income of the farmers but also help in generation of employment particularly for rural/local people. IFS provides nutritional security to family members of the farmers through various IFS components such as cereals, pulses, vegetables, fruits, milk, fish, egg, and other products, and sustains a regular income from the surplus amount of these components. IFS are found better in terms of net returns and benefit-cost ratio over the conventional method. For encountering the increasing rate of population growth, land functional shift, degradation of land resources and water, as well as environmental pollution and climate change, food diversification is one of the best policies to be implemented in achieving food independence and anticipating the food crisis. Food diversification utilizes land optimally by developing an integrated farming system. The IFS is an efficient and environment friendly agricultural system. It is a path to overcome the insecurity faced by the farmers by doubling their income. Higher profit, more yield, better utilization of resources with lesser cost of cultivation are the key features of IFS.

Keywords: *Integrated farming system (IFS), Farm income, Business opportunity.*

Introduction

The Indian economy is primarily based on agriculture. Small and marginal farmers make up the backbone of India's rural economy, accounting for 85 per cent of the overall farming population but just 44 per cent of operational land (GoI, 2014). The decreasing availability of land per capita is posing a severe threat to farming's long-term viability and profitability (Siddeswaran *et al.,* 2012). Due to the country's dwindling land resources and ever-increasing population, there is practically no scope for horizontal land development for food production. Only by combining appropriate farming components that require less land and time to ensure sufficient periodic revenue to farm households is vertical expansion viable (Gill *et al.,* 2009). Farmers have been emphasizing on single enterprise based agricultural systems since the Green Revolution, which has resulted in deterioration of soil health, higher crop failure risk, and decreasing productivity trends (Rahman and Sarkar, 2012). In developing countries like India, rapid population growth, urbanization, and income growth are driving up demand for animal-based foods, exacerbated by the competition between crops and livestock.

A system approach is urgently needed to meet the demands of an ever-increasing population while maintaining ecological balance. Integrated farming systems appear to be a viable solution to the ever-increasing demand for food production, economic stability, and nutritional security, especially for small and marginal farmers with limited resources. It is a concept of ecological soundness that leads to sustainable agriculture, as well as a reliable means of attaining a pretty high productivity with significant fertilizer economy. Furthermore, for resource-poor farmers, moderate increases in land productivity are no longer sufficient. As a result, competent resource management, including optimal resource allocation, is critical to reducing the risk of land sustainability. The Ministry of Agriculture and Farmers' Welfare, GoI has also given major emphasis on Integrated Farming System (IFS) while planning for doubling farmers' income by 2022. However, planning and implementation of different enterprises in integrated farming system in our country lacks scientific and systemic approach. Moreover, proper understanding of interactions and linkages between the components would improve food security and improve the employment opportunity in the country.

What Integration of Various Enterprises is Important?

"The agricultural system is a complex interconnected matrix of soil, plants, animals, implements, power, labor, capital, and other inputs controlled in part by farming families and influenced to varying degrees by political, economic, institutional, and social influences at multiple levels," according to Panwar *et al.* (2017) and Mahapatra (1992). 'IFS' is a widely used phrase to denote a more integrated approach to farming in compared to current monoculture. IFS is a whole-farm method that aims to increase whole-system production per unit area and time, as well as profitability, soil health, and overall environmental sustainability, by utilizing all available resources effectively. A well-balanced mix of agricultural activities, such as dairy, poultry, piggery, fisheries, and sericulture that is tailored to local agro-climatic circumstances and farmers' socioeconomic status. IFS also supplies food (cereals, legumes and oilseeds), vegetables, fruits, feed, fodder, and fibre, in addition to mushroom, honey, and other necessary household needs. Crop rotation, farm diversification, composting, vertical cropping, legume incorporation, and other basic principles of a farming system that meets the basic objectives of sustainable agriculture like food and nutritional security, sustaining yield and productivity, economic viability, and environmental improvement (Layek *et al.*, 2018 and Das *et al.*, 2021). Adoption of appropriate components in IFS improves agroforestry, input, and energy efficiency, and provides rural youths with year-round access to nutritious food, as well as income and job opportunities (Manjunatha *et al.*, 2014).

According to Behera *et al.*, 2015 an ideal IFS would satisfy following basic conditions:

a) Climatic adaptability,

b) Environmental sustainability,

c) Nutritional security,

d) Energy self-sufficiency, and

e) Economic feasibility.

Rationale of Integrated Farming

The goal of integrated farming is to reduce waste generated by the farm's numerous subsystems. Each subsystem's wastes or by-products are used as inputs to other subsystems to increase productivity and lower the cost of production of the various subsystems' outputs (Gill *et al.*, 2009). With India's shifting agrarian situation, IFS appears to be a viable option. Land fragmentation, a scarcity of agricultural inputs, and the country's changing climate necessitate the continuation of a holistic approach to agricultural use. Through crop intensification and diversification, as well as the integration of allied enterprises, an integrated farming system can boost economic yield per unit area per unit time. It also provides ample space for nutrient recycling within the system to save money and sustain the system, as well as reducing agricultural production's reliance on chemical fertilizers to increase profit. This may also result in additional employment opportunities for family members throughout the year (Kumar *et al.*, 2015).

Recycling waste products as inputs/resources in agricultural production, diversification with different crops and enterprises, which brings stability, meeting the diversified needs of the farm family, offering crop/market risk insurance, and overall soil sustainability have all been shown to be superior to monocropped situations (Manjunath and Singh, 2012). IFS is an attempt to combine various agricultural processes for a sustainable development of agriculture. Overall, an integrated farming system achieves the multiple goals of making farmers self-sufficient by ensuring a balanced diet for family members, raising the standard of living by maximizing total net returns and creating more jobs, reducing risk and uncertainty, and maintaining environmental harmony (Mali *et al.,* 2014). Simultaneous fish production in ponds combined with pig, duck, or chicken rearing in pens next or above the ponds results in continual organic fertilization of the pond by the livestock. Through the profitable exploitation of animal and feed wastes, this approach improves the efficiency and rentability of both livestock farming and fish culture (Gill *et al.,* 2005). An ideal nutrient interaction expected in an integrated farming system from the nutrient pool is depicted in Figure 2.1.

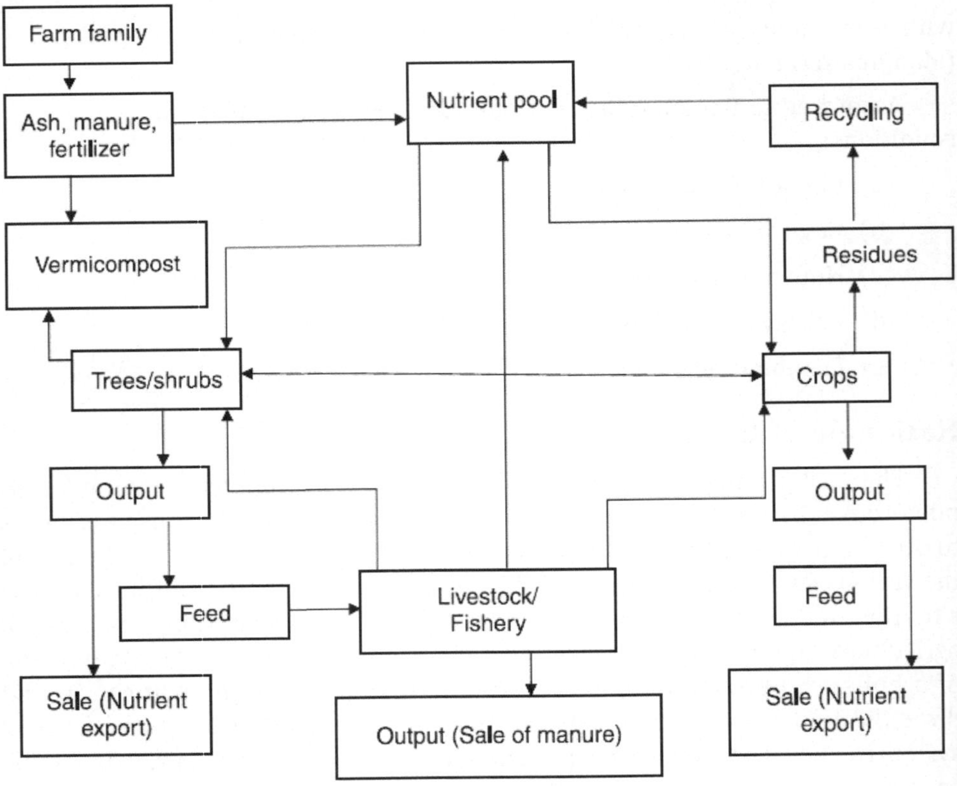

Figure 2.1: Nutrient Flow Diagram for an Integrated Farming System Model.

Advantages of Integrated Farming System

IFS is a multidisciplinary, whole-farm strategy that has proven to be particularly effective in helping small and marginal farmers solve their difficulties. Protective food and environmental safety ensures increased productivity, profitability, and sustainability. Other major benefits of an integrated farming system include waste recycling, year-round income, and energy savings, addressing fodder shortages, creating jobs, and ultimately raising the farmers' standard of living (Faroda, 2014). By integrating various farm enterprises and recycling crop residues and by products within the farm, the approach aims to increase income and employment from small-holdings (Behera and Mahapatra, 1999, Singh *et al.*, 2006). The integration of livestock with crop components has been proven to be helpful since it enhances the physical and chemical qualities of the soil in terms of N, P, K, and other mineral nutrients (Kumar *et al.*, 2012). Because of biological aeration, animal manure treatment enhances soil organic matter content, which leads to better water infiltration and water retention capacity, as well as higher cation exchange capacity. Manure and urine boost the pH of the soil, accelerating organic matter decomposition and microbial activity (Brouwer and Powell, 1998). With physical, chemical, and biological soil recovery, it aids in the improvement and conservation of soil productive capacities. It is preferable to cropping systems because it is intensive farming that provides year-round employment to small and marginal farmers, one enterprise can act as insurance for another in the event of crop failure, one enterprise's byproduct can be used in another, and it improves soil health and fertility in the long run by increasing the nutritional value of soil (Olele *et al.*, 1999; Ugwumba *et al.*, 2010). This can also help with food and nutritional security on farms, as well as rural jobs, limiting excessive migration to cities, which is a major concern in emerging countries (Singh *et al.*, 2012). The following are some of the other benefits of IFS:

- Reliance on external inputs (fertilizers, agrochemicals, feeds, electricity, *etc.*) is reduced.
- Reduce production costs to increase profitability. Poor farmers can use fertilizer from animal operations, especially when artificial fertilizers become pricey due to increased petroleum prices.
- It increases production per unit area and enhances space utilization.
- It offers a wide range of products.
- Appropriate crop rotations, as well as the use of cover crops and organic compost, improve soil fertility and physical structure.
- Crop rotation helps to control weeds, insect pests, and illnesses.
- Making proper use of crop remnants and animal waste.
- Higher net returns on the farming family's land and labour resources.

Major Components of Integrated Farming System and their Scope for Business and Income Generation in India

The integrated agricultural system has gotten a lot of press, and the concept is being advocated in a lot of places in both developing and industrialized countries. Many research based on field trials or site experiments have indicated that trees and vegetable crops can be quite profitable. These studies' financial assessments show that the systems generate net surplus revenue above and beyond the household's consumption demands. According to these studies, farmers also use their free time for other farm duties. Increased production, profit, cost-cutting through recycling, family nutrition, sustainability, ecological security, job creation, economic efficiency, and social equality are all goals that are addressed by intentional integrated systems. Small farms (up to 2 ha) are critical to India's food and nutritional security. As a result, location-specific integration of field crops, orchards, floriculture, agroforestry, livestock such as dairy, poultry, piggery, and fishery, as well as other land-intensive activities like mushroom, apiary, and boundary plantations, is being done to improve the livelihood of marginal and small households.

Crop Production

This is a fundamental farming practice that every farmer follows. Cropping methods depending on climate, soil, and water availability must adapt in order to realize potential production levels while utilizing available resources efficiently (Yadav *et al.,* 1998). In sequential cropping, the previous crop has a significant impact on the following crop. Complementary impacts include the transfer of nitrogen from earlier crop leftovers, particularly legumes, to subsequent crops (Behera and Sharma, 2008). A food crop is mostly grown in India during the most appropriate seasons, and hence a specific food crop is fundamental to the cropping system. Multitier cropping, also known as multistoried cropping, is the process of cultivating multiple crops with distinct heights, rooting patterns, and durations. Behera and Mahapatra (1999) incorporated a coconut-based multistoried cropping system in the IFS at Bhubaneswar (Odisha), which featured pumpkin, ridge gourd, and malabar spinach(*poi*) as ground storeys; ginger, turmeric, *Colocasia* (taro/ elephant-ear), and pineapple as first storeys; and coconut in the second storey. This system made the most of available space and sun energy. Proper choice of crops and high value crop in the system can help in improving the income from the system.

Livestock

Indian farming is incomplete without the integrated of livestock in it. Indian farmers have been raising crops and animals together for centuries, and livestock have always been an important part of the system (Singh *et al.,* 1998). Apart from helping farmers' food baskets, nutritional security, and household income, the sector also helps to generate meaningful employment, particularly among landless, small, and marginal farmers, and women (Tiwari *et al.,* 2013).

Poultry

Poultry rearing in a backyard system is a time-honored tradition. In India, backyard poultry is a source of income for a huge number of small and marginal farmers (Babu *et al.*, 2014). Poultry in an IFS is critical to meeting the country's ever-increasing demand for meat and eggs (Dongre, 2015). By including this component into the system, one can identify a great business opportunity.

Mushroom Cultivation

Mushroom growing is a significant sector for utilizing agro-wastes to increase income and provide a higher level of sustainability to the system. Due to its requirement for less acreage, higher year-round employment and income creation potential, and suitability for small and marginal holdings, mushroom is regarded a significant component of farming systems in India (Behera *et al.*, 2004). People in India are transitioning from non-vegetarian to vegetarian and vegan diets. The mushroom demand among these demographics can help create a lucrative business opportunity.

Apiculture and Sericulture

Apiculture is the science and culture of honeybees and their management, whereas sericulture the practice of combining mulberry cultivation, silkworm rearing and silk reeling. Though apiculture requires low investment and so it can be taken up by small, marginal and landless farmers and the educated unemployed youth, sericulture is labor intensive in all its phases, including off-farm activities such as twisting, dyeing, weaving, and printing (Behera and Sharma, 2008). Both these are considered as a good business investment because of its suitability to small and marginal farm-holdings and its low investment requirement.

Fish Culture (Aquaculture)

In India's coastal belt and other Asian countries such as Bangladesh, Vietnam, China, Indonesia, and Malaysia, this is seen as a very promising and steady component of farming systems. When completely integrated with other enterprises and household activities, fish farming has the potential to increase small farmer sustainability and income production, allowing farm families and communities to successfully manage their natural resources (Devendra, 2002).

While the aforementioned are crucial components of an IFS, small and marginal farmers also engage in a variety of additional income-generating activities (*e.g.*, vermicomposting, duck farming, rabbit rearing, floriculture, and biogas generation).

Integrated Farming Systems for Enhancing Food Security and Employment Opportunity

Farm households' dietary demands are partially or completely met by the well-integrated complementary IFS systems, which operate on a small area of land.

These systems are the future of coastal agriculture and offer the majority of the staples used by India's millions of small and marginal farmers. IFS allows for the efficient use of land and time in the production of short-duration vegetable crops, pulses, and livestock fodder. These systems are crucial for ensuring future food and nutrition for India's rapidly growing population. Kerala, India's homestead farming combined with livestock component supplies a farm family of four individuals with vegetables, milk, and eggs throughout the year on an area of 0.2 hectare (John, 2014). The constant labour demand for diverse crops and animal systems allows for increased employment generation while also keeping farm families involved in farm activities. The farm family might be productively involved in daily animal care, fodder production, agro-ecotourism activities, organic input production and marketing, and so on.

The various models proved to be effective at increasing production, marketable surplus, and profit. Table 2.1 shows an effective plan for improving sustainability and employment on the east and west coasts of the country. Reduced manufacturing costs were achieved in IFS by recycling by-products and wastes from various components within the system, resulting in improved net revenue. The implementation of IFS can lower input costs, particularly for important inputs like fertilizers, pesticides, and herbicides, by increasing resource flow and integrated pest and nutrient management (Paramesh *et al.,* 2021). The increased product diversity in IFS, particularly in livestock (dairy and poultry), has the potential to produce revenue for small and marginal farmers on a daily basis. The inclusion of high-value vegetables and spice crops in the farm is more remunerative rather than long-duration mono-cropping (Paramesh *et al.,* 2019).

Improving the Potential of Business Opportunity

The innovative features of IFS are:

a) Maximum production through proper use of resources and local materials,

b) Measures for soil and water conservation,

c) Adequate systems of water harvesting,

d) Effective recycling of farm residues and waste materials,

e) Crop production through organic farming; and

f) Generation of business opportunities through proper mixing of different components like dairy, fishery, poultry, goatry, vermicomposting, apiary, *etc.*

The lack of a systems approach is a serious limitation for the commercial potential. Much of our country's research is conducted along strict disciplinary lines, with little regard for the needs of small farmers (Devendra, 2006). The systems approach necessitates a multi-disciplinary interpretation of many system components and the biophysical environment, which are discovered through extensive evaluations of limitations, demands, and opportunities. As a result,

Table 2.1: Sustainable Integrated Farming System Models in East and West Coast of India Showing Production, Marketable Surplus, Profit and Employment

District	*Sustainable Farming System Model*	*Production (kg)*	*Marketable Surplus (kg)*	*Profit (₹)*	*Employment (Man Days)*
Srikakulam (Andhra Pradesh)	Field crops + Dairy + Goat + Poultry	5867	3600	35582	360
Bhubaneswar (Odisha)	Cropping system + Horticultural System + Dairy+ Poultry + Fishery+ Boundary plantation+Kitchen garden + Apiary	25472	21459	123167	449
Thanjavur (Tamil Nadu)	Crops + Dairy + Poultry + Horticulture + Fishery + Vermicompost + Boundary plantations	55456	50450	323167	460
24-Parganas South (West Bengal)	Field crops + Dairy + Fishery	11854	8950	38500	288
North Goa (Goa)	Field crops + Dairy + Fish + Poultry	25490	22354	126750	255
Karamana (Kerala)	Homestead Farming System	19414	14816	102545	209
Pathinamthitta (Kerala)	Field crops + Horticulture (Fruits and Vegetable crops) + Dairy + Poultry	34616	33845	34616	369
Karjat (Maharashtra)	Field crops + Horticulture + Livestock +Composting	49780	42750	192168	370
Raigad (Maharashtra)	Field Crop + Livestock + Nutritional kitchen garden	11628	9607	65708	412

Source: Ravisankar *et al.* (2016).

system perspectives are critical for improving, comprehending, and developing these farmers' contributions. Access to improved and relevant technologies for small farmers is another significant subject that needs attention. The problem is particularly acute in rain-fed areas that have escaped the Green Revolution's reach. An ideal integrated farming system (IFS) should generate adequate income and employment opportunities to support the livelihood of the farmer and his family (Behera *et al.*, 2015).

The intricacy and overall inefficiency of current marketing chains is an area that demands a lot of attention. In the current production-to-consumption cycle, insufficient access to market outlets and weak marketing arrangements are severe obstacles. Market chains include rural, peri-urban, urban, and international markets, and connecting small farmers to these markets and marketing systems is a key difficulty. Appropriate infrastructure and communication facilities, as well as collection and processing centres, are required to establish good connectivity between rural and urban markets. Before pursuing a business opportunity, all of these concerns must be thoroughly addressed.

Conclusion

Agriculture's long-term viability depends on an integrated farming system. Nutrient recycling within the system promotes system self-sufficiency, which not only reduces reliance on external inputs such as seed and fertilizers, but also provides balanced and rich nutrition to the farm family while lowering cultivation costs and increasing profit margins on the same piece of land. Each component of the system can be viewed as a potential source of money and enterprise for the kids. Without a doubt, this strategy is region specific, technically skill based, and plays a multidimensional function in meeting domestic demand, employment opportunities, resource efficiency, productivity maintenance, investment ability, and economic ability of the systems. An integrated agricultural system improves net returns, creates jobs, conserves natural resources, lowers production costs, and raises overall revenue by reducing risk.

REFERENCES

Babu, S., Avasthe, R. K., Singh, R. and Das, S. (2014). Backyard poultry farming in Sikkim with improved Vanaraja chicken. *Indian Farming*. **63**(10): 63-64.

Behera, U.K. and Mahapatra I.C. (1999). Income and employment generation of small and marginal farmers through integrated farming systems. *Indian Journal of Agronomy*. **44**(3): 431-439.

Behera, U.K., Jha, K.P. and Mahapatra, I.C. (2004). Integrated management of available resources of the small and marginal farmers for generation of income and employment in eastern India. *Crop Research*. **27**(1): 83-89.

Behera, U.K. and Sharma, A.R. (2008). Farming systems. *In*: Modern Concepts of Agriculture, National Institute of Science Communication and Information Resources (NISCAIR), CSIR, New Delhi.

Behera U.K., Babu, A., Kaechele, H. and France, J. (2015). Energy self-sufficient sustainable integrated farming systems for livelihood security under a changing climate scenario in an India context: A case-study approach. *CAB Reviews*. **10**(19):1-11.

Brouwer J. and Powell J.M. (1998). Micro-topography, water balance, millet yield and nutrient leaching in a manuring experiment on sandy soil in south-west Niger. *In*: Soil Fertility Management in West African Land Use Systems. Renard, G., Neef, A., Becker, K. and Von Oppen, M. (Eds). *Proceedings of a workshop* held during 4-8 March, 1997 in Niamey, Margraf, Weikersheim, pp. 349-360.

Das, A., Datta, D., Samajdar, T., Idapuganti, R.G., Islam, M., Choudhury, B.U., Mohapatra, K.P., Layek, J., Babu, S. and Yadav, G.S. (2021). Livelihood security of small holder farmers in eastern Himalayas, India: Pond based integrated farming system a sustainable approach. *Current Research in Environmental Sustainability*. Vol. 3: 100076.

Devendra, C. (2002). Crop-animal systems in Asia: Future Perspectives. *Agricultural Systems*. **71**(1-2): 179-186.

Devendra, C. (2006). Challenges and opportunities for research and development of small ruminants in Asia. Proceedings of the XII Asian-Australasian Animal Science Congress, 18-22 September, 2006 at Busan, pp. 1-14.

Dongre, V.B. (2015). Strategies for sustainable livestock production in rural India. *Indian Farming*, **65** (2): 29-32.

Faroda, A.S. (2014). Integrated farming system for livelihood and environmental security under rainfed conditions. *In*: Souvenir of National Symposium on Agricultural Diversification for Sustainable Livelihood and Environmental Security, held during 18-20 November 2014 at Ludhiana, Punjab, pp. 31-35.

Gill, M.S., Singh, J.P. and Gangwar, K.S. (2009). Integrated farming system and agriculture sustainability. *Indian Journal of Agronomy*. **54**(2): 128-139.

Gill, M.S., Samra, J.S. and Singh, G. (2005). Integrated farming system for realizing high productivity under shallow water-table conditions. *Research Bulletins*, Department of Agronomy, PAU, Ludhiana, pp. 1-29.

GoI. (2014). *Agricultural statistics at a glance*, Directorate of Economics and Statistics, Govt. of India, New Delhi.

John, J. (2014). Homestead farming in Kerala: A multi-faceted land-use system. Review of Agrarian Studies. **4**(1): 80-94.

Kumar, S., Bhatt, B.P, Dwivedi, S.K. and Shivani. (2015). Family Farming – A Mechanism to Contribute towards Global Food Security. *Indian Farming*. **64**(12): 13-18.

Kumar, S., Subhash, N., Shivani, Singh, S.S. and Dey, A. (2012). Evaluation of different components under Integrated Farming System (IFS) for small and marginal farmers under semi-humid climatic environment. *Experimental Agriculture.* **48**(3): 399-413.

Layek, J., Das, A., Mitran, T., Nath, C., Meena, R.S., Yadav, G.S., Shivakumar, B.G., Kumar, S. and Lal, R. (2018). Cereal + legume intercropping: An option for improving productivity and sustaining soil health. *In*: Legumes for Soil Health and Sustainable Management, Springer, Singapore, pp. 347-386.

Mahapatra, I.C. (1992). Farming systems research challenges and opportunities. *Eastern India Farming System Research and Extension Newsletter*. **6**(4): 3-10.

Mali, H., Kumar, A. and Katara, P. (2014). Integrated farming system for irrigated and rainfed conditions. *In*: Proceedings of National Symposium on Agricultural Diversification for Sustainable Livelihood and Environmental Security, held during 18-20 November 2014 at Ludhiana, Punjab, pp. 546.

Manjunath, B.L. and Singh, M.P. (2012). Farming system options for different topographical situations of west coast region. *Extended Summaries*, Vol. 3: 3rd International Agronomy Congress, held during 26-30 November 2012 at New Delhi, pp. 988-990.

Manjunatha, S.B., Shivmurthy, D.S., Satyareddi, A., Nagaraj, M.V., Basavesha, K.N. (2014). Integrated farming system – A Holistic Approach: A Review. *Research and Reviews: Journal of Agriculture and Allied Sciences*. **3**(4): 30-38.

Olele, N.F., Nweke, F.U. and Agbogidi, O.M. (1999). Role of integrated farming system in agricultural development in the delta region of Nigeria. *Delta Agriculture*. **6:** 128–134.

Panwar, A.S., Syiemlieh, H., Kalita, L., Tahashildar, M. and Layek, J. 2017. Food and Nutritional Security through Farming System Approach. *In*: *Conservation Agriculture for Advancing Food Security in Changing Climate* (Editors: Anup Das, K.P.Mohapatra, S.V.Ngachan, A.S.Panwar, D.J.Rajkhowa, G.I. Ramkrushna and JayantaLayek). Today and Tomorrow's Printers and Publishers, New Delhi - 110 002, India.

Paramesh, V., Chakurkar, E. B., Bhagat, T., Sreekanth, G. B., Kumar, H. B. C., Rajkumar, S., Gokuldas, P. P., Mahajan, G. R., Manohar, K. K. and Ravisankar, N. (2021). Impact of integrated farming system on residue recycling, nutrient budgeting and soil health. *Indian Journal of Agricultural Sciences*, **91**(1): 44-48.

Paramesh, V., Parajuli, R., Chakurkar, E.B., Sreekanth, G.B., Chetan Kumar, H.B., Gokuldas, P.P., Mahajan, G.R., Manohara, K.K., Reddy, K.V. and Ravisankar, N. (2019). Sustainability, energy budgeting, and life cycle assessment of crop-dairy-fish-poultry mixed farming system for coastal lowlands under humid tropic condition of India. *Energy*. **188:** 116101.

Rahman, F.H. and Sarkar, S. (2012). Efficient resource utilization through integrated farming system approach in the farmers' field at Burdwan district of West Bengal. *Extended Summaries*, Vol. 3: *3rd International Agronomy Congress*, held during 26-30 November 2012 at New Delhi, pp. 997-998.

Ravisankar, N., Singh, P., Mishra, R.P., Prusty, A.K., Shamir, M., Singh, R., Triatic, D. and Mohan, B. (2016). Annual Report 2016-17. ICAR-Indian Institute of Farming Systems Research (Indian Council of Agricultural Research), Modipuram, Meerut – 250110, pp. 1-258.

Siddeswaran, K., Sangetha, S.P. and Shanmugam, P.M. (2012). Integrated farming system for the small irrigated upland farmers of Tamil Nadu. *Extended Summaries*, Vol. 3: *3rd International Agronomy Congress*, held during 26-30 November 2012 at New Delhi, pp. 992-993.

Singh, D.P., Singh, K.P. and Yadavika. (1998). Integrated Farming System – A Key Issue for Research-Education-Extension Linkages to Sustain Food Security and Eco-Friendly Environment in 21st Century. *In*: Singh, P., Prasad, R. and Ahlawat, I.P.S. (Eds.), Proceedings of the 1st International Agronomy Congress: Agronomy, Environment and Food Security for the 21st Century. Indian Society of Agronomy, New Delhi, pp. 474-486.

Singh, J.P., Gangwar, B., Kochewad, S.A. and Pandey, D.K. (2012). Integrated farming system for improving livelihood of small farmers of western plain zone of Uttar Pradesh, India. *SAARC Journal of Agriculture*. **10**(1): 45–53.

Singh, K., Bohra, J.S., Singh, Y. and Singh, J.P. (2006). Development of farming system models for the north-eastern plain zone of Uttar Pradesh. *Indian Farming*. **56**(2): 5-11.

Tiwari, R., Singh, B.P. and Dutt, T. (2013). Role of Livestock Poultry and Pet Sector in sustaining food, income and employment security. *Indian Farming*. **63**(7): 73-80.

Ugwumba, C.O.A., Okoh, R.N., Ike, P.C., Nnabuife, E.L.C. and Orji, E.C. (2010). Integrated farming system and its effect on farm cash income in Awka South Agricultural Zone of Anambra State, Nigeria. *American-Eurasian Journal of Agricultural and Environmental Sciences*. **8**(1): 1-6.

Yadav, R.L., Prasad, K., Gangwar, K.S. and Dwivedi, B.S. (1998). Cropping systems and resource use efficiency. *Indian Journal of Agricultural Sciences*. **68:** 548-558.

Chapter 3

Integrated Farming System: A Promising Sphere for Entrepreneurship

Sunil Dutt[1], Nishant Kumar[2] and T.K. Mohanty[3]

[1]Ph.D. Scholar, Livestock Production Management Division,
[2]Senior Scientist and [3]Principal Scientist,
(Animal Reproduction), ICAR-National Dairy Research Institute,
Karnal – 132001, Haryana
Corresponding author: drnishantvet@yahoo.com

Abstract

Economy of India is based on agriculture with nearly 70 per cent of the population engaged with it. It is not only highest milk producer in the world with largest buffalo population but is major exporter of various agricultural products. With shrinking resources of land, increased input requirements and underemployment, conventional farming practices are topic of concern regarding sustainability. Integrated farming system ensures efficient utilisation of resources to full potential for maximum productivity. Integration of multiple enterprises with synergistic and complementary effect would not only restrain wastage of by-products but also generates source of self-employment and round the year income for the farmer. Moreover, with increasing awareness of detrimental effects of xenobiotics in food chain, demand of organic products can be easily sustained through IFS. Entrepreneurship in IFS would be golden opportunity not only for generating potential productivity through efficient resource utilisation and enterprise integration but also to exploit the uncharted market demand.

Keywords: *Agro-climatic zones, Entrepreneurship, FSR, IFS, Recycling.*

Introduction

India is a key global agricultural powerhouse. It owns the credit for highest production of milk, legumes such as pulses, and spices in the world. India possesses the highest strength of buffalo population in the world. The country also has largest cultivated share of landmass for cereal crops as wheat, rice, and cash crop such as cotton. Country's major export of agricultural products include cereal produce such as wheat and rice, cash crops such as cotton, sugarcane, vegetables, fruits, tea, farmed fish, sheep and goat meat. In past few years, concern for future resources in terms of potable water security, food security, livelihood security, preservation and conservation of natural resources and environmental preservation has become serious challenge. Developing countries are battling these difficulties while also dealing with the simultaneous burdens of climate change and globalisation.

Farming employs over 70 per cent of the Indian population. As a result, India's prosperity would be determined by the success of its farming community, which is contingent on the practical implementation of enhanced knowledge, inventions and the prudent use of assets (land, labour, capital, machinery *etc.*). Cropping systems, livestock, poultry, horticulture, aquaculture, agro-forestry, and the tools accessible to farmers which help them to gain profit are all components of farming systems. Crops, livestock, poultry, fish, and sericulture are all examples of farming enterprises. When carefully chosen, planned, and implemented, amalgamation of additional enterprises with crop farming results in higher output as compared to a sole enterprise, mainly for small and marginal farmers. Thorough examination and planning of farm must be done so as to decide the optimal incorporation of enterprises in the crop production activities. Factors influencing the amalgamation of different enterprises at farm are as follows:

a) Top soil health and weather conditions of decided area,

b) Availability of resources (capital, land, human resource, machinery),

c) Availability of resources at the moment,

d) The planned integrated farming system's economics, and

e) Farmer's managerial ability.

Farming is the process of entrapping solar energy and processing it into valuable plant and animal goods. The term "system" refers to a set of interrelated practices and processes structured into a functional entity, such as a collection of components or elements which act together to process the inputs for creation of outputs (Fresco and Westphal, 1988). The "agricultural system" is an intricate interwoven environment of topsoil, flora, animals, equipments, resource inputs that vary subjectively to different degrees by political, economic, institutional, and social forces at numerous levels (Mahapatra, 1992).

The word "integrate" derives its roots in the Latin word "integrare," which has essence "to make whole," "to complete by adding parts," or "to merge components into a whole." Keeping in view the monoculture farming, the term "integrated farming system" (integrated agriculture) is an extensively used term to illustrate a more incorporated approach to farming. It can also be synonymously termed to as integrated bio-systems, which relates to various agricultural systems that combine dairy and crop production or fish farming. The Integrated Farming system (IFS) is a coordination of systems that works together. IFS ensure the farmer for residual by-products produced from one type of agriculture would be utilised to assist the other component enterprise.

The integrated farming system (IFS) is a part of FSR (Farming System Research). IFS promote a transformation in farming methodology so as to gain highest productivity in the crop production while also ensuring optimal resource utilization. Several workers have defined the integrated farming system.

IFS is defined by Radhamani *et al.* (2003) as a part of agriculture systems that considers the notions of boosting productivity, limiting risk, and increasing returns along with enhancing the use of organic refuse and crop leftovers in the field. IFS have been described by Okigbo (1995) as an assorted agricultural system that includes minimum of two different yet conceptually interconnected crop and animal enterprise. Crops, agro-forestry, cattle, aquaculture, agri-horticulture, and sericulture are among the agricultural operations (Singh, 2004). According to Panke *et al.* (2010), incorporation is done in such a manner that the yield of one component becomes the contributing input for the other enterprise components with greater extent of complementary effects. This agricultural system exists between commodity input-output relationships which connect various activities of production, and before and after production activities, and agriculture and processing activities and services. It's work revolves around a few carefully chosen, mutually dependent, interconnected, and frequently interlinked production systems centred on a small number of crops, animals, and related sub-occupations. IFS aims to increase total productivity, sustainability, and gainful employment by using complementary and synergistic associations between various agricultural subsystems/enterprises. IFS is a set of resource-saving methods to attain satisfactory earnings with greater and stable output levels along with reducing the pessimistic consequences of comprehensive farming for the preservation the environment (Lal and Miller, 1990; Gupta *et al.*, 2012).

Evolution of Integrated Farming System in India

M.S. Swaminathan Research Foundation (MSSRF) conducted adaptive participatory research and exhibition on the Integrated Farming System at Keelamanakkudi village, Parangipettai Block, Chidambaram Taluk, Tamil Nadu, between 1996 and 2002, with a major focus on illustrating the efficient utilisation of water by integrating intermediary components into traditional paddy cultivation. According to that project, farmers could be able to reduce concerns related to water limits for the period of essential phases of crop growth if they had water storage structures on their fields. Incorporating different activities and farm ponds based

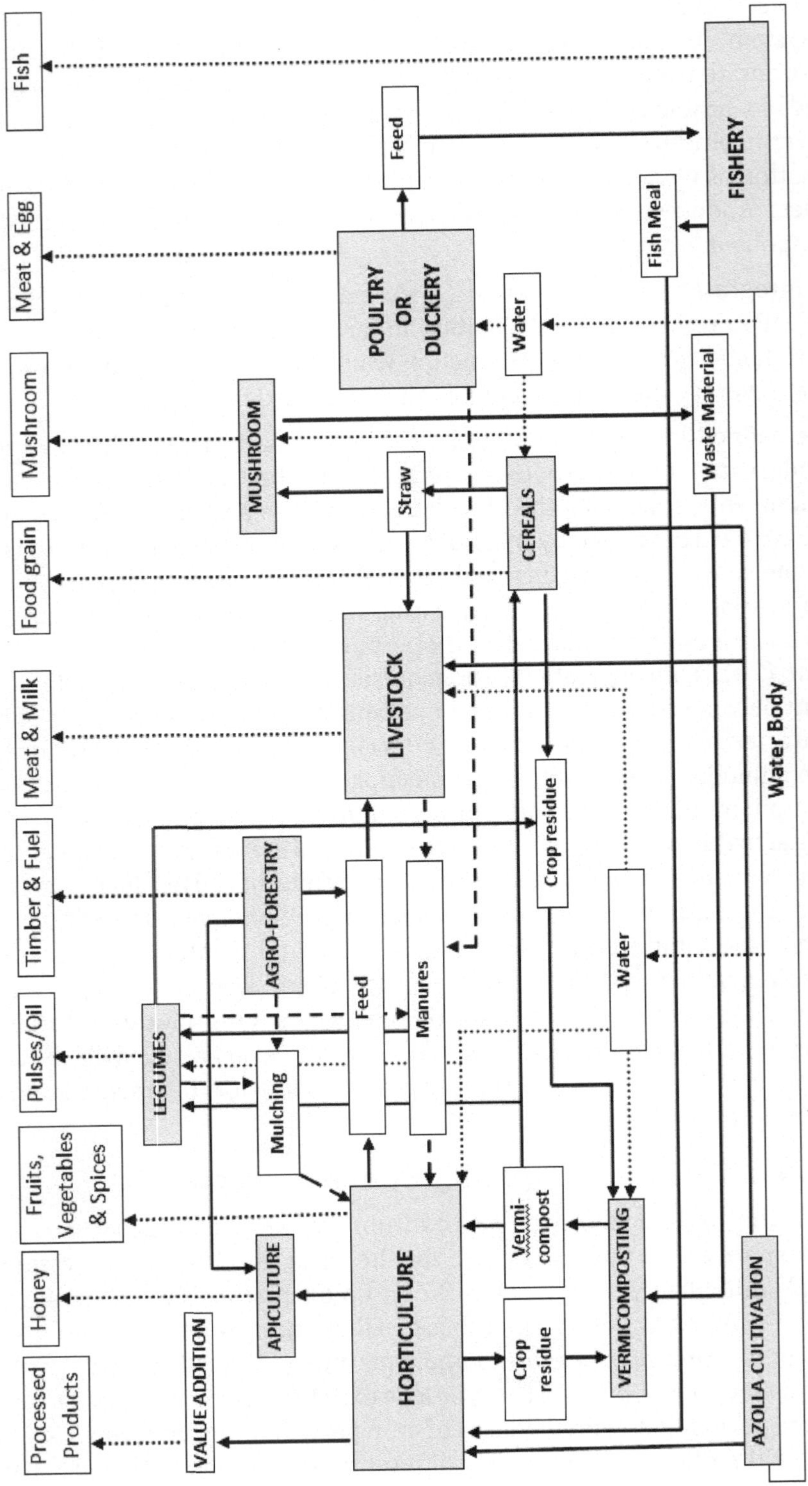

Figure 3.1: A Notional Frame of IFS to Integrate different Enterprises (Roy *et al.,* 2014).

on IFS concepts was also envisioned as providing increased livelihood possibilities in the region.

Benefits of Integrated Farming System

Merging and sharing of available resources/inputs, efficient employment of family labour, maintenance, preservation, and utilization of farm biomass, including non-traditional feed and fodder resources, efficient use of animal biomass waste, topsoil fertility and health regulation, creation of earnings and employment for many people and increased economic resources are some of the benefits of IFS. It maximizes space and offers a wide range of products.

1) **Productivity:** By intensifying crop and allied operations, the IFS give a prospect to boost economy by enhancing yield/unit area/time.

2) **Profitable:** IFS utilize least expensive waste material from one component for other. As a result, the cost of production is reduced; a link is formed between the utilisation of refuse material and the exclusion of middleman intervention in the majority of resources used. The ratio of benefit-to-cost and net profit is enhanced when working out the net profit/benefit-to-cost ratio.

3) **Potentiality or Sustainability:** The production base's potential can be maintained for considerably longer time economically and sustainably when external inputs are minimized. IFS provide this opportunity as organic by-products of various linked components can be appropriately supplemented to one another.

4) **Balanced food:** Various natural components are connected to provide various trailheads of sustenance.

5) **Safety of environment:** Leftover refuse materials are efficiently reused and recycled in IFS by connecting suitable enterprises which ultimately results in reduced pollution.

6) **Recycling:** In IFS, refuse (waste material from dairy animals, crop residues *etc.*) is effectively recycled. As a result, there is reduced dependency on more inputs like fertilisers, chemicals such as weedicide, feeds and energy.

7) **Round the year income:** It provides a year-round monetary assistance to the entrepreneur due to the interplay of different components yielding in products like cereal and cash crops, eggs, milk, honey, mushrooms, cocoons, and silk. The farming family's resources of manpower and land generate a larger net return.

8) **Implementation of latest technology:** Resourceful farmers (large farmers) make full use of technology whereas small and marginal farmers lack funds for implementing new technology. Availability of funds throughout the year encourages small/traditional farmers to adopt new technologies.

9) **Saving energy:** We need to find an alternate source of energy that will allow us to minimise our reliance on fossil fuels on time. The energy crisis can be postponed until a later day by generating energy sources like biogas. IFS promises a contribution as the bio-wastes in the system could be used for generation of clean fuel such as biogas using a practical recycling approach.

10) **Reducing fodder shortage:** Complete and efficient use of available land resource is ensured in the integrated system. Plantation of perennial legume fodder trees along field margins for nitrogen fixation can help to improve the nitrogen content of soil. These measures will significantly alleviate the lack of high-quality feed for the dairy enterprise.

11) **Eliminating shortage of timber and fuel:** When agro-forestry is linked correctly, the scale of fuel as wood-coal and industrial timber production can be increased without affecting agricultural yield. Deforestation will be considerably reduced thus maintaining the ecology of nature.

12) **Employment creation:** Combining agriculture and dairy component can not only dramatically enhance human resource demand which, as a result, greatly reduce the problem of underemployment but will allow for the year-round employment of family labours.

13) **Agro-industries:** Scaling up and market channelization results in IFS-linked products to reach commercial level where surplus value is adopted, resulting in the growth of associated agro-industries.

14) **Enhancing efficient utilisation of input:** Integrated farming gives much flexibility for making use of input resources in various enterprises to increase efficiency as well as proportion of benefit-to-cost.

15) **Agro-tourism:** IFS has the future possibilities to grow into a business that attracts tourists interested in learning about farming, rural life and to experience natural living away from hectic lifestyle in the concrete urban jungles.

Goals of Integrated Farming System

Enhancing the productivity of all enterprise businesses to full potential for ensuring the continuous and predictable returns.

a) To achieve agro-ecological equilibrium by rejuvenating/improving the system's productivity,

b) To minimize the rise of pest population, weed growth and diseases and keep them at a low intensity by using natural cropping system management, and

c) To minimize the application of agro-chemicals (fertilisers, weedicide, pesticides *etc.*), which can offer society with healthy, xenobiotic-free agricultural products and a safe environment (Manjunatha, 2014).

Principles of Integration of Components of Integrated Farming System

Crop production is the most important activity in agriculture. Crop income may not be enough to keep a farm family afloat throughout the year. A consistent cash flow may be guaranteed when the crop is integrated with other businesses. A wise combination of businesses, considering the local climate, will pay off handsomely. Simultaneously, it will encourage effective residue/waste recycling. As a result, the following integration principles must be addressed for a flourishing agricultural system.

a) The integrated agricultural system must generate an ample amount of premium food, fodder and industrial raw material for allied industries.

b) It must suit societal demands.

c) It must maintain a feasible farming enterprise.

d) It must be environmentally conscious.

e) The system must ensure that natural resources are not depleted.

Key Components of Integrated Farming System (Roy *et al.*, 2015)

A. Crop	*B. Livestock and Poultry*	*C. Pisciculture*	*D. Allied enterprises*
Cereals	Buffalo	Composite fish culture	Apiculture
Fruits	Cattle	Fingerling production	Mushroom cultivation
Vegetables	Sheep	Paddy cum fish culture	Food processing
Spices	Goat		Vermicomposting
Plantation crops	Pig		Biogas production
Flowers	Chicken		Azolla cultivation
Pulses	Duck		Sericulture
Oilseeds	Quail		Moriculture
Fodder crops			
Agro-forestry			
Sugarcane			
Fibre crops			

A system of integrated agriculture can be established by incorporating various combinations of above discussed enterprise components based on farmers' preferences, acceptable agro-climates, available technology, and marketing facilities all influence crop, livestock, fisheries, and secondary agricultural activities.

Integrated Farming System for Various Agro-Climatic Zones of India

1) **Cold deserts at high altitudes**: It included pastures integrated with agro-forestry, goats, sheep, Angora rabbits, and a few established agricultural crops such as wheat, millets, barley and fodder.

Duck Rearing

Mushroom Farming

Vermicomposting

Biogas Plant

Agroforestry

Buffalo Rearing

Aquaculture

Poultry

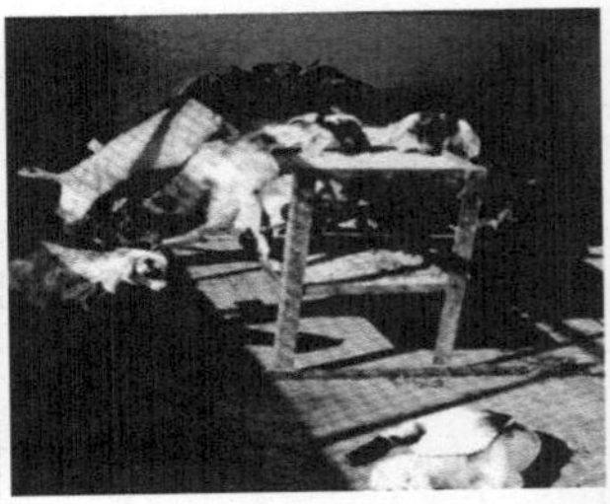
Goat Rearing

2) **Arid and desert region:** Domestic animals which include sheep, goats, and camel as well as moderate agricultural components such as fodder crops, bajra, wheat, legumes and gram.

3) **Western and central Himalayas:** Prominent component is the horticultural crops, with wheat, rice, maize, legumes, and fodder crops grown mostly on hill terraces and slopes.

4) **Eastern Himalayas:** Rice, millets, pulses, and other primitive crop husbandry is practiced. There are various agro-forestry systems in use. The main livestock activities are pigs and poultry.

5) **Indo-Gangetic Plains:** Dairy animals mainly cattle and buffaloes and crops such as rice, wheat, maize, mustard and legumes are grown intensively. In the Central and Southern peninsula, secondary enterprises include dairy cattle, poultry, sheep and goats as dairy enterprise and cotton, pulses, sorghum, millet as crop components.

6) **Western Ghats:** Plantation crops are the primary agricultural activity, with rice and pulses as supplementary crops. Dairy cattle, sheep, and goats are the major farm animal species that are usually kept in big herds and allowed to roam freely.

7) **Delta and Coastal Plains**: Paddy cultivation along with pisciculture, poultry and pig farming are mostly practiced. Marine ecosystem is also explored for capturing fisheries, which is a specialist industry that does not combine with cropping.

Table 3.1: Dominant Farming Systems in Various Agro-Climatic Regions of India

Agro-climatic Region	*States*	*Most Dominant Farming Systems in the Region*
Western Himalayas	Jammu and Kashmir, Himachal Pradesh, Uttarakhand	Crops + Dairy Agri-horticulture system
Eastern Himalayas	Assam, Meghalaya	Crops + Fish + Cattle + Piggery Mono-cropping of rice and maize + Piggery +Backyard poultry + 2-3 cows
Trans Gangetic Plain	Punjab, Haryana	Crops + Dairy
Upper Gangetic Plain	Uttar Pradesh	Crops + Dairy
Middle Gangetic Plain	Bihar, Uttar Pradesh	Crops + Dairy
Lower Gangetic Plain	West Bengal	Crops + Fishery Crops + Dairy Crops + Dairy + Poultry + Duckery
Eastern Plateau and Hills	Chhattisgarh, Jharkhand	Crops + Backyard Poultry + Fish Crops + Dairy
Central Plateau and Hills	Madhya Pradesh	Crops + Dairy

Agro-climatic Region	*States*	*Most Dominant Farming Systems in the Region*
Western Plateau and Hills	Maharashtra	Crops + Goatery Crops + Horticulture Crops + Dairy
Southern Plateau and Hills	Andhra Pradesh, Telangana, Tamil Nadu, Karnataka	Crops + Dairy Crops + Horticulture Crops + Horticulture + Dairy
East Coast Plain and Hills	Odisha	Crops + Dairy Crops + Dairy + Fish
West Coast Plain and Hills	Maharashtra, Goa, Kerala	Crops + Dairy Coconut based homestead farming Rice based farming systems
Western Dry Region	Rajasthan	Crops + Dairy
Gujarat Plain and Hills	Gujarat	Crops + Dairy
Islands	Andaman and Nicobar	Plantation crops + Piggery Crops + Cattle + Fish

Opportunities to Integrated Farming System

Indian economy is largely based on agricultural sector. Small and marginal farmers are the major contributors of the rural economy, accounting for 85 per cent of all farmers but only 44 per cent of all operational land under use (GoI, 2014). The decreasing availability of land per capita is posing a severe threat to long-term profit generation and viable sustenance of agriculture (Siddeswaran *et al.*, 2012).

As a result of country's dwindling land resources and ever-increasing population, there is practically no room for horizontal land development for food production. Combining suitable farming enterprises which require less land and time for ensuring sufficient cyclic revenue to farmer can make vertical expansion viable (Gill *et al.*, 2009).

Farmers have emphasized single enterprise based agricultural systems since advent of green revolution, which has resulted in deterioration of soil health, higher crop failure risk, and decreasing productivity trends (Rahman and Sarkar, 2012).

In emerging countries like India, rapid population expansion, urbanisation, and wealth growth are driving up demand for animal-based foods, exacerbated by the competition between crops and livestock (increasing cropping areas and reducing rangelands). A system approach is urgently needed to meet the demands of an ever-increasing population while maintaining ecological balance. Integrated farming systems appear to be a viable solution to the ever-increasing demand for food production, economic stability, and nutritional security, especially for small and marginal farmers with limited resources. It is a concept of ecological soundness that leads to sustainable agriculture and a reliable means of attaining high productivity with a significant fertilizer economy.

The existence of marginal and small farmers in India is threatened by rising agricultural production costs combined with World Trade Organization difficulties. Patra (2016) acknowledged that the scientific implementation of Integrated Farming Systems (IFS) could ensure the sustainability of Indian agriculture, particularly for small and marginal farmers. Farmers in India have always practised an integrated farming technique. However, to address India's current agricultural predicament, it is now necessary to popularise scientific IFS models among farmers. On the other hand, success is contingent on a grasp of the input-output and product-product linkages. Farmers encounter various obstacles, which have resulted in a decline in the attractiveness of agriculture as an integrated system. The restrictions differ depending on the agro-climatic zone. Extension agencies should exhibit scientific interventions in order to reduce restrictions by increasing production and income. Government initiatives are also required to fund IFS models to meet the goal of doubling farmers' income. Due to a misconception about whether the two outputs have complementary, supplementary, or competitive relationships, farmers are more vulnerable in making the best decisions.

Entrepreneurship in Integrated Farming System

Entrepreneurs and small businesses are critical to achieve economic independence. Mahatma Gandhi prioritised reviving entrepreneurship from the grass-root level so that a strong community of entrepreneurs might rebuild India. India has a great past of entrepreneurship. India's economy was based on villages, with its own entrepreneurial ecosystem that had developed over centuries. As a result, one of Gandhi's priorities was to strengthen the village economy.

The attitude of young people toward business is also shifting. Today's entrepreneurs have access to a helpful environment, institutions, a favourable business climate, and funding. An entrepreneurial and start-up ecosystem includes education and training on many aspects of management, accounting, facilitating seed finance, and incubation facilities. It was Mazzarol (2014) who first coined the term "entrepreneurial ecosystem" to illustrate the atmosphere that could collaborate individuals and encourage financial development and capital generation. The graphic below depicts the main factors that are deemed vital in the entrepreneurial ecosystem.

Currently, farming and business are two separate entities that do not work together. Farmers focus only on crop production, which is fraught with uncertainty regarding revenue and employment. In this competition, it is critical to develop a good strategy for increasing a farm's income, addressing issues in agricultural operations, and turning it into a business zone. The integrated agricultural system has transformed traditional cattle, pisciculture, horticulture and other allied

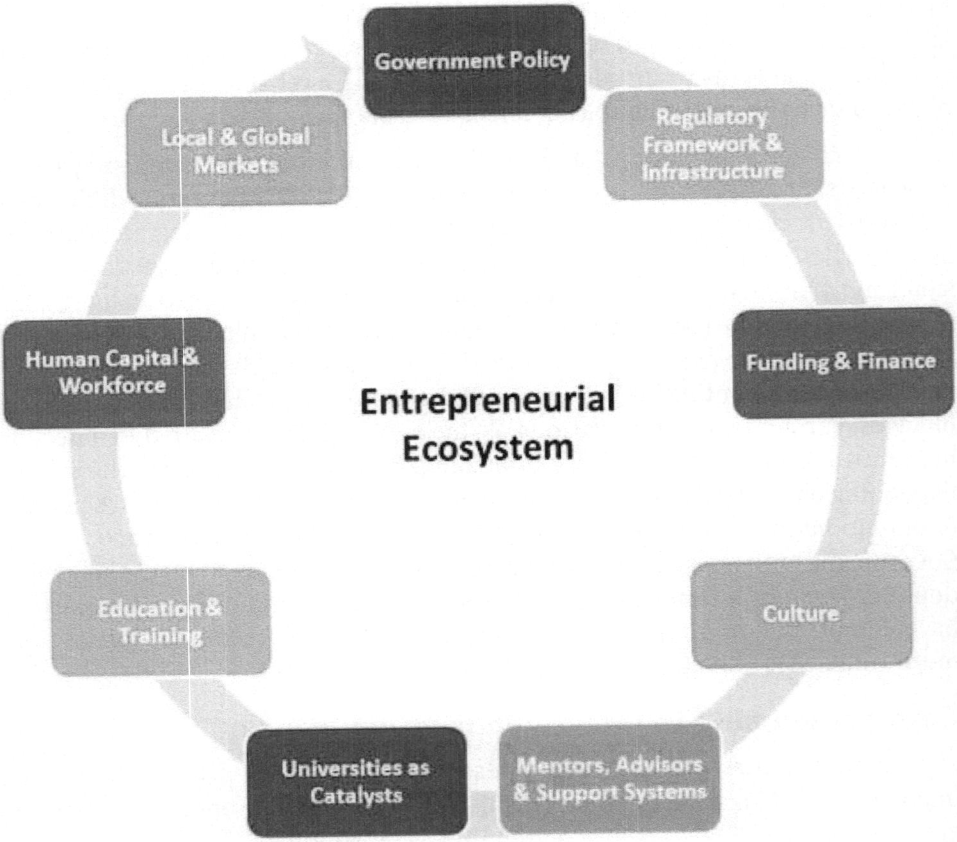

Figure 3.2: Elements of Entrepreneurship Ecosystem, Adapted from Mazzarol (2014).

activities by integration of different enterprises such as crop and fish, livestock and fish, crop, fish and livestock or a mix of crop, livestock, fish, and other components.

Steps for an Entrepreneur to Establish an Integrated Farming System

Integrated Farming System operates with a variety of agricultural activities interaction to optimize benefits through the synergistic transfer of resources, which is beneficial for soil health, weeds and pests control, and increase the efficiency of water use and its quality. The IFS needs specific characteristics in strategy, competencies and management that align with the entrepreneurship aspect. For the development of IFS, an entrepreneur has to involve 3 steps – planning, budgeting, and linear programming.

1. Planning

A plan can be defined as the pre-decided course of action to achieve objectives. Farm planning is the process of observation, appraisal, analysis and weighing the merits of new and old ideas to decide which idea is to be used in the period ahead. Planning helps the entrepreneur in the following ways:

- ☆ It helps the entrepreneur to get ideas about alternate practices.
- ☆ It helps in selecting a better combination of enterprise and most profitable practices.
- ☆ It gives reasonably expected idea about the yield, production and income.
- ☆ It helps in deciding about the input resources requirement.
- ☆ It helps avoid wastage of resources by planning their proper and efficient use.
- ☆ It gives the guide for proceeding as per by the plan.

Planning can be classified into the following seven steps:

1.1. **Survey of local conditions to explore enterprise feasibility:** This primarily includes understanding the area's meteorological records, availability of input resources, and good and broad market scope.

1.2. **Inventory of resources:** It involves the preparation of a list of various assets with their prices and listing the resources *i.e.* land, labour, and working capital, livestock, and machinery required.

1.3. **Selection of expected prices:** Assumption of future prices of the produce and the inputs will be made based on the past trend, considering that the increment or the decrease will remain the same.

1.4. **Determination of most profitable farming system:** Various enterprise components of integrated farming system can be decided based on the law of enterprise combination so that selected enterprises are joint, complementary, or supplementary.

1.5. **Determination of the technical coefficients:** To efficiently run the integrated farming system, standards are to be set to achieve the determined goals. Reasonable thresholds must be set for average production, reproduction and health parameters. Also, most efficient management practices have to be selected, which would be followed for performing routine operations.

1.6. **Estimation of the capital requirement:** Capital required for input resources, source of the capital and the rate of interest need to be planned according to the above-planned steps.

1.7. **Comparison of the old plan with the new plan:** Any new idea or improvisation will be evaluated regarding the initial basic plan for future decisions.

2. Budgeting

It is a method of analysing plans for the use of resources at the command of the decision-maker. The foremost intention of budgeting is to evaluate the component enterprises of the integrated agricultural system. Farm budgeting can also be defined as the presentation of a farm plan in economic terms by estimating cost and net income. It is of two types:

2.1. **Partial budgeting:** It is the process of estimating the returns or outcomes, or effects for a component of the business. It enables the entrepreneur to evaluate the desirability of a change in one enterprise or a single practice. It is usually done for major farm adjustments as well as the current operational problems. For instance, if the practice of hand milking is desired to be replaced by machine milking, all the expenses and net cost per animal have to be worked out for both cases before deciding on a replacement.

2.2. **Total or Complete budgeting:** It can be defined as the analysis that involves the farm business's complete re-organization. The decision taken has long-term effects. For example, it is shifting ultimately from crop farming to livestock farming after comparing the profitability of the two enterprises.

3. Linear Programming

It is the method of mathematical determination of most favourable combination of component enterprises to maximize the returns or minimize the expenditure within the limits of available resources.

Opportunities for Women Entrepreneurship in Integrated Farming System

Female farmers are the unsung heroes of agriculture, whose contributions are enormous but go unnoticed because they are largely perceived as part of their domestic chores, which are not considered a formal sector of labour. Women have a key part in household management, including agricultural operations, in India as well. This is chiefly true in high altitude and tribal locations. Men, particularly farmers with limited resources, are shifting to rural non-farm industries as well as towards urban centres, resulting in a gradual feminization of agriculture.

By utilising family labour strategically through innovative approaches, multiple applications of various household and farm resources can be assured. Empowerment of women by customised training and vital need-based assistance makes this possible. Women, in particular, require training in IFS's new permutation combinations. Women's IFS skills should also be improved because they are the major custodians of traditional knowledge. Agriculture will shift from growing productivity to enhanced affluence due to the development of different gender-friendly farming systems, which are critical for the sustainability of families with limited resources, who constitute the backbone for the agriculture.

Entrepreneurship Opportunities in an Integrated Livestock-Based Farming System

The government has implemented various projects for improvement of the farmers and raise the revenue of livestock keepers. Farmers can strengthen the livestock component of their IFS by using the following:

a) National Livestock Mission (NLM),

b) Rashtriya Gokul Mission,

c) Kisan Credit Cards (KCC) to Livestock Farmers,

d) Livestock Health and Disease Control (LH and DC) Scheme,

e) Epashuhaat portal,

f) Scheme for Integrated Development of Small Ruminants and Rabbits (IDSRR),

g) Goat Scheme under the National Mission for Protein Supplements (NMPS),

h) Rashtriya Krishi Vikas Yojana (RKVY),

i) Sheep and Wool Improvement Scheme (SWIS),

j) Bhed Palak Bima Yojna, and

k) Training programs by government institutions (*e.g.* Krishi Vigyan Kendra).

Entrepreneurship Opportunities in Organic Farming Based Integrated Farming System

India is ranked 15th in the world for organic land under cultivation (www.ifoam.org), with 5.71 million hectares certified organic (2015-16) lands. It consists of 26 per cent cultivable land (1.49 Mha) and 74 per cent forest and wild land (4.22 Mha) which serves as source of minor forest products (www.apeda.org). During 2015-16 fiscal year, production of certified organic products was approximately 1.35 million tonnes in India, which included a wide range of food items. The overall volume of exports, however, was 263687 MT. The worth of organic food exports was estimated to be around 298 million dollars. Given the rising demand for organic food items around the world, particularly in India, where the north eastern region has a huge opportunity to become a significant organic product supplier through which farmers can be benefitted from trade with comparatively high price premiums (Gopinath *et al.*, 2008). The NER of India's potential for Integrated Organic Farming can be summarised as follows –

- Land is almost undeveloped, and the crops are farmed almost entirely organically.
- The region makes limited use of xenobiotic chemicals and fertilisers.
- Because all homes keep livestock and produce enough on-farm manure, each household can be considered a little integrated agricultural system.

- ☆ In the transition period, there is no yield gap.
- ☆ The region has 47 million tonnes of organic manure potential, with 37 million tonnes coming from animal waste and 9 million tonnes coming from agricultural leftovers (Bujarbaruah, 2004).
- ☆ The North Eastern Region (NER) is abode to various cash crops with significant market demand, such as Assam lemons, medicinal rice, and passion fruits.
- ☆ In India, NER produces 45 per cent of all pineapples.
- ☆ Sikkim is the world's top producer of huge cardamom (with a 54 per cent market share).
- ☆ In India, NER is the fourth major orange grower.
- ☆ Production of finest gingers in the NER (low fibre content).
- ☆ Assam tea is renowned for its high quality all over the world.
- ☆ Diverse flora and wildlife.
- ☆ High levels of forest cover.
- ☆ Population density per square kilo-meter is lower (13–340 vs. 324 at the national level).
- ☆ Farmers at mid-and high-altitudes rely on renewable resources found on their farms.
- ☆ Tripura has established an Agri-Export Zone (AEZ) for organic pineapple growing.
- ☆ In Sikkim, another Agri-Export Zone (AEZ) has been formed for ginger.

Future Prospectives

Conventional agriculture practices seem inefficient in financially viability of the farmers. Neither has it addressed the issue of underemployment as farmer is engaged only for limited working days nor has utilised available resources of land, labour *etc.* to its full potential. With decreased possibility for horizontal extension due to constricting resources of land *etc.* and population explosion, integrated farming system offers a potential remedy for all hurdles. Moreover, with increasing awareness of detrimental effects of xenobiotics in food chain, demand of organic products can be easily sustained through IFS. Entrepreneurship in IFS would be golden opportunity not only for generating potential productivity through efficient resource utilisation and enterprise integration but also to exploit the uncharted market demand.

Conclusion

There are substantial unexplored business opportunities in the agriculture and allied sector. Compared to conventional agricultural practices, integrated farming offers better use of the farm input resources to generate round the year income for the farmer. Amalgamation of various enterprises in one farming system secures the farmer financially, reduces by-products wastage of component enterprises, generates employment and results in socio-economic upliftment of the farmer. With decreasing land resources and increasing unemployment due to population explosion, the government has taken various steps to motivate youth for entrepreneurship in integrated farming to generate self-employment and increase income of conventional farmers.

REFERENCES

Bujarbaruah, K.M. (2004). Organic farming: opportunities and challenges in North-Eastern region of India. In: *International Conference on Organic Food,* held at ICAR Research Complex for NEH Region, Umiam, Meghalaya. pp. 14-17.

Fresco, L.O. and Westphal, E. (1988). A hierarchical classification of farm systems. *Experimental Agriculture.* **24**(4): 399-419.

Gill, M.S., Singh, J.P. and Gangwar, K.S. (2009). Integrated farming system and agriculture sustainability. *Indian Journal of Agronomy.* **54**(2): 128-139.

GoI. (2014). *Agricultural statistics at a glance*, Directorate of Economics and Statistics, Govt. of India, New Delhi.

Gopinath, K.A., Saha, S., Mina, B.L., Pande, H., Kundu, S. and Gupta, H.S. (2008). Influence of organic amendments on growth, yield and quality of wheat and on soil properties during transition to organic production. *Nutrient Cycling in Agroecosystems.* **82**(1): 51-60.

Gupta, V., Rai, P.K. and Risam, K.S. (2012). Integrated crop-livestock farming systems: A strategy for resource conservation and environmental sustainability. *Indian Research Journal of Extension Education.* **2**(Special Issue**):** 49-54.

Lal, R. and Miller, F.P. (1990). Sustainable farming for tropics. *In*: Singh, R.P. (Ed.), Sustainable agriculture: Issues and Perspective. Indian Society of Agronomy, IARI, New Delhi. pp. 69-89.

Mahapatra, I.C. and Bapat, S.C. (1992). Farming systems – Research Challenges and Opportunities. *Eastern Indian Farming System Research and Extension, Newsletter.* **6**(4): 3-10.

Manjunatha, S.B., Shivmurthy, D., Sunil, A.S., Nagaraj, M.V., and Basavesha, K.N. (2014). Integrated farming system – an holistic approach: A review. *Research and Reviews: Journal of Agriculture and Allied Sciences.* **3**(4): 30-38.

Mazzarol, T.W. (2014). 6 ways governments can encourage entrepreneurship: World Economic Forum Global Agenda, World Economic Forum.

Okigbo, B.N. (1995). Major farming systems of the lowland savanna of sub-Saharan Africa and the potential for improvement. *In*: Proceedings of the IITA/FAO workshop, Ibadan, Nigeria.

Panke, S.K., Kadam, R.P. and Nakhate, C.S. (2010). Integrated Farming System for sustainable rural livelihood security. *In*: 22nd National Seminar on "Role of Extension in Integrated Farming Systems for Sustainable Rural Livelihood, Maharashtra. pp. 33-35.

Patra, A.K. (2016). Concept, scope and components of integrated farming system. *In*: Root and Tuber Crops Based Integrated Farming System: A Way Forward to Address Climate Change and Livelihood Improvement, pp. 8-13.

Prakash, N., Roy, S.S., Ansari, M.A. and Sharma, S. K. (2015). A Comprehensive Manual on Integrated Farming System: An Approach towards Livelihood Security and Natural Resource Conservation. Publication No. RCM (TM) – 08, ICAR Research Complex for NEH Region, Manipur Centre, Imphal, pp. 368.

Radhamani, S., Balasubramanian, A., Ramamootthy, K. and Geetalakshmi, V. (2003). Sustainable Integrated Farming Systems for Drylands – A review. *Agricultural Reviews*. **24**(3): 204-210.

Rahman, F.H. and Sarkar, S. (2012). Efficient resource utilization through integrated farming system approach in the farmers' field at Burdwan district of West Bengal. *In*: *Extended Summaries*, Third International Agronomy Congress, 26-30th November 2012, New Delhi, pp. 997-998.

Rana, S.S. and Chopra, P. (2013). Integrated Farming System, Department of Agronomy, College of Agriculture, CSK Himachal Pradesh Krishi Vishwavidyalaya, Palampur, p. 90

Roy, S.S., Ansari, M.A., Sharma, S.K. and Prakash, N. (2015). An Overview of Integrated Farming System. *In*: Compendium of Short Course on Integrated Farming System: An Approach towards Livelihood Security and Natural Resource Conservation. Publication No. RCM (TM) - 08, ICAR-Research Complex for NEH Region, Umiam, Meghalaya, pp. 1-18.

Roy, S.S., Sharma, S.K., Ansari, M.A., Banerjee, A., Deshmukh, N.A., Prakash, N. and Ngachan, S.V. (2014). Integrated Farming System for Sustainable Agriculture. *In*: Gupta, A. and Saxena, V.L. (*Eds*.), Integrated Farm Management, Aaviskar Publishers, Distributors, Jaipur, India, pp. 1-24,.

Siddeswaran, K., Sangetha, S.P. and Shanmugam, P.M. (2012). Integrated farming system for the small irrigated upland farmers of Tamil Nadu. *In*: Extended Summaries: Third International Agronomy Congress, 26-30th November 2012, New Delhi, pp. 992-993.

Singh, G. (2004). Farming systems options in sustainable management of national resources. *In*: Proceedings National Symposium on Alternative Farming Systems held at PDCSR, Modipuram, pp. 16-18.

Singh, J.P. (2010). Scenario of farming systems in different agroclimatic zones. *In*: Gangwar, B., Varughese, K., John, J., Rani, B. Vijayan, M. and Mathew, T. (*Eds*.) Manual on Integrated Farming Systems, KAU Cropping Systems, Thiruvananthapuram and Project Directorate of Farming Systems Research, Modipuram, Meerut, pp. 20-41.

Chapter 4

Typical Sustainable Integrated Farming System Models for Socio-Economic Upliftment of Small and Marginal Farmers

Rajeev Kumar Srivastava[1]*, Sudhanand Prasad Lal[2], Samrat Sikdar[3] and Akramul Hoque[4]

[1]*Assistant Professor (Agronomy), Directorate of Seed and Farms, Tirhut College of Agriculture Campus, Dholi (Dr. Rajendra Prasad Central Agricultural University, Pusa, Samastipur), Muzaffarpur – 843121, Bihar*

[2]*Assistant Professor cum Scientist, PG Department of Extension Education (PGCA), Dr. Rajendra Prasad Central Agricultural University Pusa, Samastipur – 848125, Bihar*

[3]*Assistant Professor at School of Agriculture, Sanskriti University, Mathura – 281401, Uttar Pradesh*

[4]*Research Scholar, Uttar Banga Krishi Viswavidyalaya, Pundibari, Coochbehar – 736165, West Bengal*

Corresponding author: raj.agronomy73@gmail.com

ABSTRACT

Green Revolution had a positive and successful impact on Indian Agriculture, thereby India had become significantly rich in the level of food grain production and Indian agriculture had seen phenomenal changes in respect to achieving "sufficiency" from "starvation" status for our population. But unfortunately, the poverty of the farmers has rather transformed into a miserable paradox of the contemporary farming scenario, even though the Government has thoroughly implemented hundreds of projects and planning. The "Green Revolution Model" of applying synthetics, rather blindly, into the field has become quite obsolete now and the farming world expects new models of farming addressing the existing controversies. It has become imperative nowadays to introduce the dimension of sustainability in farming and sustainable farming has rather transformed into a common narrative in today's farming. Integrated Farming System (IFS) has been recommended for the growth and development of small and marginal farmers all over the India in respect to doubling of the farmers' income. Basically, IFS is a multidisciplinary, whole-farm approach that has proven to be quite effective in helping the small and marginal farmers and also solve their problems in terms of profit and environmental point of view. This approach aims to increase profits and employment from small-holdings by combining diverse farm enterprises and recycling crop residues and by-products inside the farm. There are various sustainable farming system models like Crop-Livestock based farming system, Crop- Livestock-Biogas plant farming system, Crop-Livestock-Forestry based farming system, Crop-poultry based farming system, Crop-Fish-Poultry based farming system, Crop-Livestock-Poultry-Fish based farming system etc. It has been found through economic analysis that the agronomic crop + fishery + duckery/poultry + fruit crops + bee keeping + mushroom cultivation IFS model shows the highest level of profitability for 1 acre of land and the estimated annual net return is ₹ 321737.00 approximately. Besides, agronomic crop + horticultural crops + dairy + poultry + fodder crop + bee keeping + mushroom cultivation + biogas plant IFS model is highly profitable in 1 ha land area where the approximate profit level per annum is ₹ 424002.00 and in 2 ha land the approximate profit level per annum is ₹ 819446.00. These economic assessments clearly indicate that the IFS model is a profitable as well as a sustainable way of farming and holds the prospects of improving farmers' socio-economic conditions.

Keywords: *Farming system, IFS model, Profitability, Sustainability.*

Introduction

Agriculture sector is a backbone of Indian economic as it contributes 19.9 per cent share in total national Gross Domestic Product (GDP) during 2020-21 (Shagun, 2021) and on the other hand, in this aforesaid sector, the engagement of manpower of total national employment was in the tune 45.6 per cent during 2019-20 (Mahapatra, 2021). It means the majority of population of India is directly and indirectly depend on agriculture in respect to livelihood security for a majority of the population by employment generation, meeting 4Fs (Food, Fodder, Fibre and Fuel) requirement to the burgeoning population as well as supplying the raw materials to agro-industries as a major segment of our entire nation.

Beyond doubt, the Green Revolution had such a positive and successful impact on Indian Agriculture, thereby our country had become self-sufficient in food grain production and our agriculture had seen phenomenal changes in respect to achieving "sufficiency" from "starvation" status for our population. Obviously, green revolution was responsible for a massive increase in overall crop productivity of the farmland particularly rice and wheat farming. Of course, during seventies and eighties, due to green revolution, agriculture growth rate showed rapid increasing trend but whatever model and approaches we had adopted in our production system during the green revolution was rather more exploitative than supportive, thereby natural and artificial resources put under excessive pressure. Consequently, it has been observed that some issues have been raised like water table declining at an alarming rate, declining in factor productivity, degradation of crop ecosystem including entire environment as well as life supporting system altogether. These issues might have been raised due to non-holistic model we have adopted in green revolution and ignored the social issues too.

Agriculture in India and even other countries faces numerous challenging issues, which are expected to get more difficult over time. By 2020, India's population is expected to reach 1.3 billion people, who would share resources with a huge cattle population. Based on current consumption trends, total food grain demand is expected to be around 300 and 350 million tonnes by 2020 and 2030, respectively, compared to current production of around 260 million tonnes (Behera and France, 2016).

Looking into the pulse scenario of India agriculture in 2019-20, area, production and productivity were 29.03 M ha, 3.04 MT and 817 kg ha^{-1}, respectively. Though World Health Organization (WHO) has already recommended consumption of pulses 80 g^{-1} capita^{-1} day^{-1} or 29.2 kg^{-1} capita^{-1} year^{-1}, but net availability of pulses on per capita basis (kg annum^{-1}) in India has showed deficit by 32.1 g^{-1} capita^{-1} day^{-1} or 11.7 kg^{-1} capita^{-1} year on the basis of pulse availability that is about 17.5 kg^{-1} capita^{-1} year in year 2019. Moreover, it has been projected that requirement of pulses by 2050 in India will be 39 million tonnes, which can only be possible by increasing pulse production consistently 0.6 MT annum^{-1} in order to meet pulse requirement of our increased population, which will touch the level of 1.69 billion people during the year 2050 (Agri. Statistics at a Glance, 2020 and Vision 2050, IIPR, Kanpur). Keeping these facts and figures in view, intensification and diversification of cropping system with pulses in crop component of Integrated Farming System (IFS) may be the better option, which not only meet the projected pulse demand in year 2050 but pulses are also seemed to be instrumental to fill the gap between recommended requirement of pulses by WHO and its per capita per annum availability (Singh *et al.*, 2009).

Integrated Farming System (IFS) models also effectively and efficiently address the farming system issues like decline in soil organic matter, multi-nutrient deficiency, reduced biodiversity, decline in factor productivity, pests build up, receding water table at alarming rate, environmental pollution and last but not least that is climate change also, apart from low profitability and least sustainability issues. But IFS models must be formulated and validated for varying agro-climatic conditions and resource availability at the site considering the profitability, sustainability and social parameters. Considering the decline in per capita availability of land, tested IFS models may be boon for the marginal and small farmers as 85 per cent farmers in our country are categorized in the aforesaid group. Hence, IFS has been recommended for the growth and development of small and marginal farmers all over the India in respect to doubling of the farmers' income. Basically, IFS is a multidisciplinary, whole-farm approach that has proven to be quite effective in helping the small and marginal farmers and also solve their problems in terms of profit and environmental point of views. This approach aims to increase profits and employment from small-holdings by combining diverse farm enterprises and recycling crop residues and by-products inside the farm. The farming systems approach is a resource management method aimed at achieving economic and long-term productivity while sustaining the resource base and ensuring high environmental quality (Lal and Miller, 1990).

As far as Bihar state is concerned, the concrete work on Integrated Farming System is least available, but Kumar *et al.* (2012) conducted multi-locational trial at Patna, Vaishali and Munger districts with various IFS options to assess the best IFS model for Eastern India for small and marginal farmers. Results of same experiment showed that the crop + fish + duck + goat combination could be implemented in the eastern part of India than growing the crop alone on that particular land under irrigated conditions. Likewise, at regional and national level, such type of IFS model is also required which should be suitable, remunerative and sustainable for small and marginal farmers with scientifically sound approaches because some farmers of across the country have been adopted various farming system model not for profitability but for risk avoidance. Usually, the farming followed by farmer is very unscientific and the income generated through such farming is hardly sufficient to meet their livelihood. But with preference and suitability, Integrated Farming System models as per site specific need must be made available to the farmers to fulfil the all the objectivity of IFS.

Concept and Definitions

A system is a bunch of components where individual components are controlled by interconnections, allowing the system as a whole to perform specified functions in response to changing requirements. (Fresco and Westphal, 1988). A farming system has been defined in a number of ways by several people and organizations. The concept "farming system" suggests to a specific set of farming operations that are managed in response to the physical, biological, and socio-economic environment, as well as the goals, preferences, and resources of the farmers. (Shaner *et al.*, 1982). When compared to monoculture farming, the term "Integrated Farming System" (IFS) is a widely used term to describe a more integrated approach to farming.

Conclusive Advantages of Integrated Farming System

a) Augmentation of food supply with minimized cost of production in sustainable manner,

b) Increase opportunity for employment generation as well as income for many people and also included family labours by the way of intensification and diversification, thereby efficiently utilize the family labour,

c) Regular cash flow around the year,

d) Reduce the environmental pollution by effective and efficient utilization of waste of one enterprise as the input of other,

e) Pooling and sharing of inputs,

f) Preservation, conservation and utilisation of farm biomass, including nonconventional feed and fodder resources,

g) Improves space utilization and provides diversified products, and

h) Increased economic resources.

Major Constraints in the Prevailing Farming System

In spite of having several advantages, farmers are yet unable to adopt the IFS systems due to several constraints in various regions of our country, which are as follows –

a) The financial constraints are lack of required funds and high input costs,

b) Deficiency of fodder during the off-season is a major constraint in rearing livestock by the IFS farmers (Poorani *et al.*, 2011),

c) Socio-cultural constraints *viz.*, idiosyncratic character and attitude of the farmers,

d) Small and fragmented land holdings,

e) Lack of diversification,

f) Frequent crop failures due to aberrant weather condition,

g) Dependency of agriculture on monsoon rain for raising the crops,

h) Complete absence of suitable agro-processing units,

i) Lack of suitable storage facilities within the socio-economic purview of the farmers,

j) Use of low productive local breeds of livestocks, and

k) Lack of technical know-how among the farmers.

Models of Integrated Farming System

1. **Crop – Livestock Farming System:** Farming system comprises of both the crop and livestock has been significantly sustainable over centuries. Animal power is used for various farming operations. Cow dung is used as manure and fuel. Foods are obtained from the crop. Various Crop residues are used as livestock feed. Through this system, the yield becomes more sustainable because the waste of one enterprise is the input of others. Crop residues can be used for animal feed in an integrated crop livestock farming system, while manure from livestock can boost agricultural output by increasing nutrients that improve soil fertility and decreasing the usage of synthetic fertilizers (Gupta *et al.*, 2012).

2. **Crop – Livestock – Biogas Farming System:** In the conventional system, the crop residues are feed to the cattle and the cow dung and FYM are

used as manure or as fuel after making cow dung cakes. With increasing demand in rural areas most of cow dung is used for fuel purpose. But it is very good source of plant nutrients. This valuable organic source can be effectively utilized both as fuel and manure by introducing one biogas plant in the system. The biogas slurry should be incorporated into the soil that nitrogen losses may be minimized. Biogas slurry contains 1.8: 1.0: 1.0 per cent NPK, whereas FYM contains 0.5: 0.3: 0.5 per cent NPK.

3. **Crop – Livestock – Forestry Farming System:** In this system, like other two systems crop residues can be used as animal feed and dung is used as a source of FYM. The role of forest cover is to decrease runoff, soil erosion and to increase water recharge along with the preservation of environment. Forest plantation can be done as sole or on the boundaries of the farm.
4. **Crop – Poultry Farming System:** Over rice crop with recommended fertilizer doses, the intervention of green manure husk burial (profit – 24.1 per cent and net income – ₹ 32600 ha^{-1}) and vermicompost with banana pseudo stem (profit – 55.3 per cent and net income – ₹ 75000 ha^{-1}) contributed a large percentage of nutrients (NPK) in the farm. Duck droppings also increased profit percentage by 20.5 per cent, resulting in a net income of ₹ 27800 ha^{-1}. The physical qualities of soil, such as bulk density, aggregates, porosity, and infiltration rate, have also been improved by the aforesaid supplements.
5. **Crop – Fish – Poultry Farming System:** In this system fishes grown in ponded, rice field provides utilizable source of nitrogen to the crop by excreting amino-nitrogen as ammonia. So, yield of rice increased if it is cultured along with fish. To increase the productivity system, it is necessary to select appropriate varieties of rice and appropriate species of fish which are compatible and fit into the system. Fish prevents and control the certain pest, diseases and weeds. The small fish grown in the rice field is very good poultry feed. From integrated farming system experiments in Sirupura, Channabasavanna *et al.* (2002) discovered that rice-fish-poultry combinations produced the highest net profit (>₹ 157000 ha^{-1}) while also improving soil health.
6. **Crop – Livestock – Poultry – Fish Farming System:** This system will be labour intensive, economically sound and environmentally safe. The dependence of inorganic fertilizer greatly reduces. Fish gets feed from manure, through formation of protein rich zooplankton, phytoplankton *etc.* If livestock manure is added to the pond, the ratio of N: P is important because N is more likely to a limiting than P for bacterial growth and phytoplankton.

Practical Utility of Integrated Farming System

In this system, an inter-related set of enterprises is used so that the "waste" from one component becomes an input for another part of the system. This reduces costs and improves production and/or income. Since it utilizes waste as a resource, farmers not only reduce waste emissions but they also ensure an overall increase in productivity for the whole farming system. Integrated farming tries to imitate nature's principle, where not only crops but also varied types of plants, animals, birds, fish, and other aquatic flora and fauna are utilized for production. The basic principle is to enhance the ecological biodiversity:

- ☆ By choosing the appropriate cropping methodology with mixed cropping, crop rotation, crop combination and inter-cropping so that there is less competition for natural resources like water, nutrition and space and by adopting eco-friendly practices,
- ☆ By utilizing a multi-story arrangement so that the total available area is used effectively and there is a high level of interaction between biotic and abiotic components, and
- ☆ By integrating subsystems by which the various components interact positively, so that the overall farm productivity is increased.

The integrated farming system is also a sustainable agriculture system which focuses on intensification of farm productivity by increasing diversification, resource integration and creating market linkages. Thousands of small and marginal family farmers in resource-poor regions have converted their farming to this sustainable farming system to diversify farm production, increase cash income, and improve the quality and quantity of food produced and the exploitation of unutilized resources. It usually takes three to four years to establish a well-integrated farm with market linkages to ensure nutrition and the livelihood of a family.

A. Proposed IFS Models for 1 (one) Acre Area under Irrigated Conditions

1. IFS Model I: Crop + Fishery and/or Poultry

If the farmer grows the agronomic crop along with the rearing of poultry birds and fish culture in 1 acre of land, the model-based economic analysis suggests the approximate annual net return ₹ 212264.00 as indicated in the Table 4.1. The agronomic crops will include soybean, wheat, mustard, moong *etc.* This model recommends the incorporation of mushroom cultivation as an allied enterprise here in order to further enhance the profitability.

2. IFS Model II: Crop + Livestock

If the farmer grows the agronomic crop along with the rearing of livestock animals in 1 acre of land, the model-based economic analysis suggests the approximate annual net return ₹ 183359.00 as indicated in the Table 4.2. The agronomic crops will include rice, potato, mustard, moong, *Sesbania etc.* This model

Table 4.1: Economics of Crop + Fishery and/or Poultry IFS Model

IFS Model	*Enterprises Area*	*Allied Enterprises*	*Crop Component*	*Cost of Production (₹)*	*Gross Return (₹)*	*Net Return (₹ year^{-1})*
Crop + Fishery and/or Poultry	Crop – 2650 m^2	-	Soybean - Wheat + Mustard (8:1) - Moong	21399	35263	13864
	Pond – 1000 m^2	-	-	50000	150000	100000
	Birds – 50 m^2	-	-	32500 month^{-1}	37500 month^{-1}	60000
	Layout area – 300 m^2	-	-			
		Mushroom	-	4000 month^{-1}	7200 month^{-1}	38400
Net Profit by 1 Acre Land						**₹ 212264**

Table 4.2" Economics of Crop + Livestock IFS Model

IFS Model	*Enterprises Area*	*Allied Enterprises*	*Crop Component*	*Cost of Production (₹)*	*Gross Return (₹)*	*Net Return (₹ year^{-1})*
Crop + Livestock	Crop – 2700 m^2	-	Rice – Potato + Mustard – Moong – Sesbania	25292	41310	16018
	Livestock + Fodder – 900 m^2	-	-	146359	262800	116441
	Farm Layout – 400 m^2	-	-	-	-	-
	-	Vermicompost unit	-	-	46900	46900
	-	Bio-gas plant	-	5000	9000	4000
Net Profit by 1 Acre Land						**₹ 183359**

recommends the addition of vermicompost pit for the purpose of vermicompost preparation and also a biogas plant there, which would eventually enhance the net return of the stakeholder farmers.

3. IFS Model III: Crop + Fish + Duckery/Poultry

If the farmer grows the agronomic and fruit crops along with the rearing of poultry birds and fish culture in 1 acre of land, the model-based economic analysis suggests the approximate annual net return ₹ 321737.00 as indicated in the Table 4.3. The agronomic crops will include soybean, wheat, mustard, moong *etc.* and mango would be the most suitable fruit crop. This model recommends the incorporation of apiculture and mushroom cultivation as the allied enterprises here for enhancing the profitability of this model.

4. IFS Model IV: Crop + Horticulture + Goatry

If the farmer grows the agronomic and fruit crops along with the rearing of goats in 1 acre of land, the model-based economic analysis suggests the approximate annual net return of ₹ 202566.00 as indicated in the Table 4.4. The agronomic crops will include maize, toria, wheat, moong *etc.* and the fruit crops will include mango, litchi, guava *etc.* This model recommends the incorporation of apiculture as an allied enterprise here for enhancing the profitability of this model alongside the establishment of NADEP compost. Though goatry is very remunerative in this IFS model but some category of our society has social taboos. Such type of aforesaid mind set must be changed or willing to bring change.

B. Proposed IFS Models for 1 (one) Hectare Area under Irrigated Conditions

1. IFS Model I: Crop + Fish + Dairy + Poultry + Duckery

If the farmer grows the agronomic crops along with the rearing of poultry birds as well as fishes and also the dairy animals in 1 hectare of land, the model-based economic analysis suggests the approximate annual net return ₹ 409278.00 as indicated in the Table 4.5. The agronomic crops will include rice, potato, mustard, moong, *Sesbania etc.* The forage crops like maize, berseem, sorghum *etc.* would also be a part of this model. This model recommends the incorporation of mushroom cultivation as an allied enterprise here for enhancing the profitability of this model alongside the establishment of NADEP compost pit and biogas plant.

2. IFS Model II: Crop + Dairy + Poultry

If the farmer grows the agronomic crops along with the rearing of poultry birds and also the dairy animals in 1 hectare of land, the model-based economic analysis suggests the approximate annual net return ₹ 335872.00 as indicated in the Table 4.6. The agronomic crops will include soybean, wheat, mustard and moong *etc.* The forage crops like maize, berseem and sorghum would also be a part of this model. This model recommends the incorporation of mushroom cultivation as an allied

Table 4.3: Economics of Crop + Fishery + Duckery/Poultry IFS Model

IFS Model	Enterprises Area	Allied Enterprises	Crop Component	Cost of Production (₹)	Gross Return (₹)	Net Return (₹ $year^{-1}$)
Crop + Fish + Duckery/Poultry	Crop – 1500 m^2	-	Soybean - Wheat + Mustard (8:1) - Moong	12113	19960	7847
	Fish – 1950 m^2	-	-	97500	292500	195000
	Duckery/Poultry – 50 m^2	-	-	32500 $month^{-1}$	37500 $month^{-1}$	60000
	Fruit trees + Layout – 500 m^2	-	Mango	12350	25640	13290
	-	Bee keeping	-	4500	7500	3000
	-	Mushroom cultivation	-	53400	96000	42600
Net Profit by 1 Acre Land						**₹ 321737**

Table 4.4: Economics of Crop + Horticulture + Goatry IFS Model

IFS Model	Enterprises Area	Allied Enterprises	Crop Component	Cost of Production (₹)	Gross Return (₹)	Net Return (₹ $year^{-1}$)
Crop + Horticulture + Goatry	Crop – 2000 m^2	-	Maize – Toria – Wheat – Moong	20104	40600	20496
	Horticultural crop – 1500m^2	-	Guava, Mango,Litchi	37050	76920	39870
	Goat – 100 m^2	-		79800	200000	120200
	Layout – 400 m^2	-				
	-	Bee keeping		5000	9000	4000
	-	NADEP Compost		12000	30000	18000
Net Profit by 1 Acre Land						**₹ 202566**

Table 4.5: Economics of Crop + Fish + Dairy + Poultry + Duckery IFS Model

IFS Model	*Enterprises Area*	*Allied Enterprises*	*Crop Component*	*Cost of Production (₹)*	*Gross Return (₹)*	*Net Return (₹ $year^{-1}$)*
Crop + Fish + Dairy + Poultry + Duckery	Crop – 6500 m^2	-	Rice – Potato + Mustard – Moong – *Sesbania*	60888	109450	48562
	Fish - 1000 m^2	-		50000	150000	100000
	Poultry + Duckery – 50 m^2	-	500 birds	32500 $month^{-1}$	37500 $month^{-1}$	60000
	Dairy – 500 m^2	-		81311	146000	64689
	Fodder crop – 1000 m^2	-	Maize – Berseem – Sorghum	55640	108567	52927
	Farm Layout – 950 m^2	-				
	-	Bio-gas plant		8000	19000	11000
	-	NADEP Compost		12000	41500	29500
	-	Mushroom cultivation		53400	96000	42600
Net Profit by 1 Hectare Land						**₹ 409278**

Table 4.6: Economics of Crop + Dairy + Poultry IFS Model

IFS Model	*Enterprises Area*	*Allied Enterprises*	*Crop Component*	*Cost of Production (₹)*	*Gross Return (₹)*	*Net Return (₹ year^{-1})*
Crop + Dairy + Poultry	Crop – 6500 m^2	-	Soybean - Wheat + Mustard (8:1) - Moong	52490	96493	44003
	Dairy – 1000 m^2	-		162622	292000	129378
	Fodder crop – 1500m^2	-	Maize – Berseem – Sorghum	83460	162851	79391
	Farm Layout – 1000 m^2	-				
	-	Bio-gas plant		8000	19000	11000
	-	NADEP Compost		12000	41500	29500
	-	Mushroom cultivation		53400	96000	42600
Net Profit by 1 Hectare Land						**₹ 335872**

enterprise for enhancing the profitability of this model alongside the establishment of NADEP compost pit and biogas plant.

3. IFS Model III: Crop + Horticulture + Dairy + Poultry

If the farmer grows the agronomic and horticultural crops along with the rearing of poultry birds as well as the dairy animals in 1 hectare of land, the model-based economic analysis suggests the approximate net annual return of ₹ 420002.00 as indicated in the Table 4.7. The agronomic crops will include maize, toria, wheat, moong *etc.* and the horticultural crops would include the fruits like mango *etc.* The forage crops like maize, berseem, sorghum *etc.* would also be a part of this model. This model recommends the incorporation of mushroom cultivation and bee-keeping as an allied enterprise here for enhancing the profitability of this model alongside the establishment of vermicompost pit and biogas plant.

C. Proposed IFS Models for 2 (two) Hectare Area under Irrigated Conditions

1. IFS Model I: Crop + Fish + Duckery + Poultry + Dairy

If the farmer grows the agronomic crops along with the rearing of poultry birds, fishes, and also the dairy animals in 2 hectares of land, the model-based economic analysis suggests the approximate annual net return ₹ 819446.00 as indicated in the Table 4.8. The agronomic crops will include maize, toria, wheat, moong *etc.* The forage crops like maize, berseem, sorghum *etc.* would also be a part of this model. This model recommends the incorporation of mushroom cultivation as an allied enterprise here enhancing the profitability of this model alongside the establishment of NADEP compost and biogas plant.

2. IFS Model II: Crop + Dairy + Poultry + Fishery + Horticulture

If the farmer grows the agronomic and horticultural crops along with the rearing of poultry birds as well as the dairy animals with fishes in 2 hectare of land, the model-based economic analysis suggests the approximate net annual return of ₹ 767262.00 as indicated in the Table 4.9. The agronomic crops will include rice, potato, mustard, moong and *Sesbania etc.* and the horticultural crops would include the fruits like mango *etc.* The forage crops like maize, berseem and sorghum would also be a part of this model. This model recommends the incorporation of mushroom cultivation and bee-keeping as an allied enterprises here enhancing the profitability of this model alongside the establishment of NADEP compost and biogas plant.

3. IFS Model III: Crop + Horticulture + Dairy + Poultry

If the farmer grows the agronomic and horticultural crops along with the rearing of poultry birds as well as the dairy animals in 2 hectare of land, the model-based economic analysis suggests the approximate net annual return of ₹ 793024.00 as indicated in the Table 4.10. The agronomic crops will include soybean, wheat, mustard and moong *etc.* and the horticultural crops would include the

Table 4.7: Economics of Crop + Horticulture + Dairy + Poultry IFS Model

IFS Model	*Enterprises Area*	*Allied Enterprises*	*Crop Component*	*Cost of Production (₹)*	*Gross Return (₹)*	*Net Return (₹ year^{-1})*
Crop + Horticulture + Dairy + Poultry	Crop – 5000 m^2	-	Maize – Toria – Wheat – Moong	50260	101500	51240
	Horticulture – 2500 m^2	-	Mango	50000	140000	90000
	Dairy – 600 m^2	-		97573	175200	77627
	Poultry – 50 m^2		500 birds	32500 month^{-1}	37500 month^{-1}	60000
	Fodder crop – 900m^2	-	Maize – Berseem – Sorghum	50076	97711	47635
	Farm Layout – 950 m^2	-				
	-	Bio-gas plant		8000	19000	11000
	-	Vermicompost		7000	46900	39900
	-	Mushroom cultivation		53400	96000	42600
	-	Bee keeping		5000	9000	4000
Net Profit by 1 Hectare Land						**₹ 424002**

Table 4.8: Economics of Crop + Fish + Duckery + Poultry + Dairy IFS Model

IFS Model	*Enterprises Area*	*Allied Enterprises*	*Crop Component*	*Cost of Production (₹)*	*Gross Return (₹)*	*Net Return (₹ year^{-1})*
Crop + Fish + Duckery + Poultry + Dairy	Crop – 13000 m^2	-	Maize – Toria – Wheat – Moong	130676	273900	143224
	Fish–2000 m^2	-		90000	310000	220000
	Poultry + Duckery –100 m^2		1000 birds	65000 month^{-1}	76000 month^{-1}	132000
	Dairy – 750m^2	-		121966	219000	97034
	Fodder crop – 2250m^2	-	Maize – Berseem – Sorghum	125190	254278	129088
	Farm Layout – 1900 m^2	-	-	-	-	-
	-	Bio-gas plant	-	10000	29000	19000
	-	NADEP Compost	-	15000	51500	36500
	-	Mushroom cultivation	-	53400	96000	42600
Net Profit by 2 Hectare Land						**₹ 819446**

Table 4.9: Economics of Crop + Dairy + Poultry + Fishery + Horticulture IFS Model

IFS Model	*Enterprises Area*	*Allied Enterprises*	*Crop Component*	*Cost of Production (₹)*	*Gross Return (₹)*	*Net Return (₹ year^{-1})*
Crop + Dairy + Poultry + Fishery + Horticulture	Crop – 12000 m^2	-	Rice – Potato + Mustard – Moong – Sesbania	112643	202483	89840
	Dairy – 750 m^2	-		121966	219000	97034
	Fodder crop – 2500m^2	-	Maize – Berseem – Sorghum	138961	282249	143288
	Horticulture crop – 1000m^2	-	Mango	20000	56000	36000
	Fishery - 1750m^2	-		78750	271250	192500
	Poultry – 100 m^2		1000 birds	65000 month^{-1}	76000 month^{-1}	132000
	Farm Layout – 1900 m^2	-	-	-	-	-
	-	Bee keeping	-	10000	25000	15000
	-	Bio-gas plant	-	10000	29000	19000
	-	NADEP Compost	-	15000	51500	36500
	-	Mushroom cultivation	-	53400	96000	42600
Net Profit by 2 Hectare Land						**₹ 803762**

Table 4.10: Economics of Crop + Horticulture + Dairy + Poultry IFS Model

IFS Model	*Enterprises Area*	*Allied Enterprises*	*Crop Component*	*Cost of Production (₹)*	*Gross Return (₹)*	*Net Return (₹ year^{-1})*
Crop + Horticulture + Dairy + Poultry	Crop – 11000 m^2	-	Soybean - Wheat + Mustard (8:1) - Moong	88708	168863	80155
	Horticulture – 4000 m^2	-	Mango, Litchi	80000	234000	154000
	Dairy – 1000 m^2	-	-	162621	315360	152739
	Poultry – 100 m^2	-	1000 birds	65000 month^{-1}	76000 month^{-1}	132000
	Fodder crop – 2000 m^2	-	Maize – Berseem – Sorghum	111169	225799	114630
	Farm Layout – 1900 m^2	-	-	-	-	-
	-	Bee keeping	-	10000	25000	15000
	-	Mushroom cultivation	-	53400	96000	42600
	-	Bio-gas plant	-	10000	29000	19000
	-	Vermicompost	-	14000	96900	82900
Net Profit by 2 Hectare Land						**₹ 793024**

fruits like mango and litchi *etc.* The forage crops like maize, berseem, sorghum *etc.* would also be a part of this model. This model recommends the incorporation of mushroom cultivation and bee-keeping as an allied enterprise here for enhancing the profitability of this model alongside the establishment of vermicompost unit and biogas plant.

Conclusion

Farming is a dynamic business which demands for esoteric transformations and incorporations of multiple entities throughout the time scale. The enhancement in crop productivity was the principle aim of our nation primarily immediately after the independence, but the priority has been shifted to the concern of sustainability in the contemporary period. Though the "Green Revolution" as introduced in sixties was successful to a greater extent, but it did hamper the existing ecological balances in the farming landscape. Notably, there is a paradigm shift from the sole higher productivity to the enhancement of the productivity and profitability in a sustainable manner. The Food and Agricultural Organization of United Nations is thoroughly emphasizing on the implementation of the bio-diverse farms which would be agro-ecologically sustainable and also profitable for the small and marginal Indian farmers. So, the incorporation of the diverse crops, livestock, fishery, poultry birds *etc.* in the farm along with the allied enterprises like bee-keeping, mushroom cultivation *etc.* and also the production of vermicompost, biogas, compost *etc.* duly construct the domain of IFS model holistically. The IFS model is a well proved diversified farming system model as globally accepted which ensures the overall enhancement of the farmers' socio-economic standard through multiple avenues of financial inflow. The inclusion of the segments highly depends on the individual status of the farmers, mainly the land holding. As most of the Indian farmers are small or marginal in nature, they do hold small land size and accordingly, the farmers need to finalize the specific IFS model for further implementation judging their specific individual situations. Proper application of the IFS model will be a trustworthy way to improve some of the tangible constrains of the contemporary farming landscape.

REFERENCES

Agricultural Statistics at a Glance. (2020). Government of India, Ministry of Agriculture and Farmers Welfare, Department of Agriculture, Cooperation and Farmers Welfare, Directorate of Economics and Statistics.

Barbosa, F.A., Soares Filho, B.S., Merry, F.D., Azevedo, H.D.O., Costa, W.L.S., Coe, M.T., Batista, E.D.S., Maciel, T.G., Sheepers, L.C., OLIVEIRA, A.D. and Rodrigues, H.O. (2015). Cenários para a pecuária de corte amazônica. Universidade Federal de Minas Gerais: Belo Horizonte, Brazil, 154 p.

Behera, U.K. and France, J. (2016). Integrated farming systems and the livelihood security of small and marginal farmers in India and other developing countries. *Advances in Agronomy*. **138:** 235-282.

Channabasavanna, A.S., Itnal, C.J. and Patil, S.G. (2002). Productivity, economics analysis and changes in physico-chemical properties as influenced by integrated rice-based farming system. *Indian Journal of Agronomy*. **46**(1): 1-5.

Das, A., Choudhury, B.U., Ramkrushna, G.I., Tripathi, A.K., Singh, R.K., Ngachan, S.V., Patel, D.P., Layek, J. and Munda, G.C. (2013). Multiple use of pond water for enhancing water productivity and livelihood of small and marginal farmers. *Indian Journal of Hill Farming*. **26**(1): 29-36.

Devendra, C. and Thomas, D. (2002). Smallholder farming systems in Asia. *Agricultural systems*. **71**(1-2): 17-25.

Fresco, L.O. and Westphal, E. (1988). A hierarchical classification of farm systems. *Experimental Agriculture*. **24**(4): 399-419.

Gupta, V., Rai, P.K. and Risam, K.S. (2012). Integrated crop-livestock farming systems: A strategy for resource conservation and environmental sustainability. *Indian Research Journal of Extension Education*. **2**(Special Issue): 49-54.

GOI. (2014). District Census Handbook, Office of Registrar General and Census Commissioner of India, Ministry of Home Affair, Government of India, New Delhi.

Jayanti, C., Sakthivel, N., Marimuthu, S., Vivek, G. and Kandaswamy, O. S. (2004). Efficient integrated farming system for increasing the farm productivity of resource poor farmers. *In*: Singh, A.K., Gangwar, B., Sharma, G.C. and Pandey, P.S. (Eds), *Farming System Research and Development Association*, Modipuram.

Kumar, S., Singh, S.S., Meena, M.K., Shivani, and Dey, A. (2012). Resource recycling and their management under integrated farming system for lowlands of Bihar lowlands of Bihar. *Indian Journal of Agricultural Sciences*. **82**(6): 504-510.

Kumar, S., Subash, N., Shivani, Singh, S.S. and Dey, A. (2012). Evaluation of different components under integrated farming system for small and marginal farmers under semi-humid climatic environment. *Experimental Agriculture*. **48**(3): 399-413.

Lal, R. and Miller, F.P. (1990). Sustainable farming for tropics. *Sustainable Agriculture: Issues and Prospective*. **1:** 69-89.

Mahapatra, R. (2021). Agriculture: The lure of 10 million jobs a year. Down-to-earth, published online on Tuesday, 24 August, 2021.

Mahapatra, I.C. (1994). Farming system research – a key to sustainable agriculture. *Fertilizer News*. **39:** 13-25.

Mathew, T. and Varughese, K. (2007). Nutrient flow and economics of integrated farming system practiced in the coastal region of Kerala. *In*: National Symposium on Integrated Farming Systems and Its Role towards Livelihood Improvement under Indian Context, held during 26-28 October 2007 at RARS, Durgapura (Jaipur) RAU, Farming System Research and Development Association, Modipuram, pp. 26-28.

Maughan, M.W., Flores, J.P.C., Anghinoni, I., Bollero, G., Fernández, F.G. and Tracy, B.F. (2009). Soil quality and corn yield under crop-livestock integration in Illinois. *Agronomy Journal*. **101**(6): 1503-1510.

Nagram, Anake, Topark (1989). Food forage crop production systems in South East Asia. Paper Presented at the 20th Working Group Meeting, Asian Rice Farming Systems, Network, Oct 2-7, 1989, Bogor, Indonesia.

Poorani, A., Jayanthi, C. and Vennila, C. (2011). Farmer participatory research on Integrated Farming Systems. *In*: National seminar on Innovations in Farming Systems Research and Extension for Inclusive Development" 24-25th November 2011, Madras Veterinary College, Chennai, pp. 153.

Pushpa, J. (2010). Constraints in various integrated farming systems. *Agriculture Update*. **5**(3-4): 370-374.

Singh, K.K., Ali, M. and Venkatesh, M.S. (2009). Pulses in Cropping Systems. Technical Bulletin, IIPR, Kanpur, 47 p.https://iipr.icar.gov.in/pdf/pulses_in_cropping_systems.pdf

Shagun. (2021). Agriculture: Agri share in GDP hit 20 per cent after 17 years: Economic Survey. Down-to-earth, published online on Friday, 29 January, 2021.

Shaner, W.W., Philipp, P.F. and Schmehl, W.R. (1982). Farming Systems Research and Development: Guide for Developing Countries, West View Press, Colorado, USA.

Sharma, M. (2021). Agriculture: The future of Indian agriculture. Down-to-earth, published online on Thursday, 04 February, 2021.

The Economic Times. (2021). Agriculture: Food grain output to touch new record of 308.65 million tons in 2020-21 says government, published online by PTI last updated 11 August, 2021.

Chapter 5

Agri-Horticultural Systems for Doubling Farmers' Income and Climate Resilience

Narayan Lal[1], Asha Sahu[1], A.K. Vishwakarma[1], B.L. Lakaria[1], Nisha Sahu[1], D.K. Jayswal[2], A.B. Singh[1], A.K. Biswas[1] and A.K. Patra[1]*

[1]ICAR-Indian Institute of Soil Science, Bhopal – 462038, Madhya Pradesh
[2]ICAR-National Agricultural Higher Education Project, KAB-II, Pusa, New Delhi – 110012
Corresponding author: narayanlal.lal7@gmail.com

ABSTRACT

Environmental degradation, climate change and high input costs have invited new challenges in the agriculture and horticulture sector. Introducing of horticultural crops in agricultural land (agri-horti system) not only increases the income of farmers but also help in sequestration of carbon to mitigate the climate changes. The suicide of farmers and fluctuating in income has paid attention in doubling the income which is the major concern for agricultural policy maker now-a-days. Income can be enhanced by bridging the productivity gaps, intervention of latest production technology, improved land use planning, diversified crops systems towards high value crops, wasteland management, low input agriculture, best uses of available resources, climate resilient agriculture and adaptation of multi cropping systems. Agri-horticultural systems have excellent use of available natural resources such as soil, water and light that can be helpful in reduction of application of pesticides and fertilizers and thus, increasing safe food production, biodiversity and adaptation to climatic change. The levels of carbon dioxide and temperature have increased due to climate changes with changing in rainfall patterns as well as droughts and floods. The crop growth, phenology, fruit *size, quality, yield and duration are highly affected with elevated temperature but elevated CO_2 has a positive effect on the productivity of crops such as mango, citrus, grapes, guava, fig, anona, tomato, capsicum, onion, cucumber and melons. Thus, many horticultural crops can be successful grown even under changing climate.*

Keywords: *Horticultural crops, Productivity, CO_2, Farmers' income, Climate resilience.*

Introduction

Intensive agriculture has contributed to increase in the negative effects of climate change which affect agricultural production. Mono cropping systems offer low-cost food for consumers with significant negative environmental costs or externalities (water pollution, soil degradation and greenhouse gas). There is strong need to reduce agricultural inputs without significant losses in productivity may require a fundamental re-design of cropping systems (Gaba *et al.*, 2015). In modern agriculture, crop productivity has increased dramatically which accompanied by degradation in environment and other social problems (Altieri *et al.*, 2015). Fruit crop and tree based cropping system (agri-horti, agro-forestry *etc.*) can contribute to feeding a growing population in a sustainable way. The Agri-horti program intervention is successful in combating food insecurity, in response to the challenges and opportunities associated with climate change. It has attracted strong attention to the grower and policy makers in recent years because these systems have potential to reduce poverty and land degradation as well as mitigate climate changes. In agri-horti systems, fruit trees can change in micro and macro climate which is strong assets in extreme weather condition to protect the crops. Fruit trees planted at wider spacing according to the plant species which helps to develop stable and strong root systems to resist damage from wing and storms and horticultural practices like regular pruning of lower, diseased and unwanted branches also helps to avoid wind damage. Agri-horti system involves fruit crops as well as cereal, vegetable, flowers *etc.* Fruit crops are perennial in nature which gives regular fruit after a certain years. During pre-bearing stage, several vegetable, flowers and cereal crops can be grown in the space left between rows of the fruit crops. Many suitable intercrops can be grown under bearing orchard which ensures profit to the growers. Thus, integration of fruit crops with other annual crops can double or triple the farmers' income. Fruit trees crops play an important role in reducing the total amount of rainfall reaching to the soil due to interception by canopy and help in moisture conservation by increase infiltration rate of rainwater and check the runoff. Horticulture based cropping system has played a significant role in crop production and protection to the environment through carbon sequestration to mitigation going climate change problem for better well-being of mankind.

A suitable combination of fruit crops and annual needs to be identified, so that competition among the crops may be eliminated. The various components of agri-horti systems are interacting and their interactions are determined by management of the system, selection of crop species and their functional characteristics, planting geometry, stratification and fertilization regime (Niether *et al.,* 2018). At the present day, the mitigation of climate change and food security is two of the main challenges of human society (Feliciano *et al.,* 2018). There is urgent requirement to promote to agri-horti system which ensures income to the farmers and it is more profitable and stable as well as resilient to climate change (Vaast and Somarriba, 2014). Agri-horti systems ensure food security, climate change mitigation and adaptations of the crops species in extreme environmental (Mbow *et al.,* 2014). Agri-horti system must be the major climate-smart agricultural system as it ensures sustainable food production with adaptation and climate change mitigation.

Agri-Horticultural System

The horticulture sector is diversified which cover different 6 subsectors like fruits, vegetables, ornamentals, spices, plantation crops and mushrooms. The horticulture sector is highly profitable as compared to other cereal crops because it fetches good price. For example, banana gives 30 folds more remuneration than others. Flowers are seasonal and it can be more remunerative (up to 30-35 times). Considering the high potential of horticultural crops, the concept of doubling farmers' income may work magically. There is a strong need of shift from traditional crops and cropping pattern to modern/multiple cropping system. Considering potential in horticulture sector for increasing farmer's income, Government of India has allotted ₹ 2250 crores for growth and development in horticulture sector during 2021-22. Government intervention and intervention of improved technology in horticulture sector has tremendously increased horticultural production which has surpassed the agricultural production in the country. During 2019-20, the country recorded highest horticultural production 320.77 million tonnes with area of 25.66 million hectares. Horticulture has several advantages compared with agriculture crops. It is more remunerative. Horticultural crops can be easily grown on dry and hilly land where normal crops cannot be expected. Water and fertilizer requirement of many unexploited horticultural crops are lower. Horticultural crops can be grown in small area allowing marginal and small farmers to boost their earnings from small area of land. Mono-cropping is always risk system of agriculture. Thus, multiple cropping/multi storied cropping systems are being focused because it ensures income to the farmers even in extreme conditions too. The growing of season/annual crops with fruit tree at pre-bearing as well bearing stage provides more income to the grower and therefore, farmers are moving towards agri-horti system where several crops can be taken at a time from the same piece of land. Growing of seasonal crops with a variety of trees is a very old tradition in traditional farming systems.

Growing the crop in conjunction with horticultural crops is one of the most effective ways to boost agricultural production and stabilize the rural economy in the country. The suitability of different crops depends largely on the soil and climatic conditions. The competitive natures of crops are not preferred as intercrop in the system. Exhaustive crops like maize, wheat, sugarcane, cotton, are not suitable with horticultural plants to get more income. The companion and synergistic nature of crops should be considered which can ensure early income, reduce the soil erosion, facilitate better harness of solar energy, and optimize land use efficiency, increase biological efficiency both in time and space dimensions in horticulture/fruit-based production systems. Jujube plants after planting cover the entire inter-row spaces in a period of about 5 years under rainfed conditions. During this period, soil erosion is very much common from these barren interspaces.

These losses can be reduced and additional revenue can be generated by investing in the right crop as intercrops. Planting of intercrops on a newly developed jujube plantation does not have any negative effects on plant growth until 5 years. However, in the years that followed, growing plants such as pearl millet and guar were badly affected and the effect was seen at a distance of up to 2.5 m. Studies reveal that yield of both jujube and intercrops was higher than monoculture (Singh, 1997). Like intercrop, green gram, guar and sesame increase the yield of jujube (cv. Seb) to 14.8 from 5.2 kg tree^{-1}. Growing of guar provides an additional yield of 782 kg seed yield However, Gupta *et al.* (2000) reported that 3-year-old plantation of jujube at a density of 400 plant ha^{-1} in association with green gram performed well. Yield of intercropped green gram was only 160 kg ha^{-1} as against to 620 kg ha^{-1} from pure crop, but fruit yield of jujube from intercropped system increased the net profit to ₹ 2,886.00. An economic analysis of jujube and Indian goose berry based on agri-horti systems reveals that the highest profit of ₹ 13,487.00 per hectare was obtained by jujube and guar followed by ₹ 11,700.00 from the green gram with jujube (Meghwal and Henry, 2006). Jujube, as it is a multipurpose crop provides better income and the compound areas were better used for growing legume crops. The Agri-horti program reduces risk and helps to provide economic stability.

Growing with lobia (*Dolichos biflorus*) has increased water retention capacity in light soils that increase the carbon content of living carbon (Pareek and Nath, 1996). The variety of crops in these planting systems revealed that the higher plant lengths of moth beans and groundnuts were recorded under bael planting, and in the jujube area the higher crop yields were recorded with guar (Yadava *et al.*, 2006). Planting of bean with bael produced 14.6 and 27.2 per cent higher value of dry matter and grain, respectively than jujube. The total amount of revenue and profit margin was recorded with bael and groundnut followed by jujube with nuts and kinnow on groundnut planting. The highest B:C value was obtained by bael and guar followed by bael and moth. Among the fruit trees, the bael proved to be very promising with a high B:C ratio regardless of age. Research on the kinnow-based system under irrigated conditions of the northwest was conducted on pre-bearing (3-5 years) and bearing (8-10 years) orchards for two consecutive years (Bhatnagar *et al.*, 2007). During the rainy season, cotton and green gram, while in winter barley and chickpea and their various combinations were grown in both orchards. Their effects on tree growth pathways, leaf nutrient status, fruit value, fruit yield and physico-chemical parameters were studied. The major impact of different crops on tree growth in the pre-fruiting field was only seen in the first year. The nitrogen content found in pre-bearing kinnow leaves and reproduction was lower in compaction than in normal leaves and was also lower in pre-bearing trees. This may be due to the low demand and consequently the low absorption of nitrogen during the pre-fruiting phase of the orchard, which was also true of the P and K content. The effect of intercrops did not show any significant effect on the soluble content, acidity and vitamin-C content of the fruit. Singh *et al.* (1999) also found the positive effect of legumes on mandarin fruits. By incorporating cotton into the production system, a small fruit yield was recorded, which can be attributed to the competitive competition of cotton (Reddy and Reddy, 2005).

In the pomegranate sowing system, the inclusion of guar, horse gram, raw gram and henna enhances the benefits over the planting of pomegranates in the medium soils (Lal, 2008). Further, pomegranate was recorded compatible with pearl millet, green gram, isabgol, sorghum and cumin in traditional pomegranate producing areas (Gupta, 2000). Indian gooseberry fruit based multi-storey production system comprising Indian gooseberry-jujube-brinjal-moth bean-fenugreek and Indian gooseberry-bael-karonda-moth bean-chickpea showed that ground storey crops did not affect growth of over storey crops and vice versa and these systems have been found promising under arid conditions (Awasthi *et al.*, 2008). The net return obtained from above cropping through ground storey crops during first year was to the tune of ₹ 23,614.00 and ₹ 25,662.00 per hectare, respectively which increased up to 20 and 15 per cent, respectively, in the second and third years. This indicates that any type of crop can be planted during the pre-bearing stage of fruit crops that can meet the different needs of farmers. Kumar and Pandey (2004) also reported improved growth of the Indian gooseberry when planting with legumes. Saroj *et al.* (2003) have reported a positive effect of intercrop on jujube. The incorporation of plant residues into the soil improved the nutrient status of N (0.9 to 4.5 per cent), P (12.5 to 27.5 per cent) and K (2.45 to 14.7 per cent). Decomposition of crop residues and improved organisms resulted in better nutritional status (Kumar *et al.*, 2008). This study demonstrates that the agri-horti system under rainy or irrigated conditions provides a stable stability due to additional benefits from underground plants, better utilization of natural resources such as water and soil, recycling of farm waste, improving soil fertility, preventing soil erosion and land degradation. This program was found to have an additional payoff especially during the youth horticultural plants. However, for long-term sustainability, the choice of collectively grown crops is very important. Generally, legumes were found to be more compatible with horticultural plants under rainy conditions, while under irrigation conditions cumin and mustard were considered an additional source. Samadia *et al.* (2004) proposes appropriate horticultural crops for agri-horti production systems in wet and irrigated areas.

Role of Horticultural Crops in Doubling Farmers' Income

In the current context of Indian agriculture, farmers are facing the challenge of earning a living which has become an important issue and as a result rural youth are showing some skepticism in agriculture. Thus, doubling farmers' income has become the motto of the entire agricultural organization and is built to fill production gaps, using the latest production technologies, advanced processing and advanced marketing. Horticulture sector has immense potential in doubling the farmer's income. Horticulture is emerging as a core sector of agriculture which steadily growing in area coverage and production (Kumar, 2018). In the year 2018-19, horticulture covered 25.8 M ha and production was about 314.67 MT, which was 1 per cent higher than previous year. Horticulture includes fruits, vegetables, spices, plantations and ornamentals crops which lead crop diversification in agriculture. The per capita income from horticultural crops is much higher than the food grains (Table 5.1).

Table 5.1: Suitable Fruit Crops and their Cultivars for Tropical Regions

Crops	*Suitable Cultivars*
Fruit crops	
Mango	Banganapalli, Alampur Baneshan, Nelum, Mallika, Bombay Green, Amrapali, Kesar
Guava	Allahabad Safeda, Sardar, Arka Mridula
Sapota	Cricket Ball, Kalipatti
Aonla	Kanchan, Krishna, Narendra –7
Papaya	Coorg Honey Dew, Pusa Delicious, Pusa Majesty, Pusa Dwarf, Taiwan
Sweet orange	Mosambi, Kodur Sathgudi, Valencia, Blood Red, Malta
Pomegranate	Ganesh, Jyothi, P-26, Jalore seedless
Custard apple	Bala Nagar, Arka Sahan
Ber	Gola, Umran, Banarasi Karaka, Kaitli
Lime	Tenali, Promalini, Vikram
Fig	Poona, Black Ischia
Bael	Narendra Bael-5, Narendra Bael-9
Tamarind	PKM-1, Pratisthan, Yogeshwari
Passion fruit	Kaveri
Vegetable crops	
Tomato	Pusa Ruby, Pusa Early Dwarf, Swarna Mani, Vaishali, Naveen, Rupali, Rashmi
Chilli	G-5, G-3, Pusa Jwala, NP-46A, Arka Gaurav, Arka Lohit, Bharat, Sindhur.
Onion	Arka Niketan, Arka Kalyan, Pusa Red, Nasik Red, Pusa Ratnar, Pusa White Round, Pusa White Flat, Patna Red, Arka Pitambar (for export).
Brinjal	Arka Navneet, Pusa Purple Long, Pusa Purple Round, Pusa Kranthi, Arka Sheel, Arka Kusumakar, Arka Shirish, Swarna Shree, Swarna Manjari.
Cluster bean	Pusa Navbahar, Pusa Sadabahar
Cow pea	Pusa Barsati, Pusa Rituraj, Pusa Dofasali
Cauliflower	Pusa Deepali, Improved Japanese, Pusa Snowball.
Cabbage	Pusa Mukta, Pride of India, Golden Acre, Pusa Synthetic, Pusa Drumhead, Shree Ganesh Gol.

The rejuvenation of old and senile orchard, establishment of new high density planting orchard and proper marketing are key point to get remunerative income. The rootstocks in fruit crops have potential to change the architecture for efficient utilization of resources. It develops strong and deep root systems and ameliorates the soil, enhance nutrient and water use. Therefore, rootstocks have been integrated into the production system of several fruit crops such as mango, grapes, citrus, apple, peach and almond. Various cultivars are tolerant to several abiotic factors (Gora *et al.*, 2019) which can be grown even under climate changes era (Table 5.2).

Table 5.2: Cultivar/Rootstock Tolerant to Abiotic Stresses

Tolerant	*Crop*	*Cultivar/Rootstock*
Drought	Mango	Kurukkan and Neleshwar Dwarf
	Citrus	Cleopatra mandarin and Rangpur lime
	Guava	L-49
	Sapota	Khirni
	Custard apple	Arka Sahan
	Fig	Deanna, Excel and Gular (*Ficus glomerata*)
	Aonla	Goma Aishwarya, Chakiay, Francis and Kanchan
	Grape	Beauty Seedless
	Pomegranate	Ruby
	Chilli	Arka Lohit
	Onion	Arka Kalyan
	Tomato	Arka Meghali and Arka Vikas
Salinity	Mango	Kurukkan, Bappakai (Rootstock)
	Citrus	Cleopatra mandarin and Rangpur lime
	Grape	Dogridge and Salt Creek
	Okra	Pusa Sawani
	Onion	Hisar-2
High Temperature	Tomato	Pusa Hybrid-1
	Cauliflower	Pusa Meghna
	Carrot	Pusa Kesar and Pusa Vrishti
	Cucumber	Pusa Barkha
	Radish	Pusa Chetki

Genetically dwarf cultivars and dwarfing rootstocks in fruit crops are suitable for high density orchard to accommodate more number of plants and increasing income. The country has a large manufacturing base that provides growth opportunities for exports. Mangoes, banana, grape, apple, walnuts, pomegranates, okra, onion, potato, green chilli are exported from the country. These vegetable crops can be easily grown as intercrops at pre-bearing as well as bearing orchards to get more income from the same piece of land. The major Indian fruit and vegetable are exported to the UAE, Malaysia, Bangladesh, the Netherlands, Nepal, Sri Lanka, the UK, Saudi Arabia, Pakistan and Qatar. Several cut flowers are exported to Japan, Europe, Australia, the Middle East and the USA, cut leaves are exported to Europe, dried flowers to the USA, Japan, Europe, Australia, Far East and Russia. A variety of flowering plants can be grown as intercropping plants in agri-horti systems. Among the fruits, mango, guava, banana, papaya, aonla, bael, vegetables like okra, onions, potatoes, cabbage, watermelon, flowers such as rose, tuberose, gladiolus, marigold and medicinal plants and fragrances such as artemisia, mentha, citronella, ashwagandha, palmarosa, brahmi, basil, damask rose, vetiver, and lavender are the potential crops. The agri-horti program conserves natural resources through various programs under different agricultural climatic zones. Monoculture is always in risk in agriculture. Thus, multiple cropping/multi storied cropping systems as agri-horti ensure income to the farmers even in extreme cases too. In North India farming system based on fruit tree crops modules can be promoted to enhance the farmers' income. The productivity of vegetables, fruits, flowers, plantations and spice crops can be increased tremendously by supplying disease free and quality planting materials to farmers. As most of the horticultural crops are cross pollinated required pollinators and honey bee is the best pollinator for several crops Beekeeping, not only increases yield and quality of produces but also give addition income to the growers in the form honey. Bees, in addition to improving the yield of various plants and converting nectar and pollen into honey and other bee products, *viz.* royal jelly, propolis, *etc.* which provides extra income for beekeepers.

Agri-Horticultural Systems for Climate Change

Agri-horticultural can be a multifunctional tool for high food production, poverty population reduction, input reduction, water conservation, improved soil quality, biodiversity conservation, climate change mitigation and climate change adaptation. There are some ecosystem and environmental benefits such as biodiversity conservation, carbon sequestration, soil quality enrichment, and air and water quality for not only the farmers, but for society at large (Jose, 2009). The increased pollution of ground as well as surface water has provided an additional encouragement for the development and adoption of tree based farming system around the world (Kumar *et al.,* 2012). The variations in microclimate play an important role on crop environment (Singh *et al.,* 2016) and presence of trees modify site microclimate in terms of temperature, water vapor content, wind speed, temperature reductions and can help reduce heat stress of crops. Woody plants also modify the microclimate by reducing evapotranspiration and moderating extremes in soil temperatures and daily photosynthetically active radiation, Shade trees modify the interception of radiant energy by the foliage of crops, that affect the physiology of the under grown crop, photosynthetically active radiation and temperature are reduced, while the humidity is increased (Sangwan *et al.,* 2017), water use efficiency and carboxylation efficiency is better (Sangwan *et al.,* 2015), reducing vulnerability, increasing resilience of farming systems against climate related risks (Sureshbhai *et al.*, 2017). The microclimate changes will impact biological processes on insect pests of crops (directly proportional to the temperature) and the increase of humidity, can promote the development of foliar diseases like mildews and rusts. Farmers know that these diseases are more frequent near the dense hedgerows and effective management systems and new/ innovative agroforestry solutions are required and should include regional and local abiotic characteristics of climate, soil, water and nutrient balance and biotic conditions (pests, diseases and dispersal).

Climate change is the most important global environmental challenge which is facing by all living organism including humans and disturb natural ecosystems, agriculture and health. In this situation tree based cropping system will act as back bone to provide food and solve food security problem by making feeds to the growing population and mitigate malefic effects of climate change by enhancing environmental quality, sustain economic viability and enhance quality of life (Toppo and Raj, 2018). Climate change increased drought, heat, evapotranspiration and affected agriculture worldwide (Borland *et al.,* 2015). Climate change and its variability are posing the major challenges influencing the performance crops, the extreme weather events of hot and cold wave conditions have been reported to cause considerable damage to many fruit crops. The various impacts of climate changes need to be addressed in a concerted and systematic manner so that technology can be developed to face the imminent challenges (Malhotra, 2017). Climate change will increase leaf: air water vapor-pressure deficits (VPD) and will result in drought and reduce productivity (Borland *et al.,* 2015). The adoption of tree based cropping systems technologies depends on the edaphic-climatic, socioeconomic status and needs of farmers. The normally, management is influenced by physical, demographic and, institutional factors. One of the main aim of agri-horti system to grow season vegetable crops along perennial fruit crops to reduce the risk on production (Paut *et al.,* 2018). Farmers are interested in growing annual plants among young timber trees to receive the benefits of annual fertilization and weed management. However, to obtain optimal benefits, farmers must make important initial decisions about tree species, plant density and a good understanding of their advantages and disadvantages. While fruit trees have a long juvenile period (4-5 years) to come in bearing stage, the interspaces area can be used for the cultivation of agricultural crops profitably until they develop canopy (Gunaga, 2017). However, several characteristics of the trees like slow growth, long term effects, on their surroundings, long life, area over which the influence of trees extends, *etc.*, complicating the issue of assessment and adoption (Hymavathi *et al.*, 2010). Agri-horti systems can restore some of the damage from farming to improve the ecosystem services related with soil and water (Udawatta *et al.*, 2017), the primary agroforestry objective of soil conservation is to improve or maintain soil fertility aboveground and soil C stocks to mitigate climate changes (Tiwari *et al.,* 2017). Tree based cropping system improve productivity and sustainability of agricultural landscapes (Fleming *et al.,* 2019), tree presence increase C sequestration per unit of land due to the C sequestered by the tree itself

(Sureshbhai *et al.,* 2017), the potential of C sequestration is dependent on the tree component (Nair *et al.,* 2009). Finally, climate change poses a serious threat to agriculture and food security, growing land disputes require the development of land use systems that include agricultural production and the provision of more ecosystem services, including reducing climate change. A fruit tree-based planting program can be proposed as a global solution to improve land use efficiency while minimizing the negative environmental impacts and economic risks for farmers (Paul *et al.,* 2017).

Why India Needs Climate Resilient Agriculture Systems?

Temperature is increasing due to climate changes which have deleterious effects on crop growth and yields. High temperature reduces crop duration, make changes in photosynthesis, and escalate crop respiration rates. India's population reached 1.38 billion in 2020 which is 17.7 per cent of the world's population. The country's population is increased 3.35 times since Independence; by 2027, it will surpass China and become higher populated country in the world. India accounts only 2.4 per cent of the global land. Most of the farmers in Indian are marginal (with land < 1 ha) and small farmers (land holding is 1-2 ha). Most of the farmers are facing several constraints such as supply of essential agri inputs, credit availability, suitable transport, and marketing facility, *etc.* Agriculture is the main and primary source of livelihood for about 58 per cent of India's population. Food production must be doubled by 2050 to feed the country's growing populations. The small and marginal farmers, therefore, have a major role in the country's food security and meeting the sustainable development goals (SDG). Nearly 14 per cent of the population (189.2 million) in India is still undernourished. Global Hunger Index 2020 placed India at the 94th position among 107 countries. Achieving 'zero hunger' by 2030 is a big challenge, and requires an integrated and multi-dimensional approach for overall sustainable development in agriculture. Climate change is the biggest challenges for food security in the country as its impact in agriculture is extreme. The predicted 1-2.5°C temperature rise by 2030 is likely to seriously affect crop yields. Climate change increases biotic as well as abiotic stress, nutrient mineralization, reduces fertilizer use efficiency (FUE) and increases water losses though evapotranspiration. The impact of climate change is directly or indirectly related to crop, water and soil as it influences the water availability, changes the intensity and frequencies of drought, effects microbial population, soil organic matter reduction, yield reduction, depletion of soil fertility as driven by soil erosion, *etc.* An economic survey (2017-18) warned that "climate change may reduce annual agricultural income in the range of 15-18 per cent and up to 20-25 per cent for unirrigated areas". This will create shortage of food and nutrient deficiencies in human beings due to lack of healthy food and it will be vulnerable to human health. South Asian areas may face a warming effect of 2-6°C in the 21st century (Ravindranath, 2007). The concentrations of CO_2 are high, up to 410 parts per million which is the main reason for global warming (Srinivasarao *et al.,* 2019). Several states *viz.,* Madhya Pradesh, Western Rajasthan, Karnataka,

Andhra Pradesh, Haryana, Southern Gujarat and Southern Bihar are continuously facing dryness and prone to drought (Bhadwal *et al.*, 2019). Flood in 2008 in Bihar forced millions of people to stay in shelter houses and approximately 20 million populations faced the same situation in Mumbai in 2005. Kerala was also hit by the worst floods in 2018. The super cyclone in Odisha in 1999 destroyed millions of lives with heavy property losses in coastal regions. Similarly, Andhra Pradesh has experienced the same situation during Hud-Hud cyclone in 2014. India and other adjacent countries are facing heat wave during May-June. Churu, Rajasthan recorded 50.8°C, the highest temperature in 2016. Salinity is the main problem in coastal areas where most of the crops can-not be grown.

Horticulture for Carbon Sequestration

To sequester one ton of carbon from the atmosphere, it is necessary to produce about 2.2 tonnes of wood (Chaturvedi, 1994). In the tropics, fruit tress sequester carbon much faster congenial climatic conditions. In perennial horticulture and plantations, the growth rates and hence carbon sequestration potential diminish as trees approach maturity. Orchards and plantations accumulate carbon in young and middle age, but the rates approach zero as the trees mature. The reasons for this decline in the aboveground net primary production in mature orchards and plantations are –

- ☆ Altering in balance between photosynthetic process and respiring tissues,
- ☆ Decreasing soil nutrient availability, and
- ☆ Increasing stomatal limitations leading to reduced photosynthetic rates.

Changes in the below ground biomass during orchard stand development and with aging of the orchards are very poorly understood. Hence, it is difficult to speculate how this important flux may change during development of the orchards, plantations, and gardens. The long residence time of particulate organic matter from high-latitude forest soils, however, provides indirect evidence that if fluxes of carbon from vegetation to the soil increase, as a result of global change, these soils have the capacity to act as a carbon sink on decadal time scale.

Why Fruit Trees for Carbon Sequestration?

Fruit trees and plantations crops are perennial in nature that contains 25 to 100 times more biomass carbon than the agricultural lands (Ganeshamurthy, 2013). Hence, vacant and wastelands must be developed into perennial agri-horticultural system than agroforestry or agricultural crops. There are three reasons to consider horticulture as a prime area to neutralize emission of greenhouse gas and to sequester carbon:

- ☆ Fruit trees are deep-rooted and have large reserves and are less susceptible than annual crops to several biotic and abiotic stresses. Thus, fruit trees offer diversified options that can reduce production risks for smallholder farmers.

- Fruit trees are perennial resource which can be used to provide income during adverse periods, thereby reducing income risks associated with climate-related shocks.
- Fruit trees have higher potential to sequester carbon than other cropping system systems.

Tree-based systems are a convenient way of sequestering carbon from the atmosphere to reduce net emissions. Fruit trees sequester carbon dioxide through photosynthesis which remains fixed in wood and other organic matter in orchards for long time periods. This is very important for countries like India which has large areas as deforested degraded land that can be exploited for perennial horticulture. There are some properties which make fruit trees suitable to maintain soil fertility or improvement:

- The existence of deeper root system,
- Huge network of fine roots,
- Association of mycorrhizal fungi with roots,
- High rate of leafy biomass production,
- Rapid leave fall and decay which release nutrient to the soil,
- Foliage contains nutrients particularly high nitrogen and low lignin,
- Absence of toxic substances in the litter or root residues,
- Higher nitrogen fixation rate,
- Capacity to grow in marginal and wasteland,
- Absence of severe competitive effects with crops, particularly for water and nutrient, and
- Low invasiveness.

Future Perspective

Agri-horticultural system offer to improve biodiversity and it acts the backbone upon which human civilization has been established. Humans have exploited the eco system which resulted in adverse impact on environment. Land holding is shrinking day by day which demand to grow more number of crops on the same piece of land to considering the effect of climate changes. Monoculture system needs to be shifted in multi-cropping system like agri-horticultural system where combination of various characteristics of fruit crops can combat with adverse effect of global warming. The various agricultural and horticultural crops, and cultivars of each crop species needs to be screened out suitable for agri-horti systems for different agro-ecological zones. Biotic (pest and diseases) and abiotic (drought, high temperature, salinity, floods) stress tolerance cultivars needs to be identified which will be suitable for agri-horti system. Tree crops have much potential to sequester environmental CO_2 and reduce harmful effect by climate changes which needs to be promoted in agriculture.

Conclusion

Agri-horticultural systems are recommended to get more income from the same piece of land and for climatic change mitigation and adaptation. Horticulture based systems are multifunctional way for global food security, environmental protection and mitigate and climatic change adaptation. It has excellent use of soil, water and light which helps in reduction on fertilizers and pesticides load in agriculture. It ensures income even after losing one crop and high potential of carbon sequestration by fruit trees are helpful in mitigate the climate effect.

REFERENCES

Altieri, M.A., Nicholls, C.I., Henao, A. and Lana, M.A. (2015). Agroecology and the design of climate change-resilient farming systems. *Agronomy for Sustainable Development.* **35**(3): 869-890.

Awasthi, O.P., Saroj, P.L., Singh, I.S., More, T.A. (2008). Fruit based cropping system for arid regions. Central Institute for Arid Horticulture, Bikaner. Technical Publication No. 25, p. 18.

Bhadwala, S., Sharma, S., Gortia Ganesh, G. and Maya, S.S. (2019). Livelihoods, gender and climate change in the Eastern Himalayas. *Environmental Development.* **31**: 68-77.

Bhatnagar, P., Kaul, M.K. and Singh, J. (2007). Effect of intercropping in kinnow based production system. *Indian Journal of Arid Horticulture.* **2**(1): 15-17.

Borland, A.M., Wullschleger, S.D., Weston, D.J., Tuskan, G.A., Hartwell, J. and Yang, X. (2015). Climate-resilient agroforestry: physiological responses to climate change and engineering of crassulacean acid metabolism (CAM) as a mitigation strategy. *Plant, Cell and Environment.* **38**: 1833-1849.

Chaturvedi, A.N. (1994). Carbon sequestration through afforestation: role of tropical industrial plantations. *Ambio.* **25**(5): 327-330.

Feliciano, D., Ledo, A., Hillier, J., and Nayak, D.R. (2018). Which agroforestry options give the greatest soil and above ground carbon benefits in different world regions? *Agriculture, Ecosystems and Environment.* **254**: 117-129.

Fleming, A., O'Grady, A.P., Mendham, D., England, J., Mitchell, P., Moroni, M. and Lyons, A. (2019). Understanding the values behind farmer perceptions of trees on farms to increase adoption of agroforestry in Australia. *Agronomy for Sustainable Development.* **39**:9.

Gaba, S., Lescourret, F., Boudsocq, S., Enjalbert, J., Hinsinger, P., Journet, F.P., Navas, M.L., Wery, J., Louarn, G., Malézieux, E., Pelzer, E., Prudent, M., and Ozier-Lafontaine, H. (2015). Multiple cropping systems as drivers for providing multiple ecosystem services: from concepts to design. *Agronomy for Sustainable Development.* **35**: 607-623.

Ganeshamurthy, A.N. (2013). Strategies for soil carbon sequestration through horticultural crops. *In*: Climate-Resilient Horticulture: Adaptation and Mitigation Strategies (Singh, H.C.P., Krishnamurthy, N., Srinivasa Rao and Shivashankara, K.S. Eds). pp. 221-236. Springer New Delhi Heidelberg New York Dordrecht London.

Gora, J.S., Verma, A.K., Singh, J. and Choudhary, D.R. (2019). Climate Change and Production of Horticultural Crops. *In*: Agricultural Impacts of Climate Change (Kumar, R., Singh, V.P., Jhajharia, D. and Mirabbasi, R. Eds.). pp. 45-61. CRC Press, London, New York.

Gunaga, R. (2017). Crop biomass and yield patterns of dominant agroforestry systems of Navsari district, Gujarat, India. *Indian Journal of Agroforestry.* **19**: 72-78.

Gupta, J.P., Joshi, D.C. and Singh, G.B. (2000). Management of arid ecosystem. *In*: Natural Resource Management for Agriculture Production in India (Yadava, J.S.P. and Singh, G.B. Eds.). pp. 551-668. Indian Society of Soil Science, New Delhi.

Hymavathi, H., Kandya, A. and Patel, L. (2010). Beneficial effects of multiple plantation patterns in agroforestry systems. *Indian Forester.* **136**: 465-475.

Jose, S. (2009). Agroforestry for ecosystem services and environmental benefits: an overview. *Agroforestry Systems.***76**: 1-10.

Kumar, B., Gupta, R.K. and Bhandari, A.L. (2008). Soil fertility changes after long term application of organic manures and crop residues under rice–wheat system. *Journal of the Indian Society of Soil Science.* **56**(1):80-85.

Kumar, B.M., Singh, A.K. and Dhyani, S.K. (2012). South Asian agro forestry: traditions, transformations and prospects. *In*: Agro-forestry- The Future of Global Land Use (Nair, P.K.R. and Garrity, D. Eds). *Advances in Agroforestry* 9. Springer, Dordrecht.

Kumar, D. and Pandey, V. (2004). Vegetative growth of aonla as influenced by intercrops under rainfed conditions of Agra. The Odisha Journal of Horticulture. **32**: 109-111.

Kumar, S. (2018). Doubling farmers' income with production enhancement through productivity gains. *Journal of Pharmacognosy and Phytochemistry.* 7(1): 1904-1909.

Lal, G. (2008). Pomegranate-based cropping system for sustainable production under arid ecosystem. *In*: Narain P, Singh MP, Kar A, Kathju S, Kumar P (Eds.) Diversification of arid farming systems. *Arid Zone Research Association of India and Scientific Publishers,* Jodhpur, pp. 56-60.

Malhotra, S.K. (2017). Horticultural Crops and Climate Change-A Review. *Indian Journal of Agricultural Science*. **8**: 12-22.

Mbow, C., Van Noordwijk, M., Luedeling, E., Neufeldt, H., Minang, P.A. and Kowero, G. (2014). Agroforestry solutions to address food security and climate change challenges in Africa. *Current Opinion in Environmental Sustainability*. **6**: 61-67.

Meghwal, P.R. and Henry, A. (2006). Economic evaluation of different agro-horticulture systems under rainfed condition of arid zone. *In*: Kumar D, Henry A, Vittal KPR (Eds.) Legumes in dry areas. *Indian Arid Legume Society and Scientific Publishers, Jodhpur*, pp. 299-303.

Nair, P.K.R., Kumar, B.M., and Nair, V.D. (2009). Agroforestry as a strategy for carbon sequestration. *Journal of Plant Nutrition and Soil Science*. **172**: 10-23.

Niether, W., Armengot, L., Andres, C., Schneider, M. and Gerold, G. (2018). Shade trees and tree pruning alter through fall and microclimate in cocoa (*Theobroma cacao* L.) production systems. *Annals of Forest Science*. **75**: 38 (e0149949).

Pareek, O.P. and Nath, V. (1996). Coordinated fruit research in Indian arid zone – A two decade profile. National Research Centre for Arid Horticulture, Bikaner.

Paul, C., Weber, M. and Knoke, T. (2017). Agroforestry versus farm mosaic systems – Comparing land-use efficiency, economic returns and risks under climate change effects. *Science of the Total Environment*. 587-588: 22-35.

Paut, R., Sabatier R. and Tchamitchian, M. (2018). Benefits of diversified horticultural systems: assessment with the modern portfolio theory. *Proc*. International Farming Systems Association (IFSA) Symposium, Chania, Crete, Greece, pp. 1-9.

Ravindranath, N. (2007). Forests in India-Take action now. The Hindu Survey of the Environment. The Hindu, Special Issue, New Delhi.

Reddy, T.Y. and Reddy, G.H.S. (2005). Cropping system. Principles of Agronomy. Kalyani Publishers, Ludhiana, pp. 468-489.

Samadia, D.K., Vashishtha, B.B. and Raturi, G.B. (2004). Prospects of vegetables production in the arid zone of India. *In*: Kumar, S., Joshi, R.K. and Pal, S. (Eds.) Impact of Vegetable Research in India. NCAP, ICAR, New Delhi, pp. 101-114.

Sangwan, A., Dhillon, W.S., Singh, H., Chohan, S. and Gill, P. (2015). Influence of horti-silviculture combinations on pre-bearing growth and physiological parameters of pear. *Indian Journal of Horticulture*. **72**: 21-27.

Sangwan, A.K., Dhillon, W.S., Chauhan, S.K., Singh, N.P. and Bath, M.K. (2017). Performance of garlic under agri-horti-silvicultural system in relation to physiological behaviour and yield. *Indian Journal of Ecology*. **43**: 724-729.

Saroj, P.L., Dhandar, D.G., Sharma, B.D., Bhargava, R. and Purohit, C.K. (2003). Ber (*Ziziphus mauritiana* Lamk.) based agri-horti system: A sustainable land use for arid ecosystem. *Indian Journal Agroforestry*. **5**: 30-35.

Singh, R.S. (1997). Note on the effect of intercropping on growth and yield of ber (*Z. mauritiana*) in semi-arid region. *Current Agriculture.* **21**(1-2): 117-118.

Singh, S., Sharma, M. and Singh, P. (2016). Intercropping – An approach to reduce fruit drop and improve fruit quality in guava. *Journal of Chemical and Pharmaceutical Science.* **9**: 3182-3187.

Singh, S., Srivastava, A.K., Shivankar, V.J. and Singh, S. (1999). Citrus based cropping system – A review. *Agricultural Reviews.* **20**(2): 123-128.

Srinivasarao, Ch., Srinivas, T., Rao, R.V.S., Srinivasarao, N., Vinayagam, S.S., Krishnan, P. (2020). Climate Change and Indian Agriculture: Challenges and Adaptation Strategies, ICAR-National Academy of Agricultural Research Management, Hyderabad, Telangana, India, p. 584 (ISBN No. : 978-81-943090-7-9).

Sureshbhai, P.J., Thakur, N.S., Jha, S.K. and Kumar, V. (2017). Productivity and carbon sequestration under prevalent agroforestry systems in Navsari District, Gujarat, India. *International Journal of Current Microbiology and Applied Science.* **6**: 3405-3422.

Tiwari, P., Kumar, R. Thakur, L. Salve, A. and Parmar, Y. (2017). Agroforestry for Sustainable Rural Livelihood: A review. *International Journal of Pure and Applied Bioscience.* **5**: 299-309.

Toppo, P. and Raj, A. (2018). Role of agroforestry in climate change mitigation. *Journal of Pharmacognosy and Phytochemistry.* **7**(2):241-243.

Udawatta, R.P., Gantzer, C.J. and Jose, S. (2017) Agroforestry practices and soil ecosystem services. *In*: Soil Health and Intensification of Agroecosystems, Al-Kaisi, M.M., Lowry, B. (Eds.), London Academic Press, pp. 305-308.

Vaast, P. and Somarriba, E. (2014). Trade-offs between crop intensification and ecosystem services: The role of agroforestry in cocoa cultivation. *Agroforestry Systems.* **88** (6): 947-956.

Yadava, N.D., Beniwal, R.K. and Soni, M.L. (2006). Crop diversification under fruit based cropping system in arid zone of western Rajasthan. *Indian Journal of Arid Horticulture.* **1**(1): 20-22.

Chapter 6

Jute-Based Farming Systems for Sustaining Livelihood in Eastern India

Mukesh Kumar[1]*, A.K. Ghorai[2] and S. Mitra[2]

[1]Senior Scientist (Agronomy), Dr Rajendra Prasad Central Agricultural University (RPCAU), Pusa, Samastipur – 848125, Bihar
[2]Principal Scientist (Agronomy), ICAR-Central Research Institute for Jute and Allied Fibres, Barrackpore, Kolkata – 700120, West Bengal
Corresponding author: mukesh.agro@gmail.com

ABSTRACT

Jute-based farming systems is one of the great option of the self-employment for small and marginal farmers of the eastern India because it has great potential to sustain the land productivity and enhancing the income generating opportunity. Jute-based farming systems is amalgamation of different enterprises viz., poultry, apiary, fisheries, duckery, field and horticultural crops based on their compatibility within the purview of bio-physical and socio-economic environment of the farmers to make farming more profitable and sustainable. Jute is mainly grown in low rainfall. Hence, farmers should go for alternative option in such situation like jute intercropped with mung bean which may give an extra income from jute farming. This may be integrated with poultry, goatry and duckery to make jute farming more profitable. Different jute-based farming system models for one hectare can give net income of ₹ 1.90 to 2.10 lakhs per annum.

Keywords: *Jute, Farming system, Micro pond, Cropping system.*

Introduction

Indian economy is predominantly rural and agricultural based provide about 55 per cent of employment of the country and it contributes to 18.8 per cent of the GDP (at current prices in 2020-21), though it has declined from more than 50 per cent of GDP in the 1950. The agricultural growth has been buoyant for the past decade, ranged from 5.8 per cent during 2005-06 to 0.4 per cent in 2009-10 and -0.2 per cent in 2014-15 (Agricultural Statistics at a Glance, 2015). Such a volatility in agricultural growth rate influences the farm incomes and farmers scare to take risk for investing in their land holding. As the population growing continuously lead to declining land holding size from 0.5 ha in 1950-51 to 0.15 ha in 2015-16. This may further decline if population grows by about 2 per cent annually. Out of a total of 129.2 million land holders in the country, 64.8 per cent have less than 1 ha called marginal farmers and 18.5 per cent have between 1 and 2 hectare of land called small farmers. These two farmers' groups cultivate 44.5 per cent of the total cultivable land. The reduction in cultivation land, is posing a serious challenge to the sustainability and profitability of farming. On another hand, the production of food grains needs to be increased at the rate of 5.5 MT annum^{-1} in the next two decades to feed the burgeoning population (State of Agriculture, 2017 and Sharma *et al.*, 2017). Further, demand for high value commodities such as fruits, vegetables, dairy products, fish, poultry, *etc.* are increasing faster than food grains and it is expected to increase by more than 100 per cent from 2000 to 2030. It implies that there is need to produce diverse types of food in the country. Considering the gradual shrinking of land holding and growing demand of diverse food commodities it is necessary to diversify the exiting farming practice through amalgamation of field and horticultural crops with other enterprises like dairy, fishery, poultry, goatry, duckery, *etc.* within the purview of bio-physical and socio-economic environment of the farmers to make farming more diversified, sustainable and reliable (Behera *et al.*, 2004). Hence, it is imperative to develop plan and agricultural technologies that will able to provide sufficient employment and profit, especially for small and marginal farmers. The farming system approach may play the crucial role to increasing the farm productivity, round the year employment and income of small and marginal farmers. According to Mahapatra (1994), the farming system concept takes consideration of the components of soil, water, crops, livestock, labour, capital, energy and other resources available with the farm family at the centre managing agricultural and related activities. The farming system can be designated and understood by its structure and function. The structure of farming system in its broader perspective includes the size of holdings and their distribution, land use pattern, production relationships, irrigation, transport, storage, marketing, credit and financial institutions, research and education (Behera and France, 2016). Therefore, the farming system is the result of multifaceted relations among a number of interdependent components. To manage all the components, the individual farmer allotted certain quantities and qualities of four production factors *viz.* land, labour, capital and management of on- and off-farm enterprises in such manner within the skill and knowledge he has possessed, for the achieving of the goals of farming system (Mahapatra, 1994). The term farming system is a complex system which is integration of farm enterprise or enterprises resources at the control of the farmer, and farming or social environment (Behera *et al.*, 2013).

Present Scenario of Jute Crop and their Importance

Jute, the golden fibre of India, occupies only 0.42 per cent of gross cropped area but supports more than 40 lakh farm families and generates employment of about 100 lakh man-days in a season in the rural area (Kundu *et al.*, 2011). Additionally, about 2.5 lakh industrial workers and 5 lakh traders get employment in jute sector. Thus, jute farming and its related industry and trade provide livelihood support to about 50 lakh people of the country. India contributing about 60 per cent of the global production and the single largest producer of jute goods in the world. The domestic market consuming about 87 per cent of the total production and continues to be the backbone of industry (Mahapatra *et al.*, 2012; Kundu, 2016). Besides, 30 per cent of global export market share of our country and it is estimated to grow further. Presently India gets about ₹ 2050 crores annually through export of jute goods as compared to ₹ 233 crores in early sixties (Jute Price Policy, 2017). However, jute sector is undergoing a severe competition among major jute producing countries of the world as echoed in the rising demands for raw jute during last 3 to 4 years. Mismatch between demand and supply of raw jute has been causing price fluctuation in the raw jute in major jute producing countries like Bangladesh. Jute is biodegradable natural fibre whose traditional use was confined to packaging industry but its diversified use and consumer's awareness of environment have contributed to the recent development of jute sector for the last few years. In recent past non-traditional or diversified uses of jute such as jute pulp and paper, jute yarn, jute geo-textile, jute composites and other jute diversified products appears to be the main driving force for the future development of raw jute, which lead to increase the demand and consequently increasing raw jute price. The jute production is confined to in the traditional of West Bengal, Bihar, Odisha and Assam in eastern Indo-Gangetic region of India (Table 6.1).

Table 6.1: Area, Production, Productivity and Potential of Jute Crop in different Regions of India

Name of Ecological Region	*Area (000' ha)*	*Production (000' bales)*	*Productivity (kg ha^{-1})*	*Potential (kg ha^{-1})*
West Bengal	576	8454	2641	4000
Bihar	111	1637	2651	3500
Assam	75	793	1904	3000
Odisha	13	68	958	2500
Andhra Pradesh	7	51	1286	2500
All India	**810**	**11126**	**2473**	**3500**

Source: Jute Price Policy 2016-17.

During 2014-15, West Bengal is leading state which contributed about 74 per cent of total jute production in the country, thereafter Bihar (17 per cent) and Assam (6 per cent) (Table 6.1). Similarly, jute growing area is also higher in West Bengal and that is 67 per cent of total jute cultivated area followed by Bihar (17 per cent). But due to increasing availability of cheaper synthetic alternatives to jute products, jute has significantly lost their market and farmers loss their profit from jute cultivation, which influenced the jute farmers and 90 per cent of those are belong to the marginal and small categories out of which almost 65 per cent are with 1 hectare and below and 25 per cent with 2 ha and below land holdings. In the existing rice-based cropping system of eastern region, there is little scope for further increasing area under cultivation of jute. Further, mainly being a rainfed crop, its production largely depends on the vagaries of monsoon. During the last five years, area under jute cultivation in India declined from 8.6 lakh ha (2012) to 8.10 lakh ha (2018-19). However, the production of raw jute during this period hovered between 19.67 lakh MT and 19.80 lakh MT due to increase in yield during the same period. This declining trend may be attributed to stiff competition from cheaper synthetic fibres, changing climatic conditions, erratic nature of rainfall over space, time and quantity, shortage of farm labour, non-availability of quality seeds and other inputs, shrinking availability of freshwater for retting, and options of remunerative alternative crops available with the farmers. To sustain jute fibre production and meet the growing market demand, in such a situation, the only alternative left is to strive for increasing productivity, improving quality of fibre and profitability from jute cultivation. It can only be possible through intervention of the crop and cropping systems-based research which makes ways for farming systems research particularly with regard to small farmers (Jha, 2003; Kumar *et al.*, 2014). Farming system integrate of field and horticultural crops with fisheries, poultry, goatry, apiary, within the acceptability of within the purview of social and environmental condition of locality for striving the farming more profitable and interdependent (Behera *et al.*, 2004; Rautaray *et al.*, 2005). Implementation of an individual farm enterprise independently cannot sustain the farm family income, on contrary, the farming system approach could be able to fulfil the requirement of farm family and sustain the economic growth of Indian farming communities. Farming system reap benefits from synergisms among different enterprises, gives diverse type of produce and it also environmentally sound (Behera and France, 2016). Therefore, farming system model has been suggested by several workers for increasing the farm income and round the year employment of small and marginal farmers across the country (Rangaswamy *et al.*, 1996; Behera and Mahapatra, 1999; Singh *et al.*, 2006).

Existing Jute-Based Cropping Systems and their Importance in term of Productivity, Profitability and Nutrition

Jute was grown as rainfed crop during rainy season followed by mustard, potato and lentil during the initiation of First Five-Year Plan (1950-55) and hence, total acreage under jute was only 0.26 M ha with a total production of 0.70 M bales (1 bales =180 kg) of raw jute. The jute varieties available during that period were D154 (white jute) and *Chinsurah green* (*tossa* jute) of 150-160 days duration with yield potential of 15-18 q ha^{-1} only. The increase in irrigation potential and subsequent advent of intensive agriculture and introduction of multiple cropping systems exposed jute crop to stiff competition from other major crops (rice and maize). Consequently, it was necessary to fit jute crop in multiple cropping sequences and make it more remunerative through the development of short duration, high-yielding varieties with judicious manipulation of the agro-techniques like advancing sowing time; optimising planting geometry, fertilizer dose, plant-protection measures, retting techniques *etc.* Development of pre-mature flowering resistant varieties like JRO-878, JRO-7835 and JRO-524 and reducing the crop duration from 150 to 120 days not only advanced the sowing time to as early as middle of March, but also enabled jute to be fitted in multiple cropping sequence with transplanted *Aman* paddy (Das and Maiti, 1998, Mahapatra *et al.,* 2012). This led to a paradigm shift in jute agriculture as the entire jute growing area of the country was taken over by the *tossa* jute and it became essential cash crop of farming in eastern India. Beside fibre, it also provides jute stick which is being used as fuel, making hut in rural household. Jute leaf, rich in antioxidant content, is now globally recognised as a high-value vegetable and farmers can make additional income through its export (Kundu, 2016). Furthermore, benefits of growing jute in cropping system helps in build-up of soil nutrients. The amount leaf fall (senescence leaves) was about 9.3 q ha^{-1} during its crop period of 110 days which contributed about 13.5 kg N, 10.6 kg P and 11.8 kg K ha^{-1} to the soil. Moreover, the amount of leaves at harvest of about 18.6 q ha^{-1} which returned about 39 kg N, 32 kg P and 42 kg K ha^{-1} to the soil when plants were kept for 3 days in field after harvest for shedding of leaves before putting in jute stack in retting water (Saha *et al.,* 2008; Kumar *et al.,* 2018). According to land situation, diversified jute-based crop sequences are being practised in jute growing area. On rainfed uplands jute-lentil, jute-blackgram, jute-greengram, jute-winter maize and several other promising jute-based sequences are being practised West Bengal, Assam, Odisha, and Bihar (Table 6.2). Under assured irrigation, triple cropping sequence like jute-rice-lentil, jute-rice-potato, jute-rice-mustard, jute-rice-wheat was found promising in upland condition (Mitra *et al.,* 2006, Sinha *et al.,* 2009).

Important Issues Related to Jute Farming and Improvement of Existing Jute Crop Based Farming System

- ☆ Jute is an environmental friendly crop and it sequesters about 15 t CO_2 ha^{-1} during its growth period of 120 days.
- ☆ Jute is mostly grown as marginal rainfed crop, and vagaries of rainfall and scarcity of retting water and labour further limit its cultivation.
- ☆ Under irrigated condition, jute is grown in sequence with rice in *kharif* season and mustard, lentil, potato, garden pea in *Rabi* season. Such

multiple cropping systems are being practiced since long without consideration of soil health and market demands.

- As day by day price of inputs like fertilizers, pesticides, *etc.* are increasing, we should think about efficient and remunerative cropping sequence and alternative measures to maintain soil health too.

Table 6.2: Jute Based Cropping Sequence Practised under different Land Situations

Land Situation	*Crop Rotation*	*Location*
Rainfed uplands	Jute-Lentil, Jute-Blackgram, Jute-Greengram, Jute-*Toria*, Jute-Rice-*Toria*/Wheat, Jute-Rice-*Toria*, Jute-Mustard-Cowpea	North and South Bengal, Assam, Odisha, Bihar, Uttar Pradesh
Irrigated uplands	Jute-Rice-Potato/Wheat/Pea, Jute-Rice-Vegetables, Jute-Rice-Mustard, Jute-Rice-Wheat, Jute-Onion, Jute-Winter Vegetables, Jute-Berseem	North and South Bengal, Assam, Bihar
Rainfed lowlands	Jute-Rice-Lentil	West Bengal, Bihar, Odisha, Uttar Pradesh

Source: Sinha *et al.*, 2009.

Required Intervention for Jute-Based Farming Systems

There are two options for sustainable jute based cropping system – i) selection of suitable crops for fitting into jute-based cropping sequence, and ii) suitability and feasibility of intercropping/strip cropping in jute.

A. Selection of Suitable Crop for Fitting into Jute Based Cropping Sequence

The data from frontline demonstrations of three years were analyzed and it was observed that fitting legume crops in the cropping system is very important not only for improving soil health but also to get a higher profit. Jute–rice–potato cropping sequence resulted in higher net income (₹ 134868), benefit-cost ratio (2.0) and economic efficiency (₹ 369.50 ha^{-1} day^{-1}). However, the fluctuation in market price of potato leads to lower net returns in some years. Jute-rice-garden pea and Jute-rice-frenchbean recorded the second highest net return and proved to be economically efficient cropping systems for irrigated condition. In rainfed condition, lentil in jute-rice sequence gave the better results. The jute-based cropping systems, having legume in sequence, also helps in the maintenance of soil fertility.

Jute-vegetable crop rotation was also found remunerative in Nadia district of West Bengal. Jute-ridge gourd-cauliflower was the highest income generating system with benefit-cost ratio of 2.78 in case of farmers having operational holding size of 1 acre. This was followed by jute-ridge gourd-cabbage (Roy *et al.,* 2008). Sarkar *et al.* (2013) reported that strip cropping of blackgram or ricebean (in 20 per cent area) with jute fibre crop (in 80 per cent area) can enhance the jute equivalent yield up to 13.8 per cent. Inclusion of rice (*kharif*) and lentil (*Rabi*) in the jute-based system can increase the system productivity to 60 q ha^{-1} compared to jute alone with 25.0 q ha^{-1}.

System productivity, sustainability and soil quality of five jute-based cropping sequences were assessed under variable nutrient and crop residue management practices by Kumar *et al.* (2018). The jute-rice-baby corn-jute (leafy vegetable) system was recorded the higher system productivity (205 q ha^{-1}) when baby corn fit in grown in (November – January) followed by jute-rice-garden pea system (82 q ha^{-1}) and jute-rice-mustard-mung bean (76 q ha^{-1}) (Figure 6.1). Since market is one of the criteria for selecting the crop as market of baby corn is only restricted only to urban area hence, jute-rice-baby corn-jute (leafy vegetable) sequence can be adopted in sub-urban area, jute-rice-garden pea or jute-rice-mustard-mung bean (where mustard *cv.* B-54 and mung bean *cv.* TMB-37 was sown on no tillage) may be adopted in both sub-urban and rural areas. The sustainability and productivity of any cropping system depends on nutrient management strategies, 100 per cent RDF with incorporated crop residue of rice gave higher system productivity and sustainability index. Carbon management index (CMI), an indicator of soil quality, was higher (CMI 60.8) in jute-rice-mustard and mung bean cropping system. The application of 100 per cent RDF with crop residue recorded higher CMI (57.11) compared to other nutrient and crop residue practices (Figure 6.1). It indicated that multiple cropping required balance nutrients which may be obtained by recycling of crop residue with in cropping systems.

B. Suitability and Feasibility of Intercropping/Strip Cropping in Jute

1. Mixed Cropping

Growing associate/companion crops intermingled together with the main crops (not definite row arrangement in between crops) for better utilization of nutrients, water land and space and generate more income from unit land area per unit time. Mixed cropping of jute with red amaranth produced 40 to 64 q red amaranth in 21 to 40 days after sowing along with 35 to 37 q fibre ha^{-1}, besides it suppressed the weed growth by more than 50 per cent (Ghorai *et al.*, 2010). It is possible cropping under rainfed condition when jute sown after assured rain under late sowing situation. This system minimizes loss of input by weeds and more remunerative than manual weeding twice. The net return varied from ₹ 19,000 to 21,000 ha^{-1}. This system fetched ₹ 1.63 to 1.68 from each rupee invested.

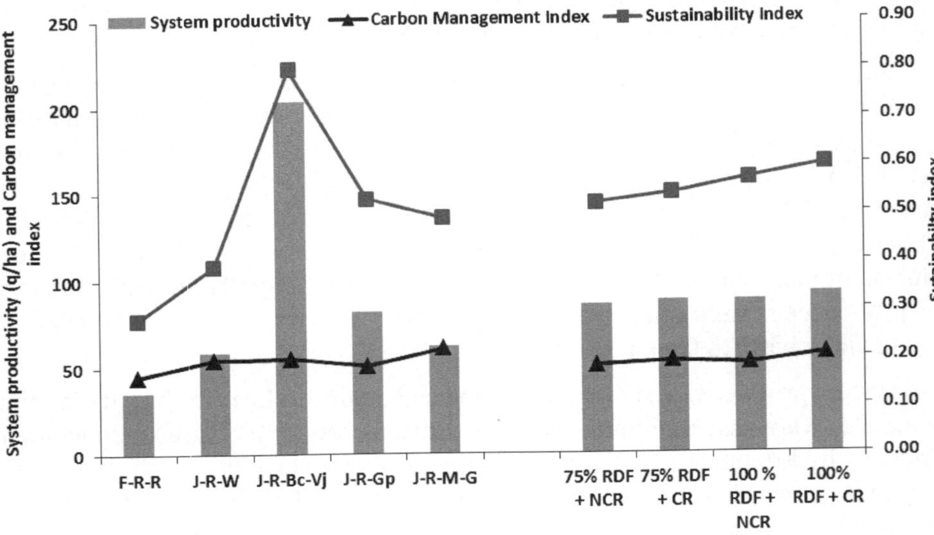

Figure 6.1: System Productivity, Sustainability Index and Carbon Management Index of different Jute based Cropping Systems under Nutrient and Crop Residue Management Practices.

(J: Jute, R: Rice, W: Wheat, Bc: Baby corn, Vj: Leafy jute vegetable, M: Mustard, G: Greengram, CR: Crop residue incorporation, NCR: Without crop residue incorporation)

2. Intercropping

Growing associate/companion crops in definite row arrangement in between main crops for better utilization of nutrients, water, land and space and generate more income from unit land area per unit time. Intercropping of jute *cv.* JRO-524 with radish *cv.* Chinese white, under rainfed upland and late sown on alluvial soil (sandy clay loam) with application of N-P-K @ 80-30-30 kg ha^{-1} was remunerative which recorded the fibre yield of 35 q ha^{-1} and radish yield of 27 to 36 q ha^{-1}. Introduction of greengram as intercrop with jute improved the system productivity (Figure 6.2) and in addition to grain yield produced 2 t ha^{-1} of greengram straw (containing 2.35 per cent N), if incorporated in soil which provide nitrogen as much as 10 tonnes of farmyard manure (FYM) and thereby, improves the soil health. It can also provide nutritious fodder to cow, goat *etc.* The comparatively short stature of "RMG-62 and TMB-37" fitted well in normal jute spacing (25 cm). Higher price grain (small shiny and polished) of these varieties increased the jute equivalent yield compared to other green gram varieties. The jute equivalent yield varied from 47 to 53 q ha^{-1} while sole jute production was around 39 q ha^{-1} only. The intercropping system recorded 27 to 34 q ha^{-1} jute fibres along with 7 to 10 q ha^{-1} of pulse grains (depending on grain size and varieties). Intercropping of jute and green gram (Pant mung 5) produced 10 q pulse grain and fibre yield of 28 q ha^{-1}, and also suppressed the growth of *Cyperus rotundus* up to 56 per cent by smothering affect (Ghorai *et al.*, 2016). In the trial conducted under All India Network Project on Jute and Allied Fibres (AINPJAF) for jute and greengram (*cv.* Pant mung 5 and RMG 62), the strip cropping of jute and greengram in 9:9 row arrangement recorded the highest benefit-cost ratio of 2.88-2.69 compared to jute alone (1.72) at Kalyani, West Bengal and 1.54 compared to sole jute (0.90) at Kendrapara, Odisha (AINPJAF, 2013). The system was not only remunerative and sustainable but it also make available protein to the poor jute farmers.

Figure 6.2: Jute-Mung Intercropping System.

Strategies for Improvement of Farm Income in Model Jute-Based Farming System

As discussed earlier, jute is mainly grown by the small and marginal farmers and about 77 per cent are having land less than 1 ha. Jute-based farming system varies from area to area and according to topographic situations. In rural areas, jute-rice-mustard/wheat + poultry + goatry (homestead farming), and jute-rice-potato/mustard/lentil + livestock in semi-urban and urban areas, jute + rice + vegetables + poultry/livestock, jute-vegetables + fisheries/livestock, jute-onion/mustard + fisheries/livestock are being practiced in unscientific manner. However, there is scope to increase the productivity as well as income of the farmers by farming in integrated model. Two jute-based integrated farming system (IFS) models for small and marginal farmers of eastern region for lowland and midland irrigated ecosystems are presented hereunder. The details of the module are given as for 1 ha land area.

IFS Model-I: Crops + Vegetables + Livestock + Goatry + Backyard Poultry

In this model, crops are allocated to 50 per cent area. In summer jute + greengram intercropping (0.5 ha), in *Kharif* rice (0.5 ha) and in *Rabi* mustard (0.15 ha)/potato (0.2 ha)/lentil (0.15 ha) will be taken in different jute-based cropping systems (Figure 6.3). Vegetable cultivation which give instant cash to the farmers is allocated to 0.25 ha area in this model. In summer/*kharif*, okra and pumpkin and in *Rabi* season cabbage and cauliflower may be taken in same piece of land. Livestock enterprise for two cows was assigned to 0.2 ha in which 0.1 ha for fodder (sudan grass/maize/berseem/oat/cowpea) and 0.05 ha for shed and 0.05 for composting/vermicomposting. Homestead goatry of black Bengal breed (four goats) and backyard poultry (15 birds) has also been taken for getting income and home consumption. The enterprises are inter linked with one or more aspects (Figure 6.4). The establishment cost under this model in ₹ 227000 and net income per annum is ₹ 190775 (component-wise cost is given in Table 6.4).

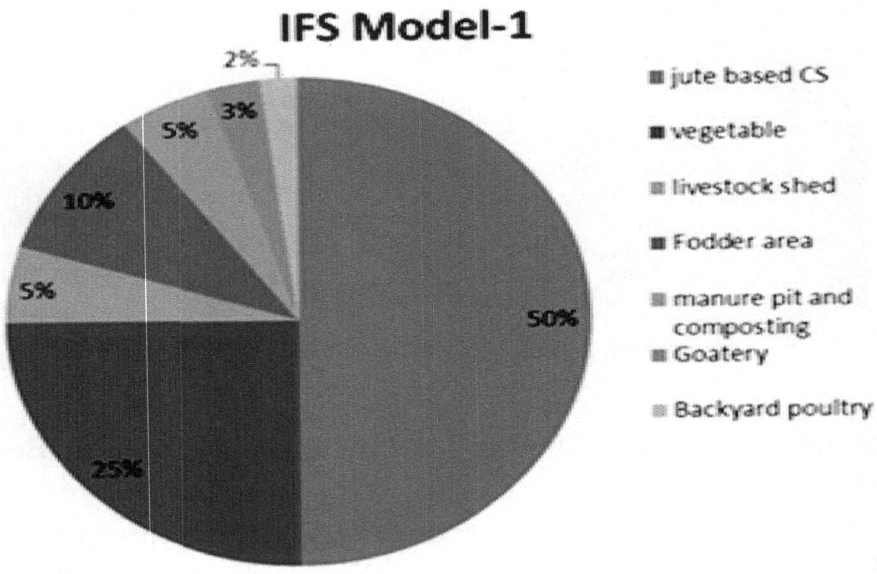

Figure 6.3: Allocation of Land for IFS Model-I: Crops + Vegetables + Livestock + Goatry + Backyard Poultry.

IFS Model-II: Crops + Vegetables + Livestock + Fishery + Duckery

In this model, the crops and vegetables are allocated to 0.6 ha of area (Figure 6.5), where the intercropping of jute + summer mung intercropping and jute+ amaranthus (leafy vegetable) are grown in summer season in 0.4 ha land area

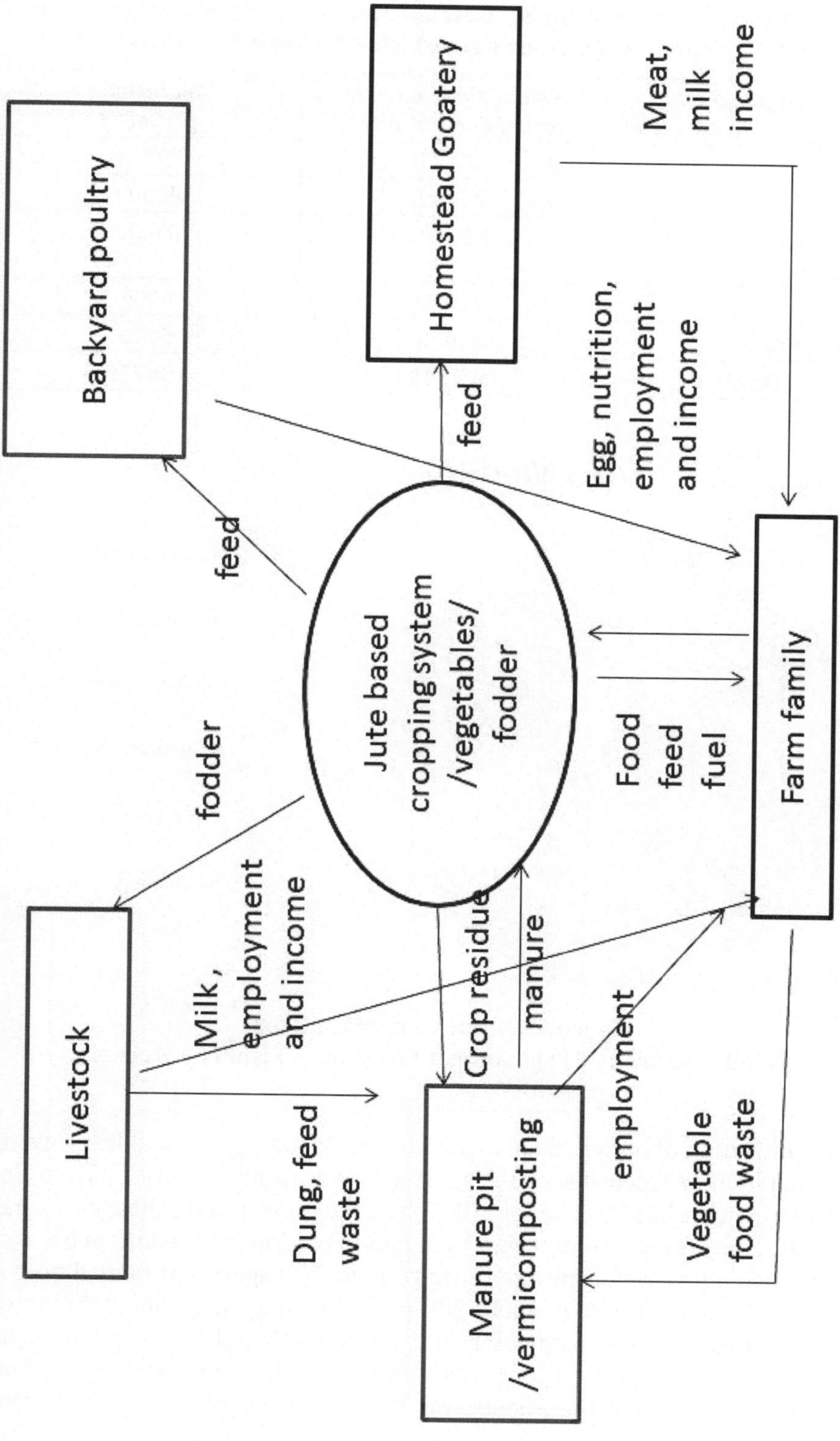

Figure 6.4: Interlinkage between different Enterprises of IFS Model-I: Crops + Vegetables + Livestock + Goatry + Backyard Poultry.

Table 6.4: Estimated Establishment Cost and Net Income/Annum by Suggested IFS Model-I: Crops + Vegetables + Livestock + Goatry + Backyard Poultry

Component	*Establishment Cost/ Cost of Cultivation (₹)*	*Net Income (₹ year^{-1})*
Crops	65625	79775
Vegetables	25000	40000
Livestock (with shed and manure pit)	125000	53000
Goat	12000	6000
Poultry 15 birds	18000	6000
Total	**227000**	**190775**

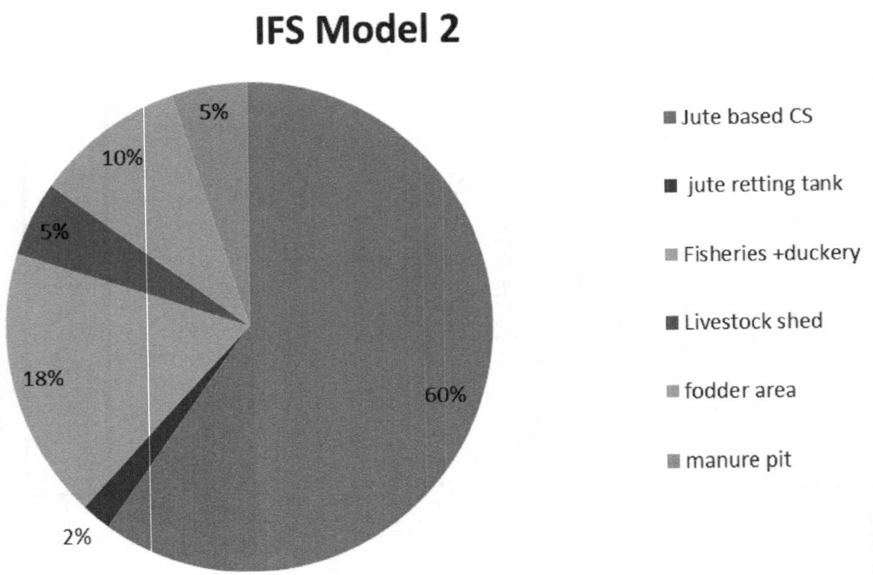

Figure 6.5: Allocation of Land for IFS Model-II: Crops + Vegetables + Livestock + Fishery + Duckery.

and the cultivation of okra is done in 0.2 ha area. In *kharif*, rice will be grown in 0.4 ha and pumpkin/cucurbits on 0.2 ha, in *Rabi* season, mustard (0.1 ha), garden pea (0.1 ha), onion (0.15 ha), potato (0.15 ha), cauliflower and cabbage (0.1 ha). Besides this, the provision of retting tank for jute of 200 m^2 has been kept because of less availability of water for jute retting due to the vagaries of rainfall now-a-days, therefore, a gross area of about 200 m^2 will be dug out to about 7.5 m top diameter and 6.5 m bottom diameter with 1.2 m depth and 1.5 m embankment (Figures 6.6 and 6.7) to collect the excess rainfall if sufficient rain is there or may be filled with tube well water, whenever deficit in rainfall, for retting of jute crop

Figure 6.6: Jute Bundle Arranged in Micro-pond.

Figure 6.7: Jute Bundle Sink by giving Weight.

Figure 6.8: Growing of Vegetables on the Embankment of Micro-pond.

Figure 6.9: Cultivation of Cauliflower Inside and Outside of the Micro-pond.

Figure 6.10: Cultivation of Cabbage Inside the Micro-pond.

Figure 6.11: Cultivation of Boro Rice Inside the Micro-pond.

(Ghorai *et al.*, 2013). This micro-pond may be utilized for growing vegetables (cabbage, cauliflower) (Figures 6.8–6.10) in winter on bund and bottom and *Boro* rice in summer in bottom (Figure 6.11). The farmers can get ₹ 1500 to 2000 from growing additional crop in and around the bunds (Ghorai *et al.*, 2013). This tank can also be used for rearing of small fish like tilapia or for rearing of the fingerlings. The integration of fisheries and duckery is allocated to 0.15 ha and livestock with fodder production area and compost pit area 0.2 ha. In this model, duckery area on the embankment of the pond. The enterprises are interlinked with one another in one or more aspect (Figure 6.12). The establishment cost under this model in ₹ 315000 and net income is about ₹ 217775 per year. The component wise cost is given in Table 6.5.

Table 6.5: Estimated Establishment Cost and Income per annum for Suggested IFS Model-II: Crops + Vegetables + Livestock + Fishery + Duckery

Component	*Establishment cost/ Cost of Cultivation (₹)*	*Net Income (₹ year^{-1})*
Crops	65625	79775
Vegetable	20000	35000
Jute retting tank	18000	8000
Fisheries	60000	25000
Duckery (pond bund)	15000	7000
Livestock	125000	53000
Total	**315400**	**217775**

Expected Outcome of Improved Jute-Based Farming System

The advantages of improved jute-based farming system is the stable and sustainable production system through diversified crops and enterprises as well as it helps in minimization of risk and resilience to climate change, diversified enterprises and nutrient for family and consumers, Diversified jute-based cropping system help in maintaining the soil and human health. Efficient utilization of resources among the different enterprises. Efficient use of family labour, conservation, preservation and utilization of farm biomass including non-conventional feed and fodder resources; effective use of manure and animal waste will helps in regulation of soil fertility and health and income and employment generation for many people.

Conclusion

There is little scope to increase the area under crop cultivation as the population and their diverse food demand increases. Simultaneously the number of small and marginal farmers is increasing due to fragmentation of land. Due to climate change and extreme weather, farming is no more remunerative. Hence, for sustaining the

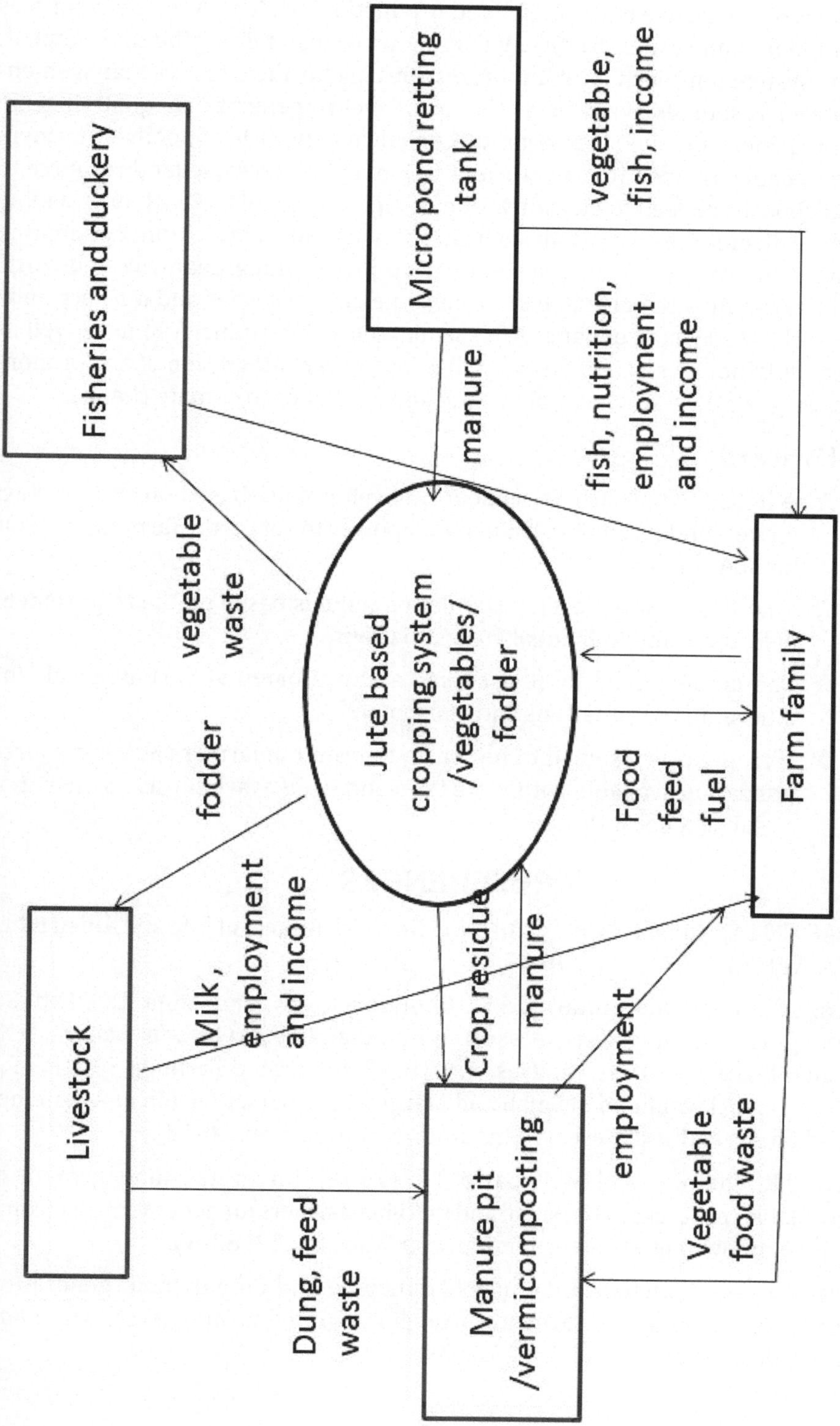

Figure 6.12: Interlinkage between different Enterprises of IFS Model-II: Crops + Vegetables + Livestock + Fishery + Duckery.

farming, it is imperative to diversify food production, while recycle the resources like land, labour and water in efficient manner while maintaining the environment. Farming system approach is an important strategy for the effective management of whatever resources available at the farm level to generate adequate income and employment for the rural poor and enhancement of livelihoods of farmers in a sustainable manner. It is an efficient way of using the same land resource to produce food, fibre, feed meat, cultivation of vegetables and fruits, thus providing a balanced diet for the farm family, reducing hunger and malnutrition. By adopting jute based integrated farming system where jute + mung bean-rice-mustard/ garden pea/lentil with vegetable crops in some area, fisheries and duckery and/ or livestock/goatry may enhance the sustainability of farming system as well as provides nutrition to soil and farm families. Moreover, integration of one or more enterprise gives diverse source of income and resilience to climate change.

Way Forward

- ✰ A holistic approach is needed to develop need-based farming system models and it requires trained manpower to set up the farming systems for specific farms.
- ✰ There is need to develop simulation models based on the requirement of farmers and individual farm situations.
- ✰ The models need to be assessed and evaluated at various levels for different land situations and consumers.
- ✰ Training centres need to be set up to train the farmers about resource utilization available within the farm and to take farming as business not as subsistence farming.

REFERENCES

AINPJAF. (2014). *Annual Report* of All India Network Project of Jute and Allied Fibre (AINPJAF) 2013-14, pp. 96-97.

Behera, U.K., Dass, A., Rautaray, S.K., Choudhary, A.K. and Rana, D.S. (2013). Integrated farming system research in India: an overview. *In*: Behera, U.K., Choudhary, A.K., Dass, A., Vyas, A.K. (Eds.), Integrated Farming Systems for Enhancing Livelihood of Small and Marginal Farmers. Division of Agronomy, Indian Agricultural Research Institute, New Delhi, pp. 40-78.

Behera, U.K., Jha, K.P. and Mahapatra, I.C. (2004). Integrated management of available resources of the small and marginal farmers for generation of income and employment in eastern India. *Crop Research*. **27:** 83-89.

Behera, U.K. and Mahapatra, I.C. (1999). Income and employment generation of small and marginal farmers through integrated farming systems. *Indian Journal of Agronomy*. **44**(3): 431-439.

Behera, U.K. and France, J. (2016). Integrated farming systems and the livelihood security of small and marginal Farmers in India and other developing countries. *Advances in Agronomy*. **138:** 235-282.

Das, B.B. and Maiti, S.N. (1998). Jute (*Corchorus* species) and allied fibres research in India. *Indian Journal of Agricultural Sciences*. **68**(8): 484-493.

Ghorai, A.K., Mitra, S. and Mahapatra, B.S. (2010). Improved production technology of Jute and Mesta in India: An overview. *Indian Farming*. **59**(12): 14-21.

Ghorai, A.K., Kumar, Mukesh and Kar, C.S. (2016). Weed smothering in jute with green gram intercropping. *Indian Journal of Weed Science*. **48**(3): 343-344.

Ghorai, A.K., Chowdhury, H., Kundu, D.K, Kumar, Mukesh, Barman D., Mahapatra, B.S. and Tripathi, M.K. (2013). *In-situ* jute retting in low volume water in polyethylene lined micro-pond using native culture and ploy culture in and around it. *Technical bulletin*, CRIJAF (6/2013). 20p.

Jha, D. (2003). An overview of farming systems research in India. *Annals of Agricultural Research*. **24**(4): 695-706.

Jute Price Policy. (2016). Jute price policy, CACP, Ministry of Agriculture and Farmers Welfare, Govt. of India, New Delhi.

Kumar, Mukesh, Singh, S.R., Jha, S.K., Shamna, A., Mazumdar, S.P., Singh, A., Kundu, D.K. and Mahapatra, B.S. (2014). System productivity, profitability and resource use efficiency of jute based cropping systems in the eastern Indo-Gangetic plain. *Indian Journal of Agricultural Sciences*. **84**(2): 209-213.

Kumar, Mukesh, Kundu, D.K., Ghorai, A.K., Mitra, S. and Singh, S.R. (2018). Carbon and nitrogen mineralization kinetics as influenced by diversified cropping systems and residue incorporation in Inceptisols of eastern Indo-Gangetic Plain. *Soil and Tillage Research*. **178:** 108-117.

Kundu, D.K. (2016). Potentials of jute in sustainable family farming for livelihood security in India and recent developments in jute production technologies. Pages 209-217 *in* Family Farming: Challenges and Opportunities (Eds. B. Mandal *et al.*). Renu Publishers, 90 Sainik Vihar, Mohan Garden, New Delhi-110059, India.

Kundu, D.K., Mitra, S., Kar, C.S., Mitra, J., Sarkar, S.K., Chowdhury, H., Meshram, J.H. and Ramasubramanian, T. (Eds. 2011. Vision 2030. Central Research Institute for Jute and Allied Fibres (ICAR), Barrackpore, Kolkata 700120. 18p.

Mahapatra, I.C. (1994). Farming systems research: A key to sustainable agriculture. *Indian J. Fertilizer Res*. **39**(11): 13-25.

Mahapatra, B.S, Mitra, S., Kumar, Mukesh, Ghorai, A.K., Sarkar, S.K., Kar, C.S., Kundu, D.K., Satpathy, S. and Karmakar, P.G. (2012). An overview of research and development in jute and allied fibre crops in India. *Indian Journal of Agronomy*. **57**(3rd IAC Special Issue): 72-82.

Mitra, S., Maiti, S.N. and Sarkar, S. (2006). Recommendations for jute and allied fibre crops: An endeavour of All India Network Project. p.28, CRIJAF, Barrackpore, Kolkata.

Rangaswamy, A., Venkitaswamy, R., Purushothaman, S. and Palaniappan, S.P. (1996). Rice poultry-fish-mushroom integrated farming systems for the lowlands of Tamil Nadu. *Indian Journal of Agronomy*. **41**(3): 344-348.

Rautaray, S.K., Das, P.C. and Sinhababu, D.P. (2005). Increasing farm income through a rice-fish based integrated farming system in rainfed lowlands of Assam. *Indian Journal of Agricultural Sciences*. **75**(2): 79-82.

Roy, B., Bera, B.K. and Nandi, A.K. (2008). An economic analysis of jute based cropping system in Nadia district of West Bengal. *Research on Crops*. **9**(1): 183-186.

Saha, M.N., Saha, A.R., Saha, S., Jana, A.K., Maji, B. and Mitra, D.N. (2008). Nutrient management, soil health and sustainable crop production. pp. 140-161. (In) : *Jute and Allied Fibre Updates,* Karmakar, P.G., Hazra, S.K., Ramasubramanian, T., Mandal, R. K., Sinha, M.K. and Sen, H. S. (Eds) CRIJAF, Barrackpore, Kolkata, India.

Sarkar, S. Majumdar, B. and Kundu D.K. (2013). Strip cropping with jute in jute-paddy-lentil sequence. *Journal of crops and weeds*. **9**(1) 207-209.

Sharma, R.L., Shalu, A., Bhagat, R. and Om Prakash. (2017). Comparative performance of integrated farming system models in Gariyaband region under rainfed and irrigated conditions. *Indian Journal of Agriculture Research*. **51**(1): 64-68.

Singh, K., Bohra, J.S., Singh, Y. and Singh, J.P. (2006). Development of farming system models for the north-eastern plain zone of Uttar Pradesh. *Indian Farming*. **56**(2): 5-11.

Sinha, M.K., Mitra, Sabyasachi, Ramasubramanain, T and Mahapatra, B.S. (2009). Crop diversification for profitability in jute and allied fibre crops. *Indian Journal of Agronomy*. **54**(3): 221-225.

Chapter 7

Integrated Livestock Farming Systems: A Cornerstone for Augmenting Farmers' Income

Amandeep Singh[1]*, Y.S. Jadoun[1], Gurpreet Kour[2] and Harpreet Kour[3]

[1]Assistant Professor, Directorate of Extension Education,
[2]Ph.D. Scholar, Department of Animal Genetics and Breeding,
[3]Ph.D. Scholar, Department of Veterinary Pharmacology and Toxicology,
Guru Angad Dev Veterinary and Animal Sciences University, Ludhiana – 141004, Punjab
Corresponding author: amandeepsinghvet@gmail.com

ABSTRACT

Integrated farming is a system of farming where poultry, livestock, fish, crops etc. are raised together in a farm. In other words, it is a type of mixed farming with cropping patterns that involves raising of animals, crops, and or trees for inter-utilitarian purposes. In India, the integration of livestock, fish and crops has proved to be a sustainable system through years of experience. The integration of different production systems have symbiotic effect on production of commodities. The diversified systems consist of components like livestock and crops that coexist independently from each other. Whereas integration minimizes risk and ensures better recyclability adding to circular economy of the farm. In the integrated system, livestock and crops interact with each other to create a synergy, with recycling allowing maximum use of available resources. The crop residues can be used as animals feed, livestock and livestock by-product production and processing can enhance agricultural productivity by increasing the nutrients that enhance soil fertility, thereby reducing the use of chemical fertilizers. For an environmentally sound integration, livestock droppings and feed waste can be poured directly into the pond to constitute feed for fish and zooplankton. Livestock manure can be used as a fertilizer for grass or other plants. Vegetables can be irrigated from the fish ponds and the by-products or residues can be fed to the livestock. This chapter deals with various combinations of livestock production systems with other cropping systems to ensure optimal production and profit for farmers which can aid in entrepreneurship development as well.

Keywords: *Integrated livestock farming system, Farmers' income, Integrated waste management.*

Introduction

India is an agricultural country and majority of its population depends directly on their own farm production for survival. Agriculture in India has a low technological and resource base, is complex and labour intensive. The agro-ecological conditions are also very complex in most parts of the country. There are several distinct land types in the country. The vast difference in the land types explains different cropping patterns in the country and make the use of land very complicated. However, the integrated system has the potential to overcome all these technical constraints in which waste of one becomes the resource for the other.

In the present times, supply and demand of food has changes due to an increase in population and shift in cropping pattern. Because of a continuous growth in population, the per capita availability of land is decreasing day by day. Therefore the need to produce more quantity and maintaining the quality of food in limited area arises to meet out the demand of increasing population arises. Adopting integrated farming is the only option left out as it requires less space and ensures higher productivity. The better economic returns from integrated system of farming have the potential to improve the economic condition of the farming community.

Integrated farming system (IFS) is a combination of different farm enterprises *viz.* livestock, horticulture, fishery, poultry, forestry and cropping systems and also the means available to the farmers to raise them and gain profit. It interacts effectively with the environment on the one hand without disturbing the ecological and socio-economic balance, and on the other hand seeks to achieve national goals. IFS will itself contribute in many ways to improving the overall agricultural efficiency and standard of living of the country's farmers.

Livestock production and rural life are inextricably linked to one another in India. The ancient sculptures carved with animal images and forms depicts the vitality of animals in Indian society. Livestock not only provides livelihood to the people but also has an integral part in the ascertaining food security and social values. Livestock can be colloquially related with the production of milk, meat, eggs and fibre but many other chores of farmers are completed by livestock in India. Fortunately India is blessed with huge livestock wealth with wondrous qualities of heat tolerance, disease resistance and ability to thrive in different agro-climatic conditions. Indian livestock industry has a colossal mark globally as India is the world's largest producer of milk, accounting for 22 per cent of the world's total milk production. India accounts for about 7.22 per cent of the world's egg production and 2.55 per cent of the world's meat production and is home to the world's largest livestock population (FAO, 2021). The annual growth rates suggests the supremacy of livestock over agriculture as livestock in India is increasing by an annual rate of 5 to 10 per cent.

Livestock waste in the form of dung and urine forms the resource base for livestock-based integrated farming systems. The livestock dung consists of ample amounts of nitrogen, phosphorous and potassium for manuring the aquatic life. The droppings from goat farms, poultry farms and pig farms serves as feed for aquatic life along with fertilizing the ponds. The dung can further be subjected to production of biogas for meeting the energy requirements of an integrated farm and can be vermicomposted for manuring agricultural crops and fertilizing fish ponds. There are tremendous utilitarian aspects of livestock based integrated farming models which can nurture farmers nutritionally as well as financially.

This chapter focuses on the association of various factors of production and their conjunction with livestock for reaping the maximum benefit. Integrated livestock offers great potential for food security and poverty alleviation not only in rural areas, but also in urban and suburban areas. This is an efficient way to produce carbohydrates, animal proteins and essential micronutrients simultaneously or continuously using the same land resources. All of these benefits can be enjoyed by farmers as well as improving soil fertility, reducing the use of chemicals and controlling pests, supporting aquatic biodiversity

Main Principles of IFS

1. **Cyclic:** This type of farming system is basically cyclical (organic resources – livestock – land – crops). Therefore, the decisions made for the management linked to one component can affect the other components.
2. **Rational:** The wise use of crop residues is an important route to get out of poverty. For resource-poor farmers, proper management of crop residues and optimal allocation of scarce resources leads to sustainable production.
3. **Ecologically Viable:** By combining ecological sustainability and economic viability, the integrated livestock-farming system maintains and improves agricultural productivity while also reducing negative impacts on the environment.

Objectives of IFS

- ☆ Identify the current farming systems in a particular area and access their relative feasibility.

- Design a farming system model that includes major and related companies for different farming situations.
- Ensure optimal use and protection of available resources and effective recycling of agricultural residues in the system.
- Maintain a viable production system without compromising resources/ environment.
- Enhance the overall productivity of farm households by complementing each other with major affiliates.
- To maintain environmental excellence and environmental stability.

Integrated Livestock Farming Systems

1. Livestock-Crop Integrated Farming System

India's agriculture is intensifying in response to both human and livestock population growth. As a result, the demand for productivity has increased, the integrated system of plants and livestock has become more important, and the role of legumes such as cowpea has become more important. Legumes serve as an important integration factor in strengthening the crop-livestock system, providing healthy plant proteins for human diets and livestock feed, and allowing nitrogen to be introduced into the agricultural system through nitrogen fixation. In this context, for example, cowpeas play a major role in food security, income generation, and the maintenance of the resource base of millions of smallholders growing cowpeas in the region. Raising goats, sheep and cattle for human consumption is the main form of livestock farming by smallholders in all developing countries. Livestock not only provide animal products, but also provide a means of saving wealth and a form of insurance in the absence of a properly functioning financial institution. Legume weeds are a valuable feed for ruminants, and their presence on or near farms can increase the profitability of legume cultivation. For this purpose, it is recommended to grow combined legumes such as peanuts, soybeans and cowpeas to provide edible grains and animal feeds. Livestock-crop integration is as old as agriculture. The livestock were fed with crop residues and the livestock dung in return was used to fertilize agricultural land for crop production.

The integrated plant-livestock system provides opportunities to promote organic farming and to facilitate the transfer of carbon and nutrients from one harvest to the next. Dual purpose/fodder bean varieties (cowpea) are cultivated in the rest of the field until the rain stops and the leaves show signs of wilting. At this stage, the plant grains are harvested and the rest are cut and rolled and stored on the roof or in tree fork. Stored leftovers are fed to ruminants during the dry season or, in some cases, sold at local markets, making a significant contribution to farmers' income during this shortage. The leaves of cereals are stripped after harvest and given to ruminants, which are used for buildings and enclosures. The ruminants in the farm are supplemented with cowpea and peanut residues, and manure is collected with household waste. At the beginning of the next cropping season, "compost" made from manure and household waste is spread in the fields before preparation of land is started.

2. Fish-Cattle Integrated Farming System

Fish farming that uses cow dung as manure is a well-established method. A cow weighing about 400 kg excretes about 5 tonnes of feces and about 8000 liters of urine each year. Cow dung particles sink at a rate of 2.6 cm minute^{-1}, while pig dung particles sink at 4.3 cm minute^{-1}. This gives the fish enough time to consume the edible portion available in the dung. The BOD of cow dung is relatively lower than other manure because it is partially degraded by microorganisms in the rumen. Manure and urine are very useful for filter feeding and omnivorous fish such as silver carp and catla. A unit of 5 to 6 cows can provide sufficient input for 1 hectare of water. These types of systems not only provide fish to farmers, but also receive milk daily, increasing income and reducing working capital for household safety. A cow shed should be built near the fishpond to make it easier to handle cow dung.

3. Fish-Duck Integrated Farming System

This system utilizes a mutually beneficial biological relationship between fish and ducks. Asia is considered a duck habitat, but the best varieties currently available for excellent egg and meat production have been developed in Europe and the United States through systematic breeding, feeding, management and disease management. In the West, duck meat is considered a delicacy. Duck eggs are an important food source in East India. These are cheap to manufacture and can play an important role in the balanced diet of Indians. In India, both the consumption and production of duck eggs are primarily driven by the socially weaker parts of the population. There are various types of ducks, such as egg-shaped, meat-shaped, and ornamental. Khaki Campbell, a production group with an average annual spawning of 310 eggs, is recommended for the integration of duck and fish.

The fish pond acts as a semi-closed biological system inhabited by multiple aquatic flora and fauna, providing an excellent disease-free environment for ducks. The ducks in turn eat young frogs, tadpoles, dragonflies, and more. This creates a safe environment for the fish. Duck droppings enter the pond and decompose. This process provides the nutrients needed for the production of natural fish feed organisms, such as carbon, nitrogen and phosphorus. The integration of ducks and fish has two advantages. One is that the availability of fertilizer is not lost, and the other is that the fertilizer is evenly applied to the pond. Ducks also aerate pond water with the effect of scraping the ground. Approximately 500 ducks can take over the fertilization plan for 1 hectare of water. This saves on fertilization costs and provides eggs and meat for human consumption in addition to fish. Duck droppings replace fish feed and pond fertilizer. These account for 60 per cent of the total input cost in aquaculture. Duck breeding costs are largely offset by income from the sale of duck eggs and duck meat. This method was able to achieve the production of 35 to 40 quintal fish per hectare of pond without resorting to additional feeding.

4. Fish-Pig Integrated Farming System

Pig farming is widespread in North-eastern and Northern India and is easier and yields more than poultry farming. Pigs are very fertile animals, and their combination with fish not only enhances economic efficiency, but also enhances their ecological effects as waste, leftovers, and aquatic plants are often used as feed for pigs. The excreta is used as an organic fertilizer in fishponds. Farmers who raise pigs can enjoy up to 28-30 per cent of the economic benefits over regular pig farming.

By building animal stalls by the pond and draining waste water from the stalls into fish ponds, farmers can successfully combine the breeding and aquaculture of small livestock such as pigs. In this integrated culture system, fish not only utilize pig spilled feed, but also fresh pig manure, which contains 70 per cent of the fish's digestible feed. Pig dung recycled in fish ponds acts as an excellent fertilizer, increasing the biological productivity of pond water and increasing fish production. This system does not require supplemental fish feed and pond fertilizer, as pig manure replaces them. This leads to a significant reduction in input costs in aquaculture. The cost of keeping a pig is largely offset by the income from the sale of pork. With proper management protocols, it is possible to achieve annual production of 30 to 35 quintal fish per hectare from a pond.

5. Fish-Poultry Integrated Farming System

Economically feasible, simple and well-tested poultry and fish integration uses poultry dung from fully constructed poultry litter and produces 45 to 50 hundred fish per hectare per year. It is recycled into a heavy fish pond. Broiler production provides immediate returns to the farmers. Before starting a business, it is important to research the market demand for the products you sell. The success of the system depends primarily on the farmer's efficiency, experience, suitability, and ability to manage herds. This includes sourcing good chicks, good breeding conditions, equipment such as feeders and water bowls, and adherence to scientific management practices, including disease prevention and management.

The leftover feed and droppings are used to increase the biological productivity of water. Poultry litter is applied to the pond in daily quantities and is postponed until the day when the pond develops blue-green algae. Adult chickens produce about 25 kg of compost (poultry manure) per year, providing enough bedding for 1000 birds to fertilize 1 hectare of water. Poultry farming is the integration of poultry such as chickens, ducks and geese into fish farming. The most common practice in India is duck or chicken poultry farming, which is widespread because of its profitability. Birds raised for eggs (laying hens) or meat (broilers) can be incorporated into fish farming. This reduces the cost of inputs such as fertilizers and feeds and maximizes profits. The poultry house can be built in the pond and raised above or next to the pond, as the case may be. Bird droppings can serve as fertilizer to fertilize the pond, or fish can feed in the pond directly. It is more ideal and better if the poultry house is raised above the pond. Vertical integration allows fertilizer to be easily transported to the pond and maximize land use. When the excreta falls directly into the pond, the poultry house becomes clean.

6. Fish-Goat Integrated Farming System

Goat farming is an ancient practice, but its integration with fish farming has not been explored much. Goat's solid excreta is many times richer in nitrogen content and phosphate than other animal excreta. Goat urine is equally rich in nitrogen and potassium. Goat droppings are coated with mucus and float in a semi-dry state, with a pellet size of about 6 to 7 mm, which has the advantage of being directly applicable to grow-out fish ponds. Feces have been observed to be consumed by fish. Goats can be kept under a house built on a raised part of the embankment to facilitate waste disposal into the pond.

7. Fish-Rabbit Integrated Farming System

Rabbits, recently considered pets, have emerged as an alternative meat source for health-conscious meat consumers due to their lower fat content than chicken, mutton, beef and pork. In addition, rabbits have the highest reproductive rates of any animal and are comparable to broilers in terms of growth. About 60 species of rabbits are popular all over the world, including Soviet chinchilla, Gray giant, White giant, New Zealand white, Russian angora and German angora. It can be bred in three ways: Cage system, Hutch system, and Floor system.

Rabbit manure fertility was assessed through studies of its composition and its effect on hydro-biological conditions compared to the traditional organic fertilizer cow dung. It has been found that nitrogen-rich rabbit excreta (10 times more than cow dung) gradually releases nutrients, thereby maintaining high plankton production over the long term. It is clearly evident that rabbit excreta with low water content and high nitrogen content is a high quality fertilizer for sustainable plankton production, so rabbit farming can be efficiently integrated with fish farming.

8. Livestock-Sericulture Integrated Farming System: "Silk and Milk"

Dairy enterprise is one of the most important activities comparable to sericulture. As is well known, the combination of '*silk and milk*' enterprises is still very popular in Kolar district of Karnataka. Many studies also confirm this fact in terms of effective combinations of enterprises. The farmers in the district are known to be highly innovative and everyday produce is combined with related businesses such as vegetables, dairy products, sericulture, poultry and pig farming, among others. Integrating dairy farms with other farms has been scientifically proven to provide greater opportunities to increase farm income and employment, especially for vulnerable sections of rural communities. When integrating livestock and sericulture, mulberry cultivation sites can be fertilized with livestock manure. Livestock can graze under mulberry trees and fertilize them in return. Mulberry leaf yields vary between 35 to 45 tonnes of fresh leaves ha^{-1} $year^{-1}$. The leaves contain almost 70 per cent water. The nutritional value of mulberry is one of the highest in plant-derived products, far exceeding the cereal crop residues and traditional feeds that can be used as feed for livestock after breeding silkworms.

9. Livestock-Tree-Crops Integrated Farming System

Among the ruminant production systems, integrated systems that use tree crops (coconut, oil palm, rubber, *etc.*) are the least developed and underestimated. Underneath the perennial tree crops of Southeast Asia an estimated area of 210,000 hectares is found. Presence of a range of perennial crops in the highlands of many countries provide a common thread for the development of integrated systems that involve ruminants. In this context, it is also appropriate to draw attention to the major tree crops that are potentially important from the perspective of developing integrated systems, and the locations of the major countries in which they are found. Trees such as cashew nuts, cocoa, teak, rubber, citrus fruits, mangoes and oil palms, can be cultivated with livestock to increase farmers' income.

10. Integrated Lowland Farming System

In Tamil Nadu, some experiments were carried on poultry farming in the lowlands, where poultry (chicken, ducks), pigs, goats, pigeons, fish and mushrooms were bred in various combinations in which residue and by-product of one combination would be recycled by another. The recycling process lowers the unit price of grains, meat, milk, eggs, mushrooms, biogas, *etc.*, thereby widening the gap between production costs and net income. On a 1-hectare farm, 0.90 hectares were used for rice cultivation. The remaining area (0.10 ha) were allocated to the pond, which is home to 1,000 young fish or fingerlings. Either 50 spawning layers or 100 pairs of productive pigeons could be linked to supplement the feed demand for fingerlings. Average mushroom production of 5 kg day^{-1} proved to be a balanced activity.

11. Integrated Upland Farming System

An integrated study of field crops, dairy products, biogas production, spawns and mushroom production, sericulture, mulberry, bee farming, and homestead gardens was conducted on a hectare of farmland under irrigated highland conditions. The dairy cow unit consisting of three cows were bred in such a way that two cows were in milk all year round. Raising three animals earned a net income of ₹ 29225.00 per year. Manure collected from three cows was sufficient to produce 2 m^3 of biogas daily. This could meet the fuel requirement of farm family apart from the preparation of gruel to the dairy unit, lighting two lamps in the farm house, and for the boiling and cooking activities of mushroom and spawn production. Economical products and by-products of their harvesting activities (corn grains, sorghum grains, cotton mills obtained from cotton mills, cakes obtained by extracting oil from sunflower and peanut seeds, soybeans) were used for preparation of concentrated for dairy animals. This has reduced the cost of concentrates to 35 per cent of the value of commercial products on the market. Similarly, millet stalks from harvesting activities, corn husks, straw, and biogas from biogas plants can be used to produce chicks and grow edible mushrooms at low cost.

The 200 m^3 homestead garden near the farmhouse having vegetables, fruit trees (guava, papaya, bananas) and greenery would supplement the nutritional needs of the family. Surplus produce can be sold, which can generate income to cover the cost of protecting seeds and crops. Vermicompost provides complete coverage of the nutrients needed in the garden. One or two honeycombs can be housed in homestead garden. This makes it easier to collect honey from flowering plants. Through these units, natural honey is available for five and a half to six months, and for the rest of the time, through artificial feeding.

12. Integrated Waste Management: A Case of Biogas Production and Vermicomposting

Livestock dung has been used to produce biogas (methane) and generate electricity for consumer use. Biogas is used in cooking, homes and street lighting. It can be used for driving self-combustion engines for a variety of purposes like running a generator to generate electricity. In livestock farms, biogas can be used for both heating purposes and conversion to electricity using a generator. The slurry (residue) could also be used to produce manure, which when applied on crops acts as an excellent fertilizer. The slurry can also be used to make vermicompost. Dung for biogas and its slurry for vermicompost can generate greater economic returns.

Vermicomposting uses earthworms to use wheat and rice straw, sugar cane bagasse, peanut shells, banana peel, cotton stalks, turmeric leaves, coconut pulp, vegetable trimming, wood chips, parthenium and water hyacinth – into rich manure that increases humus content of the soil. It is boon for sustainable agriculture. Earthworms can turn 1,000 tonnes of moist organic waste into 300 tonnes of abundant dry vermicompost. Almost all kinds of organic matter, such as bones and eggshells, can be consumed up to body weight every day. In 45-60 days, 1 kg of earthworms (about 1000 to 1250 worms) produce about 10 kg of vermicast, the nutrient-rich excreta of worms. Vermicompost (mix of vermicast and other compost) contains 5, 7, 11, 2 and 2 times more nitrogen, phosphorous, potassium, calcium and magnesium, respectively, than crop waste or animal manure. Matured vermicompost should be applied at the rate of 5 t ha^{-1} in the field.

Conclusion and Policy Implications

If farming integration with livestock species is encouraged, the farmer's income will increase and his poverty level will decrease, while at the same time more protein in the farmer's family, his employees, and the entire community, especially in urban areas will be available. Increasing food supply to cope with high population densities, especially in urban areas, requires more than increasing agricultural land. Land is a finite resource, and as more land is used for agriculture, forests are thinned at a rate that is detrimental to the environment. The cost of producing food will increase accordingly. Therefore, there is a need for ways to produce more food on existing farmlands, and integrated agriculture offers a potential solution. Integrated farming varies from region to region in terms of production combination, rate and size. It is more profitable than single system of farming as it guarantees a sharing of financial risk for its varied and diversified nature in rearing of fish, animals and crops; it has a capacity of making more food available thus enhancing food security and creating more jobs for the teeming unemployed masses in the country. Its ecological importance must be considered before this potential is fully realized. This is because it determines the pace of effective management that leads to maximum yields. Clear and explicit connections between livestock, aquaculture and the environment, develop complementary roles for livestock that contribute to environmental sustainability, and communicate to all stakeholders for optimal yields thereby enhancing the income of the farmers.

REFERENCES

Devendra, C. (2007). Perspectives on animal production systems in Asia. *Livestock Science*. **106**(1): 1-18.

Devendra, C. (2011). Integrated tree crops-ruminants systems in South East Asia: Advances in productivity enhancement and environmental sustainability. *Asian-Australasian Journal of Animal Sciences*. **24**(5): 587-602.

Ezeaku, *i.e.*, Mbah, B.N., Baiyeri, K.P. and Okechukwu, E.C. (2015). Integrated crop-livestock farming system for sustainable agricultural production in Nigeria. *African Journal of Agricultural Research*. **10**(47): 4268-4274.

Gabriel, U.U., Akinrotimi, O.A., Bekibele D.O., Anyanwu, P.E. and Onunkwo D.N. (2007). Economic benefit and ecological efficiency of integrated fish farming in Nigeria. *Scientific Research and Essay*. **2**(8): 302-308.

Gill, M.S., Singh, J.P. and Gangwar, K.S. (2009). Integrated farming system and agriculture sustainability. *Indian Journal of Agronomy*. **54**(2): 128-139.

Gupta, V., Rai, P.K. and Risam, K.S. (2012). Integrated crop-livestock farming systems: A strategy for resource conservation and environmental sustainability. *Indian Research Journal of Extension Education*. **Special Issue (2):** 49-54.

Jayanthi, C., Vennila, C., Nalini, K. and Chandrasekaran, B. (2009). Sustainable integrated management of crop with allied enterprises ensuring livelihood security of small and marginal farmers. *Special Feature: Sustainable Agriculture*, pp. 21-27.

Kumar, S., Dey, A., Kumar, U., Chandra, N. and Bhatt, B.P. (2012). Integrated Farming System for Improving Agricultural Productivity. Status of Agricultural Development in Eastern India. *Eds*: Bhatt, B.P., Sikka, A.K., Mukherjee, J., Islam, A. and Dey, A., pp. 205-230.

Mamun, S.H., Nusrat, F. and Debi, M.R. (2011). Integrated Farming System: Prospects in Bangladesh. *Journal of Environmental Science and Natural Resources*. **4**(2): 127-136.

Olukunle, O. (2014). Integrated Aquaculture: A Tool for Sustainable Development/ Food Security and Poverty Alleviation in Achieving MDG's Goals. *Report and Opinion*. **6**(6): 1-8.

Pushpa, J. (2010). Constraints in various integrated farming systems. *Agriculture Update.* **5**(3 and 4): 370-374.

Radhakrishna, P.G., Sekharappa, B.M. and Manibashetty, V.G. (2000). Silk and Milk – An economic package for rural upliftment. *Indian Silk*, pp. 11-12.

Ramrao, W.Y., Tiwari, S.P. and Singh, P. (2005). Crop-livestock integrated farming system for augmenting socio-economic status of smallholder tribal farmers of Chhattisgarh in Central India. *Livestock Research for Rural development*, **17**(8).

Saikia, A.K., Abujam, S.K., Das, D.N. and Prasad, B.S. (2015). Economics of paddy cum fish culture: A case study in Sivsagar, Assam. *International Journal of Fisheries and Aquatic Studies*. **2**(5): 198-203.

Singh, A. (2019). Development of a need-based and effective mobile app for promoting organic waste management among dairy farmers. M.V.Sc. Thesis. ICAR-Indian Veterinary Research Institute, Izatnagar. pp. 72.

Singh, A. and Rashid, M. (2017). Impact of animal waste on environment, its managemental strategies and treatment protocols to reduce environmental contamination. *Veterinary Science Research Journal*. **8**(1 and 2): 1-12.

Singh, A., Tiwari, R., Chandrahas and Dutt, T. (2021). Augmentation of farmers' income in India through sustainable waste management techniques. *Waste Management and Research*. **39**(6): 849-859.

Singh, A., Tiwari, R., Joshi, P. and Dutt, T. (2020). Insights into organic waste management practices followed by dairy farmers of Ludhiana District, Punjab: Policy challenges and solutions. *Waste Management and Research*. **38**(3): 291-299.

Singh, J.P., Gangwar, B., Pandey, D.K. and Kochewad, S.A. (2011). Integrated Farming System Model for Small Farm Holders of Western Plain Zone of Uttar Pradesh. PDFSR Bulletin No. 05, pp. 58. Project Directorate for Farming Systems Research, Modipuram, Meerut, India.

Solaiappan, U., Subramanian, V. and Sankar, G.M. (2007). Selection of suitable integrated farming system model for rainfed semi-arid vertic-inceptisols in Tamil Nadu. *Indian Journal of Agronomy*. **52**(3): 194-197.

Trivedi, S. and Sarkar, K. (2015). Comparative study on income generation through agriculture crop and sericulture at farmer's level in Murshidabad district. *Journal of Entomology and Zoology Studies*. **3**(1): 242-245.

Chapter 8

Livestock Production System with Special Reference to Northern States of India

Arunbeer Singh[1]*, Pragya Bhadauria[2] and Shabnam Sidhu[3]

[1]Assistant Professor, Directorate of Extension Education, Guru Angad Dev Veterinary and Animal Sciences University, Ludhiana – 141004, Punjab
[2]Scientist, ICAR-Agricultural Technology Application Research Institute, Zone-1, Ludhiana – 141004, Punjab
[3]Assistant Professor, Department of Veterinary Medicine, Guru Angad Dev Veterinary and Animal Sciences University, Ludhiana – 141004, Punjab
Corresponding author: brararunbir@gmail.com

ABSTRACT

Agriculture and livestock rearing under mixed farming system is the main occupation of majority of farming households in India. This system is a source of livelihood as well as nutritional security. However, for development of agriculture and livestock-based entrepreneurs, a policy needs to be devised by the central and state governments that takes the agro-ecological conditions into account. Each state has a unique blend of livestock species and livestock farmers. Species distribution varies according to the agro-climatic conditions within the state. It holds immense potential for promotion of entrepreneurship development for rural youth. Entrepreneurship development in the livestock sector needs eco-friendly and sustainable practices which can combine productivity with sustainability. It will be clear from the following review that traditional practices are equally important as the modern technology-based interventions for advancement in agricultural and livestock production.

Keywords: *Livestock, Entrepreneurship, Production, System.*

Introduction

Livestock rearing is an important economic activity along with agriculture in India. The trilateral effects of livestock rearing include income generation through milk sale and asset building; employment and food security (Shergill, 2017). Dairy farming is an important subsidiary occupation and an inseparable part of farming system. Milk production has been found to support crop production indirectly by providing for the purchase of off-farm inputs (Dhawan and Kashish, 2016). The distribution of livestock is more equitable as compared to land. It is also viewed in positive light due to opportunities for diversification and intensification. Within the livestock sector there is a huge variation in distribution of species across various geographic locations of the country (Kumar and Singh, 2008). It is important to study the distribution of livestock species across states, regions and agro-climatic zones from time to time to assess the changing dynamics in the sector, which would in turn enable policy makers to set research priorities and formulate conclusive and result oriented development programmes. It would also help to improve the functioning of the state and national animal husbandry departments.

Background

This article follows the livestock production system in the states of Himachal Pradesh (H.P), Uttarakhand (U.K.), Punjab (Pb.) and the Union Territories of Jammu-Kashmir and Leh-Ladakh. The study is based on secondary data collected from various sources.

Table 8.1: Land and Livestock Resources

Land Use	*Punjab*	*HP*	*UK*	*JK*
Total area (km^2)	50362 km^2	55673 km^2	53483 km^2	222236 km^2
Forest (per cent of total, approx.)	5	20	65	10
Permanent pasture (per cent of total, approx.)	0.10	27	3.5	0.5
Net area sown (per cent of total, approx.)	82	10	13	3
Livestock (No. in 000s)	7050355	4412846	4427089	8325324

Statistical Year Book of India (Mospi.gov.in), 20th Livestock Census.

Distribution of Livestock Species

The northern and north-western states in India are home to a wide variety of livestock species. Cattle are the most diverse species of livestock in India with a total of around 50 registered breeds followed closely by sheep (44). Eight of the registered breeds of sheep are from the hill states of H.P., Uttarakhand and the erstwhile state of Jammu and Kashmir, accounting for a little under 20 per cent of the total genetic diversity in the species. These three states are home to 5 of the 34 registered breeds of goats. The home tract of two of the most acclaimed milk breeds of buffaloes, Murrah and Nili-Ravi, is in Punjab and Haryana.

Table 8.2: Distribution of Livestock Species in Punjab, Himachal Pradesh, Uttarakhand and JK

Species	*State*	*Population (2012)*	*Population (2019)*	*Per cent change*
Cattle *	Punjab*	24,27,734 (3,63,085)	25,31,460 (4,25,873)	4.27
	HP	21,49,259 (11,65,331)	18,28,017 (7,59,082)	-14.94
	UK	20,06,053* (15,08,461*)	18,52,123 (12,75,303*)	-7.67
	JK, Leh-Ladakh	27,98,326 (13,28,639)	24,39,240 (10,96,013)	-12.83
Buffalo	Punjab*	51,59,734*	40,15,947*	-22.16
	HP	7,16,016	6,46,565	-9.69
	UK	9,87,775	8,66,318	-12.29
	JK, Leh-Ladakh	7,38,989	6,90,829	-6.51
Goat	Punjab	3,27,272	3,47,949	6.31
	HP	11,19,491	11,08,413	-0.98
	UK	13,67,413	13,71,971	0.33
	JK, Leh-Ladakh *	20,17,900*	17,30,218*	-14.25
Sheep	Punjab	1,28,534 (98,927)	85,560 (73,673)	-33.43
	HP	8,04,871 (4,99,680)	7,91,345 (7,18,524)	-1.68
	UK	3,68,756 (2,41,249)	2,84,615 (2,08,256)	-22.81
	JK, Leh-Ladakh *	33,89,485* (10,74,022)	32,47,503* (16,63,468)	-4.81
Pigs	Punjab*	32,221	52,961	64.36
	HP	5,033	2,477	-50.78
	UK	19,907	17,659	-11.29
	JK	2,421	1,215	-49.81

*Values in brackets () represent the indigenous breeds.

The region in focus boasts of some of the most productive breeds of livestock. With an aggregate of 19,34,62,871 heads, cattle are the largest species by number, followed by buffaloes (10,98,51,678), goats (14,88,84,786), sheep (7,42,60,615) and pigs (90,55,488). The largest number of cattle (25,31,460) are present in Punjab. The largest number of goats (17,30,218) and sheep (32,47,503) are present in Jammu-Kashmir. Buffaloes (40,15,947) are the main livestock species in Punjab. The population of bovines (cattle and buffalo), sheep, goats and pigs is presented in Table 8.2.

The results of the latest census how negative growth rate of the main livestock species in all the states except a few exceptions. The cattle population has declined in H.P, Uttarakhand and Jammu and Kashmir. However, the cattle population in Punjab has increased by over one lakh heads as compared to 2012. The increase in cattle population may be attributed mainly to the increase in the population of indigenous cattle in the state. Jammu and Kashmir, which ranked first among the four states in terms of the number of cattle has slipped to second behind Punjab. The population of buffaloes has spiralled downwards between the two censuses. Despite the contrasting growth rate of cattle and buffaloes in Punjab, the state continues to Punjab and Uttarakhand have achieved a positive change in the sheep population. Punjab, though, has a more favourable growth rate as compared to Uttarakhand. The sheep population has dwindled in all the states under consideration. However, the number of indigenous sheep has increased in H.P as well as J&K. Indigenous breeds which contributed a little over 50 per cent of the total sheep population in HP, now stands at a staggering 90 per cent. The contribution of indigenous breeds has increased from around one-third to over half of the total in J&K. The decline in the population of exotic sheep in Uttarakhand is steeper as compared to indigenous. Punjab is the only state that has recorded a positive growth in the pig population.

Livestock Production System

Livestock farming is an integral component of mixed farming system in Uttarakhand (Sati and Singh, 2010). There is wide variation in the composition of livestock species with difference in altitude. An average household rears more animals at higher altitude. These households also possessed more goats and sheep as compared to villages at lower altitude. Cattle and buffaloes are the main source of milk. The contribution of small ruminants to milk was negligible till 2010. Pasturelands, locally known as *bugyals* or *kharaks* are an important resource for the feeding of animals. Tiwari and Pratap (2017) reported that village communities in tribal regions of hills are highly dependent on forest for obtaining fodder for their animals. Sati (2016) observed that there is an inclination towards raising milch animals and poultry in Uttarakhand, owing to subsidies and easy marketing, among other factors. The small ruminants were mainly concentrated in the hills in the past. Increased human migration towards the valleys may have resulted in decrease in the population of sheep.

The cattle population is characterized by the presence of a larger proportion (54.5 per cent) of exotic/crossbred cows in the states under consideration. The state of Uttarakhand is an exception with indigenous cattle contributing 75.1 per cent to the total population of cattle (Census, 2019). Milk is the most important product of livestock farming in Uttarakhand. The average income was estimated at Rs. 1,25,000 per annum from a milch cow (crossbred) and Rs. 75,000 per milch buffalo. The contribution of small ruminants to milk production has increased in the past couple of years. Sharma *et al.* (2007) reported on the various channels of marketing of milk. The producers got the best price spread through self-marketing followed by sale through the state co-operative.

Jithendran (2000) noted that the rural population in Himachal Pradesh, especially the northern border districts are majorly dependent on livestock farming for livelihood. Animal husbandry is distributed over a range of systems including transhumant migration, local grazing in community and forest lands, as well as stall feeding during winters (Thakur *et al.*, 2012). In 2017-18, the total milk production of the state was over 13 lakh tonnes. The production of meat, wool and eggs was 4492 tonnes, 1484 tonnes and 98.14 million respectively (Anonymous, 2018). Anupam (2016) reported that around 61 per cent of the milk production comes from cows, followed by buffalo (35 per cent) and goats (4 per cent). Sharma *et al.* (2006) concluded that whereas both, cross bred cows and buffaloes were profitable in the low hills, cross bred cows were more suitable in the mid hills.

It has been estimated that 70 per cent of the total goat population in Himachal Pradesh is reared under nomadic system (Dogra *et al.*, 2018). These nomads migrate to higher altitude during the summers along with their herds and migrate back during the winters. These livestock keepers sell 40 per cent of their animals on an average every year to generate income. The contribution of goats to milk and meat has increased marginally over the years. The share of sheep and pigs in meat production has declined. The livestock sector in Himachal Pradesh recorded a growth of 2.37 per cent against 1.59 per cent growth rate in the agriculture sector between 2004-05 and 2010-11 (Kumar *et al.*, 2012). Wool production has increased to over 1600 tonnes at the rate of 0.21 per cent per annum. Milk and egg production has increased by about 300 per cent since the 1980s despite the fact that there was a negative growth rate of 0.40 per cent in livestock population between 1992-2012 (Anupam, 2016). The increase in egg production is reflected by the increase in the number of improved breeds of poultry.

The erstwhile state of Jammu and Kashmir is endowed with natural resources favourable to livestock rearing in abundance. The favourable agro-climatic conditions are matched by an unparalleled richness in the native germplasms of farm animals. Besides providing livelihood to the locals, the animal species are an integral part of the ecosystem of the area. There are a variety of breeds of cattle, sheep and goat native to Jammu, Kashmir and Ladakh area. Some of these breeds have not been registered yet. But these animal genetic resources are well adapted to the harsh climate and low feed and fodder availability. Besides, these local breeds are known to possess fecundity gene (Rather *et al.,* 2020). There is huge variation in the distribution of livestock species across the geographically diverse regions of the arid Ladakh, temperate Kashmir and sub-tropical Jammu (Kour *et al.,* 2018). Recent trends suggest that Jersey crosses are being preferred in the temperate regions such as Kashmir. It was postulated that there was an increase in the influx of crossbred cattle from Punjab and Haryana to replace the low yielding local breeds (Ishfaq *et al.,* 2017).

Livestock farming involves semi-pastoralism, intensive farming as well as mixed farming system. Buffaloes are mainly reared under migratory system in the Jammu region. There are no recognized breeds of cows native to this part of the country. Around half of the non-descript/indigenous cattle have been crossed with exotic germplasms of Holstein Friesian and Jersey. But, the region is home to pashmina yielding goat breeds which are considered an important genetic resource. Sheep breeding has shifted focus from wool to mutton production mainly. Sheep and goat rearing is an important occupation for tribal population. Jammu and Kashmir was perhaps the only state to have an independent directorate dedicated to small ruminants. Only 1 to 2 per cent of the total milk is handled by the organized sector (Anonymous 2016).

A unique trend in livestock population in Punjab is that the number of buffaloes exceeds cows by about 1.5 million (20th Livestock Census). This is perhaps due to the general preference for buffalo milk as well as the higher productivity of buffaloes as compared to indigenous/non-descript cows in the state. Nili-Ravi buffalo is a registered breed of buffalo native to Punjab. Dairy farming and agriculture are closely related in Punjab. Punjab farmers have a long running tradition of rearing cattle and buffaloes to produce milk for the household as well as the agricultural labour. Dairy farmers retain 45 per cent of the total milk produced, while the rest is sold. A number of commercial dairy farms rearing cross bred cattle are coming up. A Progressive Dairy Farmers' Association, established with technical guidance of the Veterinary University, has emerged as the centre of a mini-revolution for dairy farmers in the state. There is a strong infrastructure in the state devoted to healthcare and breeding. (Anonymous, 2014). Another important trend is the small proportion of small ruminants in the state. Whereas goats contribute 27 per cent of the combined population of buffaloes, cows, small ruminants and swine at the national level, their contribution to the combined population in Punjab is a meagre 4 per cent. Pig farming is an upcoming business venture.

The potential of pig farming to supplement the income of small farmers is well recognized (Agriculture Policy for Punjab, 2013). Nearly 80 per cent of the exotic population is being reared in the rural areas. The increasing population of exotic pigs is indicative of adoption of organized pig rearing in Punjab. Majority of the farmers have small land holding (Singh, 2014), rearing of small groups of pigs, is a characteristic feature of pig farming in Punjab. A combination of kitchen waste and industrial by products along with feeding of concentrate ration is common. A higher prevalence of Gastro-Intestinal parasites is common as compared to organized farms. Marketing of pigs is mainly through middlemen (Sidhu *et al.*, 2019).

Goats are mainly reared by landless 'Sansi' tribals in the breeding tract of Beetal, the only recognized breed from Punjab. The average flock size of goats is 5.27 (Tantia *et al.*, 2001). In the previous census, goat population was largely concentrated in Bathinda, Sangrur, Fazilka, Muktsar and Ludhiana in that order (District wise goat population, 2019). Goat farming is mainly an occupation of landless in Punjab (Singh *et al.*, 2020). Over 50 per cent of the surveyed goat farmers lacked formal education. Goat keepers are largely dependent on common/ public lands for grazing (Rath, 1992). The use of protected forests is an unresolved issue between the farmers and administration (NAIP report). Intensive housing is uncommon in south-western districts (Sandhu, 2016). Marketing channels for the sale of goats in south-western Punjab were limited to Middlemen and local market. Adoption of scientific management practices was low. These findings hint at the unorganized nature of the enterprise.

So far, there are no registered breeds of sheep in Punjab. A study on the phenotypic traits and performance of Kajali sheep revealed that these sheep were being reared in average flock size of 56 animals. The herds in Punjab and Haryana are relatively stationary, except for grazing activity within the district (Bhatia and Arora, 2005). The sheep were being maintained mainly on grazing with some farmers providing concentrate during breeding season. Majority (67.12 per cent) provided fodder during lean season. In India, sheep are largely dependent on monsoon grasses (Shinde and Bhatta, 2002). Yadav *et al.* (2011) reported some aspects of marketing in Munjal sheep native to Rajasthan, Haryana and Punjab. The wool yield per animal is worth ₹ 80625 on an average. The marketing of lambs is based on the body weight at the age of 3 to 6 months. Three-month lambs are sold for an average price of ₹ 1100.00. The price of adult animals was in the range of ₹ 2500 to 12000 depending on the gender.

Constraints

Kale *et al.* (2016) observed wide inter-state disparities in milk production between the states. The study categorized Punjab as the most progressive state in dairy farming, whereas Himachal was among the least progressive. Himachal was also the worst hit in terms of resource availability despite having the largest area under pasture. Despite the advances in dairy farming, the organised marketing in Punjab fared poorly. Jammu-Kashmir and Uttarakhand were not included in the study. Manhas (2005) found that the major marketing constraints encountered by the farmers in Jammu included nascent co-operatives and an underwhelming infrastructure for processing of milk. Further, discrepancies in the functioning of the cooperative sector have been reported from Punjab perceived shortcomings in milk procurement drive the farmers towards unorganized market, an already flourishing sector (Gupta and Sharma, 2010).

Singh *et al.* (2006) reported that migratory sheep and goat rearing was not profitable in small flocks. Migratory system also suffered due to poor veterinary facilities at higher altitude. Medical aid for the animals was found to be satisfactory in the foothills and plains by half of the respondents. Other constraints included poor marketing and processing facilities, hostility of forest officials and wild animal attacks. The geography of hill states pre-disposes the livestock keepers, especially tribal, to conflict with wildlife (Bargali and Ahmed, 2018). The economic losses to the livestock keepers were estimated to the tune of ₹ 15719200.00 per year.

Singh *et al.* (2004) found that the main challenges to livestock farming in the hills included feed shortage, poor management practices and poor knowledge of scientific techniques among other factors. The low awareness of scientific practice was attributed to the poor extension services. The need for training has an inverse relationship with extension contacts (Shahjar *et al.*, 2018). Frequency of trainings falls short of the requirement in many regions. There is a need for imparting need based and skill-oriented trainings (Raina *et al.*, 2017), demonstrations, exhibitions, discussions *etc.* (Minhaj *et al.*, 2017) for the farmers.

Another constraint in the dairy sector that may be considered is the large numbers of non-descript animals and a huge population of male animals in the country (Dairy sector statistics; Breed Survey, 2013) coupled with constraints in the adoption of recommended feeding practices (Singh *et al.*, 2015). The area under fodder crops was reported to be 2.91 lakh hectare. Yet the fodder deficit was in the range of 70 per cent (Singh *et al.*, 2012). Over the years, area under fodder has been shrinking in the face of increased cultivation of wheat and paddy (Grover and Kumar, 2012). The intensification of dairy farming sector is bound to add pressure on the feed and fodder resources (Singh and Datta, 2013). Considering fodder deficit is all the more important given the steady rise in milk yield of dairy animals since 1990s.

Future Strategy

Based on the overview of the livestock sector and the constraints holding back production, it is evident that a multi-pronged approach would be required to boost production of livestock sector across the states. Some strategies have been reviewed in this section for adoption at national level. Policy decisions must be based on the needs of the farming community which differ across states, location and agro-climatic zones.

1. **Strengthening Traditional Livestock Rearing:** Dogra *et al.* (2018) concluded that the low input migratory system of livestock rearing is a sustainable avenue with huge scope for expansion. Timely interventions and supply of critical inputs has resulted in increase in the herd strength as well as income of the nomadic shepherds registered under AICRP. Sankhyan and Thakur (2019) identified that selective breeding among native breeds of sheep, veterinary aid during migration and organization of wool procurement system was essential for the survival of sheep

farming especially the migratory communities. Mittal *et al.* (2008) suggested the use of forest area for grazing of animals where pasture land is not available. This is particularly significant for the hill states which are abundant in forest and pasture-based resources as compared to Punjab. Similar strategies could be extended to indigenous breeds of bovines. An urgent need for grasses and fodder development programmes has long been proposed in the hills through emphasis on farm forestry, agro-forestry, silvi-pastures, horti-pastures *etc.* (Dev *et al.*, 2006)

2. **Sustainable Development of Intensive Farming:** Baba *et al.* (2011) found a positive association of bovine density with geographical area (excluding net sown area) and net sown area in Jammu-Kashmir. The dependence of bovine population on the geographical area could prove detrimental to the environment in the future. It was suggested that certain measures must be introduced for the protection of common resources. Further, diversification of cross-bred animals must be complemented with strengthening of 'livestock-crop association', if the livestock sector were to be sustainable. Subash and Kaur (2017) ruled in favour of increasing the forest cover to lend ecological sustainability to livestock sector. Improving productivity to reduce pressure on the resources was also recommended. Singh and Kaur (2018) calculated that milk yield below 6.8 kg and 8.5 kg in buffaloes and cows, respectively and was not viable in Punjab. Kumar (2007) observed that a commercial goat unit must have 50 breeding goats to be self-sufficient. Advancement in dairy farming calls for improved nutrition and improved practices in fodder cultivation to support the large number of crossbred cattle. The advantage of a healthy infrastructural support in the dairy sector has already been discussed.
3. **Conservation of Indigenous Animal Genetic Resources:** The focus on cross breeding has had its share of criticism. Intensive cross breeding has resulted not only in the marginalization of valuable indigenous breeds; it has caused great amount of strain on the local communities of farmers as well as a disruption in the crop-livestock system (Dey, 2017). An alternative strategy would be to encourage conservation of local breeds in their native tracts in the face of low availability of concentrate feeds (Kale *et al.*, 2016). Rather *et al.* (2020) also stressed upon the need to conserve the native breeds citing their unique production traits and importance to the ecosystem. It is imperative to employ a *sui* generis system to protect the local breeds and indigenous knowledge associated with livestock for conservation efforts. Small and marginal livestock farmers can benefit from a legal system to manage the use of livestock biodiversity (Ramesha, 2011).
4. **Research Component:** Research priorities for different species must be based on the relative significance of the output from each (Birthal *et al.*, 2002). Specific research and development programmes

based on specific problems must be given priority (Dixit *et al.*, 2015). Breed characterization, molecular studies, performance evaluation, documentation, registration and establishment of breeding farms in niche area were suggested as possible interventions to promote conservation of animal genetic resources. Thiruvenkadan and Rajendran (2015) stressed upon the need for coordination in information sharing and exchange of germplasm, diagnostics and immunization in research on goats.

5. **Training Facilities and ICT:** Ishfaq *et al.* (2017) emphasized that knowledge of scientific practices and nutritional interventions were needed to boost production and health of the farm animals. Gangil (2004) found that the farmers were inclined towards receiving training on various aspects of dairy farming. It was suggested that need based training must be provided at village level for better participation. Need based training have the potential to raise production through better management of livestocks. ICT tools including web modules will also strengthen information dissemination channels for the livestock farmers (Singh *et al.*, 2019). At the same time, stakeholders have to focus on improving the core competencies of extension professionals to brace them for the challenges in the sector (Anonymous, 2019). Vocational training has a positive influence on the adoption behaviour of participants. In the face of processing and marketing as a major component of business model, training programmes must be designed for maximizing self-employment through dairy farming (Bhattu *et al.*, 2015).

6. **Marketing and Processing Facilities:** Reliability and profitability in market opportunities are important motivating factors for the farmers (Sirohi *et al.*, 2017). Traditional marketing system tends to be exploitative in nature when organized marketing channels are weak or non-existent. Replicating the Anand cooperative model to organize the dairy farmers along with promoting private processing sector could open up new avenues for producers. The dairy sector can profit from linking milk procurement with value addition and processing (Wani and Wani, 2010). Modern milk supply chains have been found to be inclusive for all types of land holders. Development of milk supply chain would depend on infrastructural support and incentives on quality of production Pathania (2015) suggested that formation of co-operative societies was necessary in the breeding tract of pashmina goats in the face of poor marketing facilities. Formation of co-operatives could boost the processing as well.

7. **Organic Farming:** Hamdani and Khan (2015) advocated the promotion of organic farming in the states of Himachal Pradesh, Uttaranchal and the twin UTs. It was argued that the agro-climatic conditions in these regions are better suited for organic farming. Production of organic milk, eggs and meat can be effectively used to attract youth towards livestock farming. Organic farming products from livestock sector are in great

demand in the country. The situation is also favourable owing to the positive attitude of media (Ribadiya *et al.,* 2016). Any efforts to promote organic farming would necessitate the promotion of native breeds which are comparatively resistant to climatic variations, disease and parasitic infestations.

REFERENCES

Anonymous. (2014). Dairying in Punjab: A Statistical Profile. National Dairy Development Board.

Anonymous. (2016). Handbook of Right to Information, Animal and Sheep Husbandry Department, Government of Jammu and Kashmir.

Anonymous. (2018). Economic survey of Himachal Pradesh 2018-19. Economic and statistics department.

Anonymous. (2019). Assessment of core competencies of livestock extension professionals of Jammu and Kashmir state of India. Discussion paper 8. National Institute of Agricultural Extension Management.

Anupam, K. (2016). Livestock sector: growth, production and contribution in the economy of Himachal Pradesh. *International Journal of Multidisciplinary Research and Development.* 3(5): 213-216.

Baba, S.H., Wani, M.H. and Zargar, B.A. (2011). Dynamics and sustainability of livestock sector in Jammu and Kashmir. *Agricultural Economics Research Review.* **24:** 119-132.

Bargali, H.S. and Ahmed, T. (2018). Patterns of livestock depredation by Tiger (*Panthera tigris*) and Leopard (*Panthera pardus*) in and around Corbett Tiger Reserve, Uttarakhand, India. *PloS ONE.* **13**(5): e0195612.

Bhattu, B.S., Dhaliwal, A.P.S., Singh, S. and Singh, G. (2015). Impact of vocational training on adoption of dairy enterprise in south-western districts of Punjab. *Journal of Krishi Vigyan.* **3**(2): 62-65.

Birthal, P.S., Joshi, P.K. and Kumar, A. (2002). Assessment of research priorities for livestock sector in India. Policy paper 15. National Centre for Agricultural Economics and Policy Research (ICAR).

Dev, I., Misri, B. and Pathania, M.S. (2006). Forage demand and supply in western Himalaya: A Balance Sheet for Himachal Pradesh. *Indian Journal of Animal Sciences.* **76**(9): 720-726.

Dey, D. (2017). Sahiwal- A Policy Pointer in Indian context. *Journal of Livestock Sciences.* **8:** 88-91.

Dhawan, V. and Kashish. (2016). Transforming livestock economy in India with special reference to Punjab: a review. *Economic Affairs.* **61**(2): 259-271.

Dixit, A.K., Singh, S.K., Tripathi, M.K., Singh, M.K. and Kumar, V. (2015). Economic gains from technological and marketing interventions in goat production in

India: An *ex ante* assessment. *Agricultural Economics and Research Review.* **28**(2): 285-292.

Dogra, P.K., Sankhyan, V., Thakur, A. and Kumar, N. (2018). Enhancing profitability of nomadic Gaddi goat production system for augmenting farmer's income. *Indian Journal of Animal Production Management.* **34**(3-4): 36-39.

Grover, D.K. and Kumar, S. (2012). Economics of production, processing and marketing of fodder crops in India. AERC Study No. 29. Punjab Agricultural University, Ludhiana.

Gulati, A., Roy, R. and Hussain, S. (2017). Getting Punjab agriculture Back on High Growth Path: Sources, Drivers and Policy Lessons. Indian Council for Research on International Economic Relations, Core 6A, 4th Floor, Habitat Centre, Lodhi Road, New Delhi, pp. 1-51.

Gupta, M. and Sharma, V. (2010). Seasonal variations in dairying – A study of milk producers in Punjab. *International Journal of Applied Agricultural Research.* **5**(4): 419-427.

Hamdani, H. and Khan, A.A. (2015). Organic dairy farming – An overview. *Journal of Livestock Science.* **6:** 4-9.

Ishfaq, A., Gani, A.M., Ahmed, H.A., Beigh, S.A., Khan, H.M. and Ahmad, S.B. (2017). Rearing practices, production performance and reproductive problems of cattle of Budgam district in Kashmir valley. *Journal of Animal Health and Production.* **5**(2): 68-73.

Jithendran, K.P. (2000). Helminth parasites – A constraint in animal health management in Himachal Pradesh. *ENVIS Bulletin.* **8**(2): 7-20.

Kale, R.B., Ponnusamy, K., Chakravarty, A.K., Sendhil, R. and Mohammad, A. (2016). Assessing resource and infrastructure disparities to strengthen Indian dairy sector. *Indian Journal of Animal Sciences.* **86**(6): 720-725.

Kour, G., Singh, A., Kumar, P. and Kumar, D. (2018). An overview of diversified animal genetic resources in the Indian state of Jammu and Kashmir. *International Journal of Current Microbiology and Applied Sciences.* **7**(10): 3113-3121.

Kumar, A. and Singh, D.K. (2008). Livestock production systems in India: An appraisal across agro-ecological regions. *Indian Journal of Agricultural Economics.* **63**(4): 577-597.

Kumar, A., Staal, S.J. and Singh, D.K. (2011). Smallholder dairy farmers' access to modern milk marketing chains in India. *Agricultural Economics Research Review.* **24:** 243-253.

Kumar, S. (2007). Commercial goat farming in India: An emerging agri-business opportunity. *Agricultural Economics and Research Review.* **20:** 503-520.

Kumar, V., Lal, H. and Priyanka. (2012). Livestock economy of Himachal Pradesh: Need to strengthen animal health and veterinary services. *Agricultural Economics Research Review.* **25**(Conference Number): 501-506.

Manhas, J.S. (2005). An analytical study of dairy entrepreneurship and its future strategy in Jammu district of Jammu and Kashmir, Thesis. MPUAT, Udaipur.

Minhaj, S.U., Khandi, S.A., Bafanda, R.A. and Bhushan, B. (2017). Adoption of improved animal husbandry practices by dairy farmers of Doda district of Jammu and Kashmir, India. *International Journal of Current Microbiology and Applied Sciences.* **6**(12): 2703-2707.

Mittal, S., Tripathi, G. and Sethi, D. (2008). Development strategy for the hill districts of Uttarakhand. ICRIER, Working paper No. 217.

Pathania, M.S. (2015). Impact of National Agricultural Innovation Project on socio-Economic Status of Pashmina Goat Keepers in Himachal Pradesh. *Himachal Journal of Agricultural Research.* **41**(1): 13-19.

Raina, V., Sharma, N., Khajuria, S., Kumar, K., Choudhary, S., Syed, Z. and Hussain, K. (2017). Training needs of dairy farmers. *International Journal of Agriculture, Environment and Biotechnology.* **10**(2): 245-251.

Ramesha, K.P. (2011). Intellectual property rights regime for livestock agriculture in India – Present status and future prospects. *Journal of Intellectual Property Rights.* **16:** 154-162.

Rather, M.A., Shanaz, S., Ganai, N.A. and Hamadani, A. (2020). Status of farm animal genetic resources of Jammu and Kashmir – A review. *International Journal of Livestock Research.* **10**(4): 39-48.

Ribadiya, N.K., Savsani, H.H., Patil, S.S., Dutta, K.S., Garg, D.D. and Karangiya, V.K. (2016). Strategic nutritional management for organic livestock and poultry farming – A review. *Agricultural Reviews.* **37**(1): 42-48.

Sankhyan, V. and Thakur, Y.P. (2019). Migratory sheep husbandry in Himalayan state of Himachal Pradesh, India: Status and headway options for improvement and sustainability. *International Journal of Livestock Research.* **9**(8): 71-77.

Sati, V. (2016). Livestock farming in the Uttarakhand Himalaya, India: Use pattern and potentiality. *Current Science.* **111**(12): 1955-1960.

Sati, V. and Singh, R. (2010). Prospects of sustainable livestock farming in the Uttarakhand Himalaya, India. *Journal of Livestock Science.* **1**(1): 9-16.

Shahjar, F., Khandi, S.A., Bafanda, R.A., Bhushan, B. and Khateeb, A.M. (2018). Study of relationship between socio-economic profile and perceived training needs of dairy farmers in Jammu district of Jammu and Kashmir, India. *The Pharma Innovation Journal.* **7**(11): 71-76.

Sharma, K.D., Pathania, M.S. and Lal, H. (2006). Farming system approach for sustainable development of agriculture in mountain regions – A case of Himachal Pradesh. *Agricultural Economics Research Review.* **19**(Conference Issue): 101-112.

Sharma, M.L., Saxena, R., Mahato, T. and Das, D. (2007). Potential and Prospects of Dairy Business in Uttarakhand: A case study of Uttaranchal Cooperative Dairy Federation Limited. *Agricultural Economics Research and Review.* **20**(Conference Issue): 489-502.

Shergill, B.S. (2017). Livestock economy of Punjab: Theory and empirical evidence from farm households. Conference paper. National Seminar on agribusiness potential of Punjab state at CRRID, Chandigarh.

Singh, D.R., Kaul, S. and Sivaramane, N. (2006). Migratory sheep and goat production system: The mainstay of tribal hill economy in Himachal Pradesh. *Agricultural Economics Research Review.* **19:** 387-398.

Singh, J. Grover, D.K. and Dhaliwal, T.K. (2012). State Agricultural Profile – Punjab. AERC Study No. 30. Punjab Agricultural University, Ludhiana.

Singh, J., Kumar, P. and Singh, A. (2019). Dissemination of information to dairy farmers in Jammu and Kashmir: Developing a web module. *Information Development.* **36**(4): 546-558.

Singh, P., Bhatti, J.S., Hundal, J.S. and Kansal, S.K. (2015). Constraints faced by farmers in adoption of dairy as entrepreneurship. *Haryana Veterinarian.* **54**(1): 67-69.

Singh, P.R., Singh, M. and Jaiswal, R.S. (2004). Constraints and strategies in rural livestock farming in Almora district of hilly Uttaranchal. *Indian Journal of Animal Research.* **38**(2): 91-96.

Singh, S.R. and Datta, K.K. (2013). Future of Smallholders in the Indian Dairy Sector – Some Anecdotal Evidence. *Indian Journal of Agricultural Economics.* **68**(2): 182-194.

Singh, V.P. and Kaur, I. (2018). Economics of Buffalo in Livestock Production System in Punjab: Current Status and Future Prospectus. *International Journal of Current Microbiology and Applied Sciences.* **7**(10): 2702-2708.

Sirohi, S., Sridhar, V., Srivastava, A.K., Kalamkar, S.S., Sharma, D. and Boyal, V. (2017). Ration balancing: Promising option for doubling income from dairying. *Agricultural Economics Research Review.* **30:** 193-203.

Subash, S.P. and Kaur, P. (2017). Sustainability of livestock sector in Punjab. *Indian Journal of Economics and Development.* **13**(1): 769-776.

Thakur, D., Sharma, A., Katoch, S., Chander, M., Mane, B.G. and Sharma, P. (2012). High altitude livestock farming: A participatory appraisal in Himachal Pradesh, India. *Indian Journal of Animal Sciences.* **82**(6): 644-648.

Thiruvenkadan, A.K. and Rajendran, R. (2015). Strategies to improve goat production in India. *Proceeding* of International Seminar "*Improving Tropical Animal Production for Food Security*", 3-5 November 2015, Southeast Sulawesi, Indonesia.

Tiwari, N. and Pratap, D. (2017). Use and management of forest in Chakrata tribal region of Uttarakhand Himalaya: A case study of Koruwa village. *International Journal of Multidisciplinary Research and Development.* **4**(5): 204-209.

Wani, S.A. and Wani, M.H. (2010). Marketing of Milk in various agro-climatic zones of Jammu and Kashmir. *Agricultural Economics Research Review.* **23**(1): 83-90.

Chapter 9

Rice-Fish Culture Integration in Agro-climatic Condition of Assam, India

Avinash Talukdar[1], Dharitri Baruah[2], Imtiaz Ahmed[1], Kankana Borah[1] and Chinmoy Kashyap[3]*

[1]Department of Fisheries Resource Management,
[3]Department of Aquatic Environment Management,
College of Fisheries, Assam Agricultural University (AAU), Raha – 782103, Assam
[2]World Bank, Guwahati, Assam
Corresponding author: avinashicarfish@gmail.com

ABSTRACT

Rice is a key agricultural component in Assam and grown both in flood plains and uplands covering various agro-climatic ecosystems. Around 85 per cent of cultivators in Assam are small and marginal farmers having 1 to 2 hectares land holdings and depends entirely on rice monoculture. Scientific fish culture along with the rice could be a promising approach to boost the rural economy and sustainable livelihood – two basic staple items in normal local diets. Majority of farmers in Assam have low-income sources and rice-cum-fish culture provides avenues to fully utilize the resources and economic earnings. Besides, rice-fish culture is also a zero-wastage system complimenting organic agriculture and harmonizing with the nature. Assam experiences heavy monsoon rainfall which forms two wet ecosystem including shallow water rice fields (below 50 cm water depth) and deep-water rice fields (above 50 cm water depth). These rice fields provide ample opportunity to harness productivity viz., rice and fish through technical and indigenous innovations. Several tribal communities in Assam have developed indigenous practices and have been efficiently utilizing the resources through concurrent rice-fish venture. Yet, the prospects of rice-fish culture in not commercialized in large scale or potentiality is not realized by the small and marginal farmers due to several socio-economic factors. Nevertheless, integrated rice-fish culture is a way forward towards organic farming and augments the optimize use of land and water resources through synergies between rice, fish and environment.

Keywords: *Rice-fish farming, Integrated farming, Organic farming, Assam.*

Introduction

The global population is set to touch the 9.8 billion mark by 2050 and the production of surplus food to feed the growing population is becoming inevitable to support the nutritional demands. The demand for food can only be fulfilled through utilizing the existing resources to their full potential where the quality of the product and sustainability goals are certainly compromised to meet the requirements. Considering the present scenario, certain strategies need to be established for efficient food production with a focus on resource sustainability. In this context, we have to produce sufficient food from utilizing the same resources without horizontal expansion of agriculture or its allied ventures. Further, the growing use of antibiotics, chemicals, and fertilizers to increase the per unit area agricultural production has residual effects which created health concerns for the consumers. India is an agriculture-based country where the life span of any individual is significantly depends on agricultural products directly or indirectly. Therefore, considerable attention is required to address the issues regarding environmental and human health safety through environment-friendly approaches by sustainable scaling the agricultural production. When sustainability comes, organic farming is considered as an environment-friendly culture practice that mainly relies upon the natural or ecological cycles with no chemical inputs (Das *et al.,* 2017). Organic farming mainly focused on promoting biological activity by assuring sustainability, nutritional security and health safety for the upcoming generation (Deepitha *et al.,* 2019).

Generally, nature works on the principle of mutual relationships between the environment and living organisms. Most of the scientific techniques and concepts were developed or understand based upon natural interactions. The concept of integrated farming also emerges from the natural integration where two or more related farming activities are integrated to better utilize the available resources. Integrated farming can be defined as "*sequential linkages between two or more agricultural related farming activities with one farming as major component*" (Handbook of Fisheries, 2018). The integrated farming system ensures sustainability of natural resources, efficient resource utilization, risk reduction and provides supplementary income for small and marginal farmers. This system mainly works on the principle of nutrient recycling where the waste of one system is an input for the other system and maximizing the farm production per unit area. When we consider combining rice culture with fish, then, is called as 'rice-fish integrated culture system', where rice is the main farming component. The culture of fish with rice usually augment the production of rice through fish movement resulting in good aeration and tilling, increased fertility of the soil by the fish excreta waste generated and feeding on harmful insect and weeds pest which is detrimental for paddy. Rice-Fish culture thus enables efficient use of land to produce two basic food items by complementing each other. Therefore, the concept of organic rice farming with fish culture will not only delivers the farmer's surplus income but also promote holistic rural development.

Rice Culture in Assam

Rice certainly the most important and staple food item for humans especially in Asia with global annual production in 2021-22 of 510.8 million metric tons (USDA). Unlike other cereals, rice contributes significantly toward global food security and human nutrition. In Assam, rice is the most widely consumable crop cultivated in over 2.4 million ha of the area and accounts for 96 per cent of state food grain production. The annual production of rice in Assam in 2019-2020 is 5.21 million tonnes with the area under a high yielding variety of rice being highest in Lakhimpur followed by Golaghat and Chachar districts (Statistical Handbook of Assam, 2020). However, the productivity of the rice is either declined or stagnant in the last decade possibly due to the inappropriate adaptation of the technology paradigm which is mainly specific to certain favourable farming situations. Further, high rainfall, flood and water-logged rice growing situations call for an alternate approach to commensurate the farm production in such areas. Nevertheless, the state has three rice growing seasons based on different agro-climatic conditions, seasonal variation and rainfall (Ahmed *et al.*, 2011).

Seasonal Class	***Growing Season***	***Duration (Days)***	***Remarks***
1) Ahu or Autumn rice	March-April to June-July	80-130	Photoperiod insensitive, early maturing usually broadcast, grown under variable water depth (0-25 cm). In Barak valley, ahu rice is further classified as Dumai, Murali and Chengri.
Dumai		80-90	Generally grown broadcast, Red kernelled.
Murali		90-100	
Chengri		90-100	
Ahu (transplanted)		>100	Grown transplanted.
2) Sali or Winter rice	June-July to Nov.-Dec.	150-180	Photoperiod sensitive, long duration, grown transplanted under variable water depth (0-30cm) There are sub-classes like sali, lahi and joha based on the grain characteristics.
Sali			Coarse grained called sail in Barak valley.
Lahi			Medium grained.

Seasonal Class	Growing Season	Duration (Days)	Remarks
Joha			Scented, fine grained.
Boro			Glutinous or sticky rice, called birain in Barak valley.
Chakuwa			Soft rice with low amylose content.
Asra	April-May to Dec.-Jan.	240-270	Medium deep water rice, grown broadcast or transplanted in the low lying areas, can endure water depth<100 cm. Sown at the time of ahu rice is sown and harvested at the time when sali rice is harvested.
Bao	April-May to Jan.-Feb.	270-300	Deep water or floating rice, normally grown broadcast, can endure water depth>100 cm. Sown at the time when ahu rice is sown and harvested at the time when sali rice is harvested.
3) Boro or Summer rice	Nov.-Dec. to May-June	180-200	Photoperiod insensitive, cold tolerant at the vegetative stage, grown transplanted, traditionally in the beel and marshy land situation with minimal or no tillage.
4) Hill or Jhum rice	Mar.-Apr. to Sep.-Oct.	210-250	Photoperiod insensitive. Grown in hills slope as direct seeded.

The different varieties of rice-based on seasonal adaptation are:

a) **Semi dwarf:** IR-36U, Jaya, Pankaj, Lakhimi, Bahadur, Piolee, Kushal, Moniram, Ranjit, Kmj 10-2-2 and TTB 101-15.

b) **Glutinous:** Rangalee, Bhogalee, Aghoni, Kmj 3-144 and Kmj 2-9-2.

c) **Multiple cropping:** Satya and Basundhara.

d) **Tall:** Monoharsali, Mahsuri and Swarnaprova.

e) **Scented Rice:** Keteki jaha.

f) **Post flood transplanted (late Sali):** Manohar Sali, Biraj, Andrewasali, Solpona, Prasad bhog, Govindabahog and Kmj-1-19-1.

g) **Unfavorable rainfed lowlands:** Prafulla and Gitesh.

h) **Rainfed, lowland flash flood-affected areas:** Jalashree and Jalkunwari

Rice-Fish Integrated Farming – A Harnessing Complementarities

Paddy cultivation and wild fish collection from water bodies encompass the major traditional economic activity for the majority of rural households in India for earning their livelihood. The state of Assam is bestowed with two major river systems namely Brahmaputra and Barak Rivers and receives monsoon rainfall (150-

200 cm) during the major rice-growing season (June to November). The different rice-growing fields covering floodplains to hilly regions in Assam, therefore, receive water from both rain and neighbouring rivers in the late month of July/August. The aquatic phase in the rice field remained for 4 to 6 months with varying water depths until the decline of water at the end of November. This allows wild fishes to enter the low-lying paddy fields and eventually grow in the habitat. These water-logged rice fields provide a natural habitat for varieties of fishes and are the most common fishing grounds for the adjacent rural people during June to November (wet season). The wet rice fields provide a productive biological system for rice where different organisms and plants colonize to form a stable ecosystem. Thus, simultaneous fishing of auto-stocked fishes in wet fields during the rainy season is an old-age practice in North-East India. Gradually, this traditional method is tuned more scientifically and has developed different systems of rearing fish in paddy fields which can be called "*Rice cum Fish culture*" by integrating rice and fish at the same time (Figure 9.1). Further, culturing fish with rice greatly reduces the application of chemicals and pesticides in the rice crop which could be a way forward towards organic farming. The benefits of culturing fish with rice are as follows:

a) Rice-Fish integrated farming system yield more income because the occurrence of fishes did not jeopardize the production of rice,

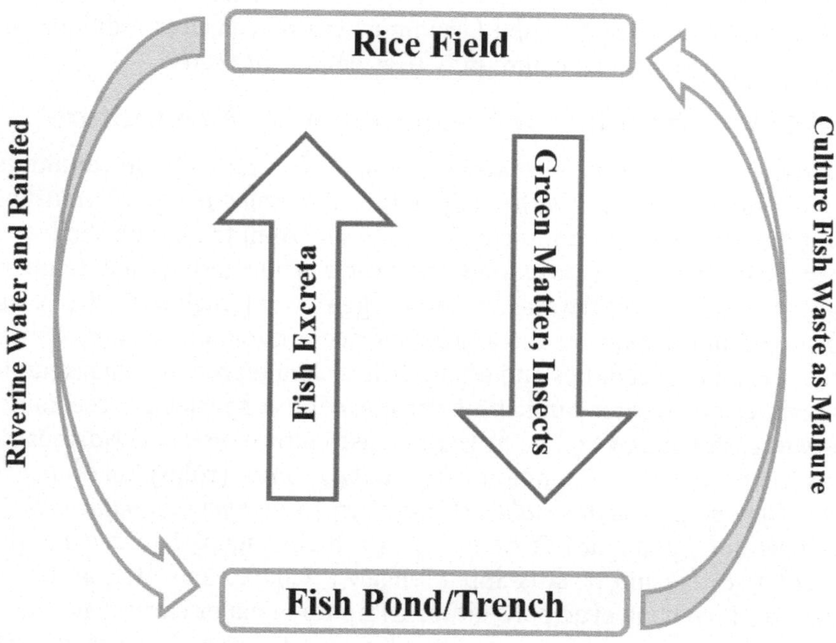

Figure 9.1: Biological Interaction in the Integrated Rice-Fish Farming System.

b) Presence of fish in wet rice fields increases soil fertility through the mineralization of fish excreta which also enhanced the availability of nutrients for rice crops,

c) Fish also graze on unwanted algal blooms and weed plants which otherwise could compete for nutrients,

d) Continuous movement of fish increases the oxygen content in water and soil,

e) Fish helps in releasing nutrients from soil to water by dispersing the soil-water interface, and

f) Fish act as a biological control agent in rice fields as it feeds on insect pests and snails which otherwise is harmful to the rice crop.

Around 80-85 per cent of the farming population of Assam are considered to be small and marginal farmers having very low-income generating sources to run their livelihood. Their basic source of income relies upon rice cultivation and fishing from the adjacent rivers or from the rice field where fishes take shelter during the monsoon season. In this context, scientific integration of rice and fish culture could augment their farm production with greater economic benefits and fully utilize the available resources. Fish act as a biological control for pest and weed management in paddy fields that could substantially reduce paddy growth. Further, this integrated system increases soil productivity by partially eliminating the application of any inorganic nutrient or fertilizers. Rice-Fish culture is a low-cost, ecologically sound and reliable farming activity that ensures additional income for the low-income farming community (Goswami *et al.*, 2004).

Suitable Fish Species for Integration with Rice Culture

Certain considerations are necessary for selecting suitable candidate fish species for integration along with paddy culture. Generally, fish species must be able to resist unfavourable environmental conditions prompted by the shallow water depth of the rice field. Fish must be able to survive in fluctuating water temperature (up to 10-40°C), high turbid water and low oxygen levels (Singh *et al.*, 2011). Further, fishes should not jeopardize the efficacy of rice production as some fishes may dislodge the paddy seedlings and finally, fishes should grow to a marketable size in a short period. Several compatible varieties of fish species are recommended for the successful integration of the rice and fish culture system in Northeast India such as *Cyprinus carpio* (common carp), *Labeo rohita* (rohu), *Cirrhinus mrigal* (mrigal), *Hypopthalmichyths molitrix* (Silver carp), *Ctenophyryngodon idella* (grass carp), *Labeo bata* (bhangun), *Cirrhinus reba* (lachit bhangun), *Labeo gonius* (kurhi), *Labeo catla* (catla) and Mystus Spp. (Tengara). Tillapia and African magur can also be introduced however, polyculture with Indian major carps is not practical. Other than these, some catfishes, small indigenous fishes and air breathing fishes are naturally found in the wet rice fields namely *Anabas testudineus, Sperata aor, Channa marulias, C. punctatus, C. striatus, Clarias magur, Heteropneustes fossilis, L.*

calbasu, *Macrognathus aral*, *Mastacembelus puncalus*, *Nandus nandus*, Puntius Spp. and *Notopterus notopterus* (Das, 2018).

Types of Rice-Fish Farming Based on System and Practices

1. Concurrent Rice-Fish Culture System

This type of rice-fish farming practice is most common in Assam which involves the culture of fish with rice at the same period simultaneously by utilizing the same resource. This type of culture system enables the biological interaction in the rice-fish ecosystem and at once has economic benefits from the two agricultural produces. The suitable candidate species for culture take less time to attain the marketable size. Typically, around 10-20 per cent of the rice field area is used to construct a fish pond or trench with a side slope of 1:1, 10-20 cm field ridge than the highest water level and 1-1.5 m water depth. This type of culture system usually yields 300 to 1000 kg ha^{-1} of fish harvest with 2500 to 3600 kg ha^{-1} of paddy and 4250 to 5000 kg ha^{-1} of straw (APART, 2018). Generally, fry fishes can be stocked at the rate of 60,000 fry ha^{-1} during the early weeding period of paddy (21-28 days) and fingerlings can be stocked at the rate of 1,000 to 2,000 fish ha^{-1} after the subsequent weeding of paddy (40 to 50 days) (Halwart and Gupta, 2004). In Assam, a suitably tall variety of paddy mainly *Ahu* and *Sali* dhan are mostly cultured along with fish covering both seasons. The recommended varieties of *Ahu* are Rangadoria, Fapori, Govind, Lachit and CH-63. Whereas, recommended varieties of *Sali* are Pankaj, Manohar, Ranga Sali, Gudumoni, Badshah bhog, Ranjit, Jalashree *etc.* Based on the system, three models of concurrent rice-fish farming were developed *viz.*, perennial system, synchronous refuge system and enclosure system.

In the perennial rice-fish farming system, one crop of fish is cultured along with two crops of paddy namely *Ahu* and *Sali* during the winter and autumn seasons. All the species of Indian Major Carps except grass carp are generally reared to table size. The perennial system of raising fish along with rice is more suitable for low-lying areas where fish ponds are constructed covering $1/3^{rd}$ of the total rice cultivating area. In a synchronous refuge system, the fish crop is reared simultaneously with the *Sali* paddy field during the monsoon season period where the cultivating plot is renovated to provide shelter for fish. This system is mainly suitable for medium low-lying areas with poor drainage and plot having naturally existing small ponds. The inner embankments are deferred to allow free passes of the fish from ponds to the paddy field. In an enclosure system, the fish crop is raised with *Boro* paddy (deep water paddy variety) in deep water areas by enclosing the plot with bamboo pegged screens (kept 1 m above the highest flood level). This system is simple and advantageous as compared to others as no perpetual loss of land due to pond construction is required.

There are several designs of rice-fish farming depending on the topographical situations of the paddy cultivating plot (Figure 9.2). However, the perimeter trench model, lateral trench model and central trench model are more suitable for the topographic situation of Assam. The perimeter trench is constructed in fields

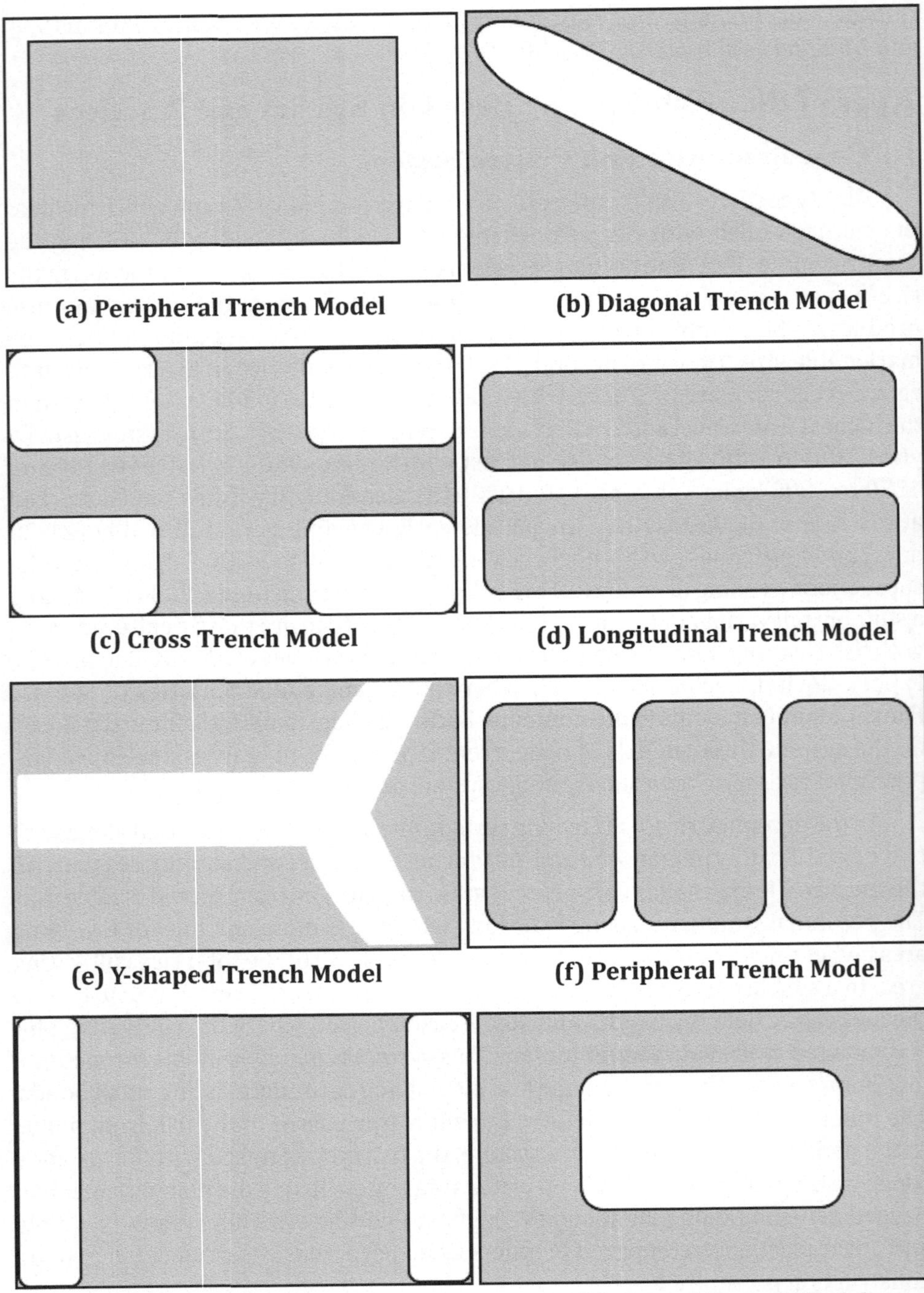

(a) Peripheral Trench Model

(b) Diagonal Trench Model

(c) Cross Trench Model

(d) Longitudinal Trench Model

(e) Y-shaped Trench Model

(f) Peripheral Trench Model

(g) Lateral Trench Model

(h) Central Trench Model

Figure 9.2: Diagrammatic Representation of Various Models Adopted for Rice-Fish Farming System. *Green colour depicts rice culture and white colour depicts fish culture. (*Source*: Halwart and Gupta, 2004).

having moderate raise towards the middle and ground sloping on all sides to allow drainage of water. Whereas, the lateral trench is constructed on one or both sides of the rice-growing fields and the central trench is made in the middle of the rice field.

2. Alternative Rice-Fish Culture System

In this type of culture system, rice cultivation and culture of fish are commenced one after another in the identical field following crop or annually. Shuffling between rice-fish culture and only fish culture can generate higher income for the farmers. However, adopting a combination of two rice crops followed by fish crops or two rice-fish crops followed by only fish crops would be more practicable in getting higher economic benefits (Das, 2018). But these combinations of alternative culture systems are only meant for growing the fish to a table size usually for 40-45 days. The most common method followed for this type of culture system is having single crop of paddy followed by single crop of fish within a year. Besides, the biological interaction between fish and paddy is not present in the alternative culture system as compared to the concurrent culture system. Although, unlike the concurrent culture system the alternative culture system does not require any extra earthwork to construct a trench or pond. Alternative culture systems allow fish to grow for a longer period and are appropriate for fish species that can adjust to deep water. Generally, the fishes are stocked after the harvesting of paddy in October and harvested before transplanting the paddy seedlings in July. The fish production from this type of culture system ranges between 1500 to 3000 kg ha^{-1} (Singh *et al.*, 2011).

3. Rice-Field Capture-Based Fishery and Wild Fishery Farming System

The rice-field capture-based fishery is usually unmanaged where wild fishes or fish seeds found their entry to rice fields along with floodwater, thus forming a capture or capture cum culture fishery. The vast waterlogged rice environment provides favourable habitat for fishes to grow where farmers drive their innovations and practices to harness maximum output from the available resources. This is more common along with the perennial or flooded rice fields of the Brahmaputra and Barak valley of Assam. The fishing practices start right from the inundation of water into the rice fields during June and continue till the decline of water level till November. As this practice relies entirely on the natural stocking of fishes, the average rate of fish capture is 3 to 5 kg ha^{-1} yr^{-1} (Bhuyan, 2014) and about 72.5 per cent of farmers do fishing to earn only the average household income (Pegu *et al.*, 2019). Further, these waterlogged flooded rice fields are common property thus community fishing is often practiced by the tribal folks of Assam. The rice field capture fishery is mostly common in deep water rice fields in the districts of Dhemaji, Lakhimpur, Barpeta, Nalbari, Bongaingaon, Dhubri and Cachar districts of Assam and among which Lakhimpur had the highest rice area (58430 ha) followed by Barpeta (47895 ha) and Nalbari (29564 ha). The average production of fish from this farming practice ranges between 78 to 138 kg ha^{-1} $season^{-1}$ (Das, 2018).

The wild fishery cropping system is practiced in seasonally flooded rice fields of Assam and culture of fish is linked with the paddy cropping cycles. In this system, farmers allow fish to pass in the whole rice fields or deeper areas and grow it till the recession of the flood water *i.e.* late November. The water level in such fields may rise up to 2.5 m where a tall variety of rice such as *Sali* is sown during April-May. These fields are then inundated with rainwater during late July/August and have connectivity to the adjacent water courses allowing entry to the wild fish and seeds. Farmers usually dig small trenches or canals (8 per cent of the total area and 0.6 to 1.5 m depth) inside the rice field to provide shelter for fishes and varieties of fishes enter these ditches till the water level goes below the entry level. The fishes are harvested after complete removal of water after the harvesting of rice crop and the production of fish varies from 200 to 300 kg ha^{-1} season^{-1} (Das, 2018). For example, in the Barpeta, Bongaingaon and Kokrajhar districts of Assam, farmers construct canals connected to the perennial water bodies locally known as 'Dong' which is guarded with bamboo screens to avoid the escape of fishes that already finds its way to the rice fields. The classic example for this type of farming is *Jungal Balahu Garh* (37 ha) in Nagaon district of Assam where horizontal deep ditches with high perimeter dyke are constructed along with the rice farming (Ghosh, 1992). Therefore, fish farming in flooded fields of Assam is generally culturing of auto-stocked fishes from riverine sources with no such management practices being followed. The concurrent rice and fish culture activity is merely followed in the state of Assam which could be due to the lack of inputs, appropriate training, knowledge and small landholdings.

Economic Profitability

Several socio-economic surveys concluded that integrated rice-fish culture offers high profitability to the farmers. An obvious increase in revenue is possible because farmers can produce two marketable agricultural products by optimizing the utilization of available resources. Further, culturing fish along with paddy can substantially surge the rice yield that to a monoculture of only paddy (Frei and Becker, 2005). The cost of insecticides, pesticides and inorganic fertilization is also greatly compensated in the rice-fish culture system which is a way forward towards organic and sustainable agriculture. Although, capital and operational cost incurred in the rice-fish culture practice cannot be denied and must be taken into consideration. Capital costs include costs involved in the excavation of ponds or trenches and the construction of dykes. The operational cost involved labour cost, buying of fish seed or fingerlings (if not depends on auto-stoking), supplementary feed and fertilization. However, despite of having additional initial cost, the profitability of rice and fish culture is incontestable and allows greater fish availability to the rural farming community (Gupta *et al.*, 1998; Goswami *et al.*, 2004) (Table 9.1). The potentiality of the rice-fish culture system in Assam is about 501.1 kg ha^{-1} in 4 to 5 months of fish culture with 17.5 per cent increase in rice production with just 10-20 per cent area used for ponds or trenches and dykes (Baruah *et al.*, 2000).

Table 9.1: Economics of Perennial System of Rice-Fish Farming (0.28 ha)

Sl.No.	Particulars	Quantity	Rate (₹)	Cost (₹)
A.	VARIABLE COST			
1	Earthwork	-	-	8000.00
2	Bamboo Bridge	-	-	1000.00
3	Ahu crop	-	-	2000.00
4	Sali crop	-	-	2000.00
5	Agricultural lime	95 kg	7 kg^{-1}	665.00
6	Fish seed	2160 Nos.	5 each	10,800.00
7	Cow dung	259 kg	0.6 kg^{-1}	156.00
8	Urea	5.30 kg	10 kg^{-1}	53.00
9	SSP	4.35 kg	10 kg^{-1}	43.00
10	Mustard oil cake	482 kg	20 kg^{-1}	9640.00
11	Rice Polish	482 kg	10 kg^{-1}	4820.00
12	Harvesting cost	-	-	1500.00
13	Misc. expenditure	-	-	1000.00
	Total			**₹ 39,520.00**
	Interest @ 10 per cent			**₹ 3952.00**
	Total Variable Cost			**₹ 43,472.00**
Sl.No.	**Product**	**Quantity (kg)**	**Rate (₹ kg^{-1})**	**Sale**
B.	RETURN			
1	Paddy	1053	11.25 kg^{-1}	₹ 11,846.00
2	Hay	1215	2.00 kg^{-1}	₹ 2430.00
3	Fish	405	120 kg^{-1}	₹ 48,600.00
	Total Return			**₹ 62,876.00**
	Area for Paddy Cultivation			0.182 ha
	Area for Trench and Dyke			0.094 ha
	Income from Paddy Cultivation			₹ 14,276.00
	Income from Fish Culture			₹ 48,600.00
	Gross Income from Perennial System of Rice-Fish Farming			₹ 62,876.00
	Net Income (Profit) from Perennial System of Rice-Fish Farming			₹ 19,404.00
	Per cent Profit to Turnover			30.86 per cent
	Per cent Return on Variable Cost			44.64 per cent
	BCR			1.45

Source: Package of Practice of Perennial System of Rice-Fish Farming, APART (2018).

Table 9.2: Annual Work Calendar for a Perennial System of Rice-Fish Farming

Months	*Works for Paddy Cultivation*	*Works for Fish Cultivation*
Jan.-Feb.	☆ Plot renovation	☆ Excavation of trench/pond ☆ Dyke construction
Feb.-Mar.	☆ Plot preparation for *Ahu* ☆ Ploughing and manuring ☆ Preparation of seed bed (for transplanted variety)	☆ Cleaning of trench/pond, muck removal ☆ Eradication of weed and unwanted fishes (if necessary)
Mar.-Apr.	☆ Sowing of seeds ☆ Initial plant protection measures	☆ Liming ☆ Manuring 7 days after liming
Apr.-May	☆ Transplantation ☆ Water management ☆ Weeding	☆ Fertilization 7 days after manuring
May-Jun.	☆ Weeding ☆ Top dressing ☆ Plant protection measures ☆ Water management	☆ Stocking of fish seed 15 days after transplantation/ fertilization ☆ Supplementary feeding
Jun.-Jul.	☆ Harvesting of *Ahu* crop ☆ Seed bed preparation for *Sali* crop ☆ Plot preparation for *Sali* crop	☆ Manuring and fertilization ☆ Liming ☆ Supplementary feeding ☆ Health monitoring
Jul.-Aug.	☆ Sowing of seed bed ☆ Plot preparation for *Sali* crops ☆ Manuring/fertilization ☆ Initial plant protection measures	☆ Liming ☆ Supplementary feeding ☆ Manuring and fertilization ☆ Health monitoring
Aug.-Sep.	☆ Transplantation of *Sali* ☆ Plant protection measures ☆ Water management	☆ Raising of water level 15 days after transplantation ☆ Liming ☆ Supplementary feeding ☆ Manuring and fertilization
Sep.-Oct.	☆ Plant protection measures ☆ Water management ☆ Top dressing	☆ Manuring and fertilization ☆ Supplementary feeding

Months	Works for Paddy Cultivation	Works for Fish Cultivation
Oct.-Nov.	☆ Plant protection measures ☆ Water management	☆ Manuring and fertilization ☆ Supplementary feeding
Nov.-Dec.	☆ Harvesting	☆ Partial harvest followed by final harvest.

Constraints in Adoption of Integrated Rice-Fish Culture

Nevertheless, the integrated rice-fish culture practice is a profitable venture for the small and marginal farmers, in reality, the adaptation of such practice is marginal due to socio-economic, environmental and technological constraints (Nabi, 2008). Several factors are also responsible for limiting the popularization and adaptation of such activity:

a) Large number of farmers in Assam are small and marginal farmers with less own land, therefore, they are reluctant to take risks associated with losing the rice crop or to adopt unfamiliar activities,

b) Lack of technical knowledge and extension services among farmers, as the rice-fish culture needs skills and knowledge,

c) Unlinked policy frameworks which give focus to only intensive monoculture of rice,

d) The capital requirement is also high in a rice-fish culture which becomes a major drawback for the marginal and small farmers,

e) Lack of proper water disposal facilities, low water holding capacity of soil, unfavourable topography and loss of fish due to flood,

f) Ambiguous use of chemicals in intensive rice culture to increase the rice production,

g) Prone to natural calamities such as high flooding, poaching and poisoning, and

h) Farmers are eventually more interested towards short term benefits and reluctant to change their existing traditional rice culture practice.

Measures taken in Adoption of Integrated Rice-Fish System

The measures to preserve the present practice of rice-fish farming system is as follows:

a) Location-specific technology refinement is necessary before recommending to farmers of any agro-climatic zones of the region,

b) Training and awareness camps are necessary to promote the culture system among the rural community through people participation,

c) Financial support through implementing various governmental schemes for promoting the culture system to incur the cost including labour, fish seed, feed and fertilizer,

d) Better irrigation facilities to enable rice and fish culture integration that is sali rice-fish culture followed by *Boro* rice-fish culture depending on agro-climatic condition,

e) Present practice of rice-fish culture system in the sector required SWOT analysis for its sustainable promotion and expansion,

f) Precedence should be on the organic farming approach and negligible use of agrochemicals,

g) Adaptation of community-based management practice in large wet rice fields, proper monitoring of entry and exit point of water channel which could led to harness better output from the culture system, and

h) Construction of strong dykes at considerable height, bamboo barriers, or pens are necessary to prevent the escape of fish during flooding.

Conclusion

The tremendous growth in agriculture is mainly driven by 'Green Revolution' aimed at higher crop production per unit area through extensive use of chemicals and pesticides. On contrast, the agriculture production no uncertainty increased but at the cost of degrading the environment sustainability. Organic farming provides excellent benefits in terms of produce quality, health and environmental concerns. The integration of rice-fish also ensures the nutritional security to the growing population as both are cheap source of quality protein and carbohydrates. Further, it also enhances the efficiency of farm productivity, judicious use of resources, diversified harvest, increase rice production and greater economic benefits to the farmers. However, despite of huge potentiality of rice-fish culture in Assam, majority of farmers are inclined towards unintentional culture of fish in rice field where they do fishing only to meet there daily food requirement. There is an urgent need of extension services and technical demonstration of rice-fish culture to make them realize the potentiality of such culture system in Assam. In Assam, the prospects of rice-fish culture system are very congenial but needs detailed research outcomes by integrating the policy makers, researchers and rice cultivators to develop a package of practices which have to be location specific. Although, rice-fish culture will remain an ideal way to bring up the concept of organic agriculture and moving towards a greener earth.

REFERENCES

Ahmed, T., Chetiam S.K., Chowdhury, R. and Ali, S. (2011). Status paper on rice in Assam. Rice knowledge management portal. Directorate of rice research, Hyderabad: Rice State Wise, pp. 1-49.

APART. (2018). Package of Practices on Fisheries and Aquaculture in Assam. Directorate of Fisheries, Government of Assam, pp.76-93.

Baruah, U.K., A.K. Bhagowati and R.K. Talukdar. (2000). Prospects of fish farming in rice fields in Assam. Indian Journal of Fisheries, **47**:149-153.

Bhuyan, D. (2014). Paddy cum fish farming - a sustainable way of agriculture, practiced in Jorhat district of Assam, with special reference to Hatigarh village. *Journal of Aquatic Biology and Fisheries*, **2**:126-132.

Das, A., Patel, D. P., Kumar, M., Ramkrushna, G.I., Mukherjee, A., Layek, J., Ngachan, S.V. and Buragohain J. (2017). Impact of seven years of organic farming on soil and produce quality and crop yields in eastern Himalayas, India. *Agriculture Ecosystem and Environment*, **236**:142-53.

Das, D. N. (2018). Farming of fishes in ricefields of Northeast India: A review. *Coldwater Fisheries Society of India*, **1(1)**: 27-41.

Deepitha, R.P., Roy, S.D., Sankar, R.K., Saravanan, K., Haridas, H. and Praveenraj, J. (2019). Rice-Fish Culture for Nutritional and Livelihood Security in Coastal India. An Overview of Integrated Farming Systems of Coastal India, pp.113-121.

Frei, M. and Becker, K. (2005). Integrated rice]fish culture: Coupled production saves resources. In *Natural Resources Forum*, Oxford, UK: Blackwell Publishing, Ltd, Vol. 29, pp. 135-143.

Ghosh, A. (1992). Rice-fish farming development in India: past, present and future. *In* Rice-fish research and development in Asia (eds. C. R. dela Cruz, C. Lightfoot, B. A. Costa-Pierce, V.R. carangal and M.P. Bimbao). ICLARM Conference Proceedings, 24: pp. 27-43.

Goswami, M., Biradar, R. S. and Sathiadhas, R. (2004). Techno]economic viability of rice]fish culture in Assam. *Aquaculture Economics and Management*, **8 (5-6)**:309-317.

Gupta, M. V., Sollows, J. D., Mazid, M. A., Rahman, A., Hussain, M. G. and Dey, M. M. (1998). Integrating aquaculture with rice farming in Bangladesh: feasibility and economic viability, its adoption and impact. ICLARM.

Halwart, M. and Gupta, M. V. (2004). *Culture of fish in rice fields*. FAO, World Fish Center, Malaysia.

Handbook of Fisheries and Aquaculture (2006). Integrated Farming, 3rd and Revised Edition. Indian Council of Agricultural Research, New Delhi, India.

Nabi, R. (2008). Constraints to the adoption of rice-fish farming by smallholders in Bangladesh: a farming systems analysis. *Aquaculture Economics and Management*, **12(2)**:145-153.

Pegu, R., Singh, Y. J., Pal, P., Upadhyay, A. D., Sarkar, A., Khuman, O. N. and Bharati, H. (2019). Paddy cum Fish Farming: A Case Study in Assam. *International Journal of Current Microbiology and Applied Science*, **8(4)**:373-380.

Singh, A. S., Devi, M. S., Mandal, S. C. and Barman, D. (2011). Paddy cum fish culture in north-eastern states of India: A better scope in the exploitation of resources. *World Aquaculture*, 58-60.

Statistical Handbook of Assam (2020). Directorate of Economics and Statistics of Assam, Beltola, Guwahati-28. pp. 77.

United States Department of Agriculture (USDA, 2021). Rice Outlook, Monthly report, December, 2021, Washington DC.

Chapter 10

Horticulture-cum-Fish Culture: An Integrated Approach

R. Somu Sunder Lingam[1]*, S. Aanand[2] and P. Velmurugan[3]

[1]*Assistant Professor, Krishnagiri-Barur Centre for Sustainable Aquaculture, Tamil Nadu Dr. J. Jayalalithaa Fisheries University, Barur, Krishnagiri – 635201, Tamil Nadu*

[2]*Assistant Professor, Erode-Bhavanisagar Centre for Sustainable Aquaculture, Tamil Nadu Dr. J. Jayalalithaa Fisheries University, Bhavanisagar, Erode – 638451, Tamil Nadu*

[3]*Assistant Professor, Kanyakumari-Parakkai Centre for Sustainable Aquaculture, Tamil Nadu Dr. J. Jayalalithaa Fisheries University, Parakkai, Kanyakumari – 629601, Tamil Nadu*

Corresponding author: somusundarlingam@tnfu.ac.in

ABSTRACT

In future, in the context of lack of natural resources and malnutrition, integrated farming practices are going to be well accepted by farming communities to address the contemporary challenges. Presently, there are two major integrated farming practices, such as animal to animal integration and animal to plant integration, available to the farming communities to achieve the local needs. However, the farmer prefers to adopt the one which requires less technical knowledge, lower investment and higher returns. Among the various integrated systems, integration of fish with agriculture plants is considered as a highly sustainable production model. Horticulture cum fish culture is a traditional type of farming practice in which fish culture is integrated with horticulture plants like vegetables, fruits and flowers.

Apart from the water spread area, fish culture pond offers a very good productive area, especially the dyke area, for horticultural plants cultivation. The wise exploitation of pond dyke results in lower operational cost and better management of farm which in turn increase the overall output of the farm. Traditional horticulture cum fish culture system can be sub-divided into vegetables cum fish culture, fruits cum fish culture and flowers cum fish culture. The aquaponics system, a modern type of horticulture cum fish farming, is also evolving as more potential rearing system to tackle the modern challenges faced by the urban farmers. In general, integration is going to occupy an important role in future to fulfil the nutritional requirement of the population, in a more sustainable way. However, compare to other integrated food production systems, horticulture cum fish system produces both animal and plant based nutritious food which will play a dynamic role in sustaining the production and supply of balanced diet to the human population.

Keywords: *Aquaculture, Aquaponics, Integration, Food, Fruits, Vegetables.*

Introduction

All animals, including human beings, require a balanced diet for a healthy life in this planet. However, production and distribution of food determine the food consumption pattern which in turn influences the health status of the population (NIN, 2011). Besides supplying valuable nutrients, food plays a crucial role in the existence of human population. The production and distribution of food, either plant based or animal based, through agriculture makes more employer in the world. In addition to employment, it also involves in fulfilling the local nutritional requirement as well as development. However, demand for nutritious food, rural upliftment and poverty mitigation are going to be the need of present as well as future. On the other side, in the context of climate change, lack of natural resources, low productive soil, poor water quality and other natural calamities may significantly hinder the progress of inclusive growth in future. At this juncture, integrated farming is considered as the only solution for subsistence farming by farming communities.

In the integrated farming system (IFS), one or two separate inter-dependent systems will be connected, where one system output was used as input for the other production system and it is also called as mixed farming (Mishra *et al.*, 2020). Unlike present modern agriculture systems, IFS is an age-old subsistent farming practices which mainly concentrating on the welfare of farming community. Compare to monoculture intensive farming practices, integrated farming system have more added advantages like high productivity, profitability, sustainability, recycling of wastes, nutritious food supply, *etc.* Among the various IFS systems, combination of fish and agricultural crops or livestock (Integrated Fish Farming -IFF) is reflected as the most efficient farming system. The IFF system can be broadly classified into agri-based fish farming and livestock-based fish farming. However, apart from energy efficiency, the wider expansion of fish consumption pattern and the demand for fruits and vegetables making the aqua-agri integration, horticulture -cum fish, as an inevitable IFF system of future.

In India, 80-85 per cent of farming lands are located in the rural area and those who practice the farming are mainly small or marginal farmers (Laxmi *et al.*, 2015). On the other side, Indian Council of Medical Research has recommended to consume around 300 g of vegetable and 85 g of fruits per day for a healthy life. In this context, horticulture-cum fish culture, cultivation of horticulture plants like vegetables, fruits and flowers with fish, is a profitable and eco-friendly farming practice to the small and marginal farmers. In this method, the embankment and dyke of fish pond and other adjoining areas can be well exploited for growing of vegetables, fruits and flowers. This system has many advantages like;

a) No need of any additional infrastructure creation and input addition,

b) Accurate utilization of empty/vacant spaces,

c) Availability of horticulture seed and other inputs,

d) Fertile fish culture water used for irrigation and bottom silt as manure,

e) Less technical and small and marginal farmers are more familiar with this practice,

f) Easy access to local market and 20-25 per cent more return,

g) Lower risk of disease and natural hazards, and

h) Recycling of crop residues.

Therefore, it is the right time to promote this horticulture-cum fish culture at commercial level for sustaining the production and supply of balanced diet to the human population.

Present Scenario of Horticulture-cum-Fish Culture

Around 80 per cent of the country's current water, a major agriculture input, has been used by agriculture and allied sectors. However, the post green revolution era has been subjected to sustainability issues due to its over-exploitation of natural resources. This was the turning point of integrated based agriculture in India. As per the latest Economy Survey, the growth of agriculture sector, including livestock dairy and fisheries, was 3.6 per cent during 2020-21 and it was resilient during 2021-22 with a growth rate of 3.9 per cent. Agriculture and allied sectors alone contributed a GVA (Gross Value Added) of 20.2 per cent to the overall economy during 2020-21. India achieved a record production of around 308.65 million tonnes of food grain in 2020-21. On the other side, India stands 2nd position in the fish production at global level, with net contribution of 7.56 per cent. Fisheries alone contribute 1.24 per cent of the country's GVA and 7.28 per cent to the agriculture sector with an estimated net fish production of 14.5 MT during 2020-21. Additionally, in the past few years, from 2014-25 onwards, the sector registered a double digit annual average growth rate of 10.87 per cent. Further to promote the integrated fish farming among small scale farmers, the government has recently announced Kishan Credit Card (KCC) to meet out their working capitals and launched a flagship scheme of Pradhan Mantri Matsya Sampada Yojana (PMMSY) worth of Rs. 20050 crores.

The rapid change in lifestyle of population had significantly altered our food consumption pattern. However, the evolution of health-conscious population had increased the demand of nutritive food in the market. Horticulture, a branch of agriculture, has supplied the agri-products like fruits, vegetable and flowers to the market through their sub-culture practices like pomology, olericulture and floriculture, respectively. The present production of horticulture products are 320.48 MT which alone contributes around 33 per cent of the country's GVA. The implementation of various government schemes, during 2014-2020, had significantly improved the horticulture crop cultivation area from 9 per cent to 14 per cent. For further boost up of this sector, the government has launched a Mission on Integrated Development of Horticulture (MIDH) for an amount of Rs. 2250 crore. Therefore, with an immense financial support from the Govt. sector, the integrated horticulture-cum fish culture can travel a long way in a more secured condition.

Types of Horticulture-cum-Fish Culture

The horticulture-cum-fish culture has three different types of culture systems based on the type of plant species integrated. These three systems are highly traditional and do not integrate any modern management/technical practices. These systems are: i) fruits-cum-fish culture, ii) vegetables-cum-fish culture and iii) flowers-cum-fish culture.

1) Fruits-cum-Fish Culture

The horticulture-cum-fruit plants can be practiced either as small orchard – an intentional planting of fruit bearing trees in a line or bund of the fish pond – or as boundary planting – fruit trees are planted along the fence of the farm and make it as live fence or barrier plants. Fruit trees like banana, papaya, coconut and pineapple can be planted serially along the bundth of fish pond. Otherwise, ditches or deeper channels can be constructed between the fruit plants in which fish can be reared. However, this kind of channel system can be suitable only for hardy air breathing larvivore fish like murrels and catfishes. Other fruit plants like mango, guava, sapota, lime and litchi can also be planted in this kind of integrated system.

A very limited study has been carried out in the fruit-cum-fish integrated culture. However, those limited studies can clearly explain the economic advantage of integrated farming systems over the monoculture-based fish culture or agriculture practices. For example, a study by Sharma *et al.* (2016) revealed that papaya based integrated fish culture system produced a net quantity of 650 kg of papaya which resulted in higher cost benefit ratio (4.52) compare to the existing farmer's monoculture farming system. Even among the integrated horticulture system, an improved horticulture-cum fish culture indicated 66.67 per cent and 27.58 per cent higher production of fish and fruits, respectively, compared to the traditional integrated system (Yadav *et al.*, 2013).

The following are some of the key points need to be adopted in a good fruit-cum fish culture system:

Sl.No.	Type of Integration	Key features
1.	Banana-Fish Integration	☆ Avoid plantation in low-lying bundth area. ☆ Spacing between the trees varies. Ideal spacing of banana is either 3m × 3m or 3m × 2m. ☆ Avoid saline and calcareous soil area. ☆ Provide a drip irrigation system.
2.	Coconut-Fish Integration	☆ Avoid low-lying area with highly clayey soil. ☆ In serial plantation, the minimum ideal space between two trees is 20 feet or 6.5 m. ☆ Plantation can be done in bundth area but ensure at least 1m height between plantation and water level. ☆ Intercropping can be done as per local need.
3.	Papaya-Fish Integration	☆ Highly suitable one. Grows well in 25-30°C temperature. ☆ Ideal spacing between two papaya plants is 1.5 to 2.0 m. ☆ Buds and old leaves should be removed at regular interval for optimal growth. ☆ Provide a stake for support.

Figure 10.1: Fruit Plants in the Bundth of Nursery Fish Pond.

Figure 10.3: Banana Planted in the Bundth of Fish Pond.

2. Vegetables-cum-Fish Culture

The integration fish and vegetable increase nutritive food production, as both of they contain high number of nutritive elements. In this system, farmer can adopt either intensive or semi-intensive vegetable cultivation along with fish culture. In the intensive type, 2/3rd of bundth area will be covered/utilized for vegetable cultivation. On the other side, semi-intensive type of farming is practiced where the fish pond bundth is not good. However, success of each system depends on the plant species selected for integration. In general, seasonal plants with highly regenerative capacity and greenery dwarf plants with less shady nature are more preferred. In addition to that, the selected fish and plant species need to have local demand for ease of marketing. Vegetables which are used in day today life like tomato, chilly, brinjal, cabbage, cauliflower, ladies finger and green pea are highly recommended.

Compare to the existing monoculture-based fish/horticulture practices, integration of nutritious fish with valuable vegetables ensured the higher net returns. An integration study conducted in Karnataka revealed that integration of vegetables with livestock increased net returns (*i.e.* 243.3 per cent) over their regular cropping pattern of millet followed by groundnut cultivation (Kulkarni *et al.*, 2014). Another study by Laxmi *et al.* (2015) found better production of fish and vegetables (like pumpkin, bottle guard, bitter guard and ridge guard) in integrated farming practice rather than monoculture technique. Additionally, the study found better food security, employment, income generation and crop diversity among the rural farming community. Integration of Chinese cabbage (*Brassica rapa chinensis*) with African sharp tooth catfish (*Clarias gariepinus*) and Nile tilapia (*Oreochromis niloticus*) in a polyculture-based system produced an increased yield, net return and cash flow among the farmers compare to fish culture and cabbage farming (Limbu *et al.*, 2016). Further, field level experiment and data are need to be generated on this vegetable-cum fish culture practice, with various combination of vegetables and fish varieties, to understand the economic efficiency of this system.

Figure 10.3: Preparation of Bundth for Vegetable Plantation.

Figure 10.4: Vegetable (brinjal) Planted around the Fish Pond.

3. Flowers-cum-Fish Culture

It is the least practiced horticulture-cum-fish integration method, in which flowers are integrated with the fish production ponds. Flowers are considered as symbol of purity, kindness and strength. Moreover, integration of flowers in fish ponds brings fragrance and beauty to the farm. It has a ready market value as it occupies important place in the Indian tradition and spiritual occasions. Flower varieties like jasmine, rose, marigold, gladiolus, chrysanthemum and poppy can be planted in the fish pond dykes and farm areas. To till date, there has been no any extensive study on flower cum fish integration system, however, literature says it can generate an additional income of 20-25 per cent compare to the fish culture alone.

Aquaponics – An Integration of Horticulture and Fish Culture

The evolution of aquaponic system, an advanced integration system in which fish culture is integrated with the horticulture plants using hydroponic unit, is giving a reassurance to the combined production of fish and horticultural products. The lack of responsible use of natural resources forcing the human beings to adopt the sustainable intensive production systems, in a limited area for augmenting the food production. In this context, popularizing a technology like aquaponic system, which continuously recirculate the water and utilize the fish waste as nutrients for plants with a help of microbial communities, is indeed a viable option. Aquaponics is completely indoor based closed food production system and use fish waste water for horticultural plant production which ensures efficient use of resources. Presently, smaller aquaponic units have been built by various research institutes and private sectors in several places which are giving additional popularity to this system. However, on the other hand, big lobbies are searching for a profitable venture out of this eco-friendly system.

The following key features make the aquaponic system as future integration model for fish and vegetable cultivation;

- ☆ Highly suitable for land and water scare area.
- ☆ Ideal food production system for urban area.
- ☆ Lower operational cost and reasonable returns.
- ☆ Organic way of nutritive food production.
- ☆ Compare to traditional vegetable production, the system utilizes only 2-10 per cent of water due to its continuous water flow and reuse.

Figure 10.5: Preparation of Hydroponic Unit for Plantation.

Figure 10.6: Mint and Tomato Plants grown in Aquaponic Unit.

Economic Analysis of Integrated Aquaponics System

Particulars	*Unit Price*	*Quantity*	*Total (₹)*
A. Capital Cost			
Fish Tank Construction (100 Cum)	100000	1	2,00,000.00
Grow Bed for Plantation (12x4 ft)	20000	10	2,00,000.00
MBBR Filtration Set (De-coupled and Anaerobic)	1,08,000	2	2,16,000.00
Pond Pump 15000 LPH	10000	4	40,000.00
Air oxytube	1000	12	12,000.00
Air oxytube Base	1500	8	12,000.00
Aerators (Air pump for oxygen supply) 120 LPM	11500	6	69,000.00
Aerators (Air pump for oxygen supply) 60 LPM	8500	2	17,000.00
Water Testing kit, Pressure Washer	240000	1	24,000.00
Biosecurity with Power Backup	200000	1	2,00,000.00
Plumbing items	60000	1	60,000.00
Consultation, Installation, Supply and Misc. Expenses	1,50,000	1	1,50,000.00
Total (A)			**₹ 12,00,000.00**
B. Operational Cost (for 1 Year)			
Gift Tilapia Seeds	2.5	24000	60,000.00
Feed Cost	42	8400	3,52,800.00
Plant Samplings	5000	1	5,000.00
Transportation Cost	50000	1	50,000.00
Electricity Cost Annual (per unit cost × consumption)	12	4000	48,000.00
Total (B)			**₹ 12,00,000.00**
C. Cost-Benefit Analysis			
C1. Annual Fixed Cost –			
a. Depreciation on capital investment @ 15 per cent			₹ 1,80,000.00
b. Insurance premium @ 2 per cent of the capital investment			₹ 24,000.00
c. Interest on 75 per cent of the capital investment @ 7.8 per cent per annum			₹ 70,200.00
d. Administrative/Other expenses @ 2 per cent of 75 per cent capital investment			₹ 18,000.00
Total Annual Fixed Cost (C1 = a + Cb + c + d)			**₹ 2,92,200.00**

Particulars	*Unit Price*	*Quantity*	*Total (₹)*
C2. Annual Variable Cost –			
a. Annual operational cost (B)			₹ 5,15,800.00
b. Interest on operational cost @ 7.8 per cent			₹ 40,232.40
Total Annual Variable Cost (C2 = a + b)			**₹ 5,56,032.40**
Total Operational Cost (C1+C2)			**₹ 8,48,232.40**
D. INCOME (2 Cycles in a year)			
a. Income from GIFT Tilapia production (500g) @ Rs. 150/- for 10,000 kg			₹ 15,00,000.00
b. Income from Vegetable @ Rs. 40/- for 5000 kg			₹ 2,00,000.00
Total Income (D)			**₹ 17,00,000.00**
E. Financial Analysis			
a. Operating cost			₹ 5,15,800.00
b. Total cost			₹ 5,56,032.40
c. Gross Revenue			₹ 17,00,000.00
d. Net Operating income (C-a)			**₹ 11,84,200.00**
e. Net profit (C-b)			**₹ 11,43,967.60**
f. B:C ratio			**2.05**

Conditions to Achieve the Above-said Income from Integrated Aquaponics System

- ☆ Farmer should strictly follow standard procedure of aquaponic system,
- ☆ Farmer should procure seed and feed from the certified/registered hatchery and feed companies,
- ☆ Farmer should take necessary steps to market the produce for the rates mentioned,
- ☆ Vegetable production is calculated at given average and the actual may vary depending upon the crop type, climatic conditions and quality of produce at any given point in time and may be much higher than what is quoted above, and
- ☆ All the above quoted rates are in Indian Rupees (INR) and are subjected to change as per geographical location and other economic factors.

Conclusion

The production of nutritive food is going to play a crucial role in supporting the existence of human population in this planet. The scarcity of natural resources and uncontrolled exploitation of available resources are augmenting a pressure on the food production systems. In addition to this, the introduction of chemicals and fertilizers for enhancing the food production has opened the gate for questioning of quality of food is being produced. On the other side, the shift of human kind towards the production and consumption of organic food is escalating which further fuels the pressure on food production sectors for quality food production. In this context, popularizing the integrated horticulture-cum fish culture could be a viable option for the production of original organic food with real social and economic welfares.

REFERENCES

Kulkarni, S.R., Upperi, S.N., Sujay, H, Zaheer, A. and Goudar, S.B. (2014). Sustainable livelihood improvement of farmer through integrated farming system. *In*: Proceedings of National Symposium on Agricultural Diversification for Sustainable Livelihood and Environmental Security, Ludhiana, Punjab, pp. 574.

Laxmi, Gupta, N., Singh, M. and Sharma, R.P. (2015). Evaluation of Integrated Horticulture-cum-Fish Farming in Malwa Region of Madhya Pradesh, India. *Current World Environment.* **10**(2): 667-671.

Limbu, S.M., Shoko, A.P., Lamtane, H.A., Kishe-Machumu, M.A., Joram, M.C., Mbonde, A.S., Mgana, H.F. and Mgaya, Y.D. (2016). Fish polyculture system integrated with vegetable farming improves yield and economic benefits of small-scale farmers. *Aquaculture Research.* **48**(7): 3631-3644.

Mishra, P., Pandey, B.K., Shrivastva, A., Mishra, V., Shukla, S.K. and Dikshit, A. (2020). Advancement in prospects of integrated agriculture: A new era of sustainable rural livelihood. *Journal of Emerging Technologies and Innovative Research.* **7**(12): 1346-1362.

NIN. (2011). Dietary guidelines for Indians – A manual (2nd Edition). National Institute of Nutrition, Hyderabad. p. 1-2.

Sharma, S.K., Nainwal, G.C., Kumar, P. and Ujjania, N.C. (2016). Integrated fish-cum-poultry-cum-horticulture farming system: An approach to uplift the socio-economic status of fish farmers. *International Journal Progressive Research.* **11**(1): 411-414.

Yadav, G.S, Chandan, D., Datta, M., Ngachan, S. V., Yadav, J.S. and Babu S. (2013). Comparative evaluation of traditional and improved farming practices in Tripura. *Indian Journal of Agricultural Sciences.* **83**(3): 70-74.

Chapter 11

Integrated Farming of Fish with Cattle: An Entrepreneurial Approach

S. Sangavi[1]*, C. Judith Betsy[2] and J. Stephen Sampath Kumar[3]

[1]Assistant Professor, Trichy Centre for Sustainable Aquaculture, Directorate of Sustainable Aquaculture, Tamil Nadu Dr. J Jayalalitha Fisheries University, Trichy – 639101, Tamil Nadu
[2]Assistant Professor, Fisheries College and Research Institute, Tamil Nadu Dr. J Jayalalitha Fisheries University, Tuticorin – 628008, Tamil Nadu
[3]Professor, Director- Directorate of Sustainable Aquaculture, TamilNadu Dr. J Jayalalitha Fisheries University, Nagapattinam – 611002, Tamil Nadu
Corresponding author: sangavi@tnfu.ac.in

ABSTRACT

Integrated fish-cattle culture approach allows the utilization of wastes/by-product from one system to be used as inputs in another system; by aiming conservation of resources and in turn increasing farm returns. So far, fish-cattle integration is one of the most popular fish integrated farming across Asian countries, especially in India. Cattle manure as a source of organic fertilizer for fish culture in ponds and biogas production paved the way to reduce capital costs in purchasing inorganic fertilizers for primary production. Due to the increase in pellet fish feed cost, integrated fish farming shows an option for decreasing feed cost and concurrently cumulating the yield, leading to high economic returns. This type of integrated farming stands as a sustainable way of fish production. It is considered a forgotten asset that guarantees the fish farmers more production and yield from a small-scale farm, better employment opportunities, improving the socio-economic status and livelihood of fisher folk communities, and finally, the progress of the rural economy.

Keywords: *Cattle integration, Fish culture, Husbandry practices, Livestock farming, Sustainable aquaculture.*

Introduction

The principle of the Integrated Farming System (IFS) involves farming fish integrated with livestock or/and agricultural/horticulture crops along the fish culture pond area. This type of farming provides a huge efficacy in the usage of resources in which the waste/by-product from one system forms the input to other systems; hence, the wastes are effectively recycled. In addition, this system enables the effective use of available farming space to maximize production (Kumar and Ayyapan, 1998). Rearing of fishes with agriculture or livestock such as cattle, poultry, goats rearing *etc.* is a unique lucrative venture. In addition to fish, rearing of animals in this integrated system offers meat, milk, eggs, *etc.* and farming of crops offer fruits, vegetables, mushroom, fodder and grains. Hence this system gives us restored production, more employment, improving the socio-economic status of fish farmers and, finally, the betterment of the rural economy.

The IFS is most suitable model for sustainable food production in terms of wastes utilization in a biodegradable way. For instance, in carp farming, animal manure can be efficiently employed as potential nutrients as organic manure in culture ponds for the development of primary producers. Likewise, plants and fish integration results in a polyculture that enhances diversity and multiple yields (Trivedi, 2020). In this system, water is re-used by bio-filtration and recirculation methods and also involves the production of healthy foods and improvement of the local economy. Further, the detailed description and perspectives of fish-cum-cattle integrated farming are discussed in the following chapter.

Method of Fish-cum-Cattle Integrated Farming

The method of fish farming involves using manure in fish ponds and therefore, it reduces the need to purchase and use fertilisers and fish feeds in the rearing ponds. This ultimately helps the farmers to generate more income by avoiding the procurement of fertilisers for the farm.

This method had been usually managed by building animal sheds near fish ponds, thus simplifies manure handling and saving time and labour (Figure 11.1). The organic wastes from animal sheds can be collected separately and then mixed and flushed directly into the ponds. In the other type, if the shed floor is higher than the pond dyke, a manure ditch can be dug near for collection of organic wastes together. Simultaneously, the area of fish pond has to be directly proportional with the number of livestock and wastes produced from shed; the species ratio and target output of fish *etc.* Kumar and Ayyapan (1998) reported that the frequency of manuring depends on the seasonal conversion of manure and development of fish food organisms in culture ponds.

The organic wastes produced from cow is advantageous omnivorous and filter feeders (*e.g.* silver carp and catla) and typically assorted with omnivorous (*e.g.* common carp) and herbivorous species of 15-20 per cent. The optimal output from herbivorous fishes in this integrated farming will be around 12 per cent of the total pond output. Stocking of excess herbivores species in pond necessarily requires supplement feeds.

Figure 11.1: Fish-cum-Cattle Integrated Farming System.

Global and Indian Scenario of Livestock Integrated Farming

Integrated fish farming is also called as Agro-pisciculture or Integrated agriculture aquaculture and history dated back above 1500 years in Asia, especially in India (Coche, 1967) and 2400 years in China (Wiliman *et al.*, 1998). This type of mutually beneficial integrated system in China started as a compound farming of poultry, livestock and crops with fish polyculture and used crops, grass and manure as feed and fertiliser. This long-term practice of IFS has led to higher production more than 60 per cent of the work against total aquaculture production (Rana *et al.*, 1998). Certainly, Asia is the world's leading continent with vast areas of land for livestock rearing and especially rice fields in China, India, Japan, Indonesia, Thailand, Vietnam, Philippines and Bangladesh.

Brown (2012) reported that in African countries like Nigeria, Benin, Madagascar, Zambia, Cameroon and Malawi practised this type of farming at a subsistence level and use of larger areas in Hungary. Cattle-fish farming had been predominantly practised in Czech-republic and Slovakia and a little attention in Spain and Italy.

India is enriched with ~150 lakh hectares of freshwater resources (DoF, 2020) and being an agricultural economy produces large quantities of animal and plant residues of 322 and 1,000 million metric tonnes, respectively. In addition, the country comprises a larger cattle population over 307 million heads (Trivedi, 2020), providing enormous quantities of organic material for aquaculture. Thus, fish-cattle integrated farming offers a solution to the energy and food crisis problem globally.

Fish Husbandry

Aquaculture or fish culture is defined as the farming of aquatic food organisms in a controlled environment; farming implies the intervention of rearing processes such as pre-stocking management, stocking and post-stocking management to enhance pond production.

A. Preparation and Management of Nursery, Rearing and Grow-out Ponds

- ☆ In *Nursery ponds* spawn are reared to fry stage, *i.e.* carp spawn takes 15-20 days to grow fry size.
- ☆ In *Rearing ponds* fry are grown to fingerling size, *i.e.* carp fry takes 2-3 months to reach fingerlings size.
- ☆ In *Grow-out ponds* fingerlings are stocked and are grown to harvestable size, *i.e.* carp fingerlings reaches marketable size in about 10 to 12 months.

B. Pre-Stocking Management

The preparation of fish culture ponds is very necessary as the pond environment offers an optimum condition for fish growth. The steps involved in pre-stocking (Figure 11.2) and post-stocking management are similar in the nursery, rearing and grow-out ponds and are explained briefly below:

Preparation of Culture Ponds

(i) Draining and Drying

Fish culture ponds have to be dried before starting of the rearing operations for a period of 7-10 days till the soil cracks. This facilitates and helps in oxidation of organic matter, degassing of toxic gases (ammonia and hydrogen sulphide), killing of pathogenic microorganisms, predatory and weed fish and unwanted aquatic plants along the dyke margins and bottom of the pond.

(ii) Ploughing

Ploughing of ponds using wooden ploughs/power tillers/tractors plays a vital role in soil mixing up and helps in oxidation of organic matter, proper degassing of soil from toxic gases and mineralization of nutrients.

(iii) Liming

Liming corrects the soil pH, thus improves the quality of the pond soil and so enhances the primary productivity. Therefore, the desirable pH to be maintained in the pond is 6.5 to 7.0, by using of liming materials such as agricultural lime or calcite ($CaCO_3$), dolomite [$CaMg(CO_3)$], calcium hydroxide/slaked line $Ca(OH)_2$ and calcium oxide/quicklime (CaO) at a rate of 200-500 kg ha^{-1}. Apart from correcting the soil pH, liming has several advantages such as mineralization of organic matter, release of soil phosphorous in water, and disinfection of pond bottom.

Manuring of Fish Pond

Liming of Nursery Pond

Spraying of Oil Emulsion

Watering of Nursery Pond

Figure 11.2: Pre-stocking Management Practices in Nursery Ponds.

(iv) Fertilization of Pond

The fry and fingerlings of most freshwater fish species (*e.g.* carps) feed on zooplankton. Therefore, sustained zooplankton production in ponds depends on adequate availability of nutrients added to water (in the form of organic manure and inorganic fertilizers such as nitrogen, phosphorous, carbon and micronutrients), good phytoplankton growth and bacterial base.

Organic manures are rich in carbon element and nutrients such as N and P in small amounts. They decompose and release the nutrients slowly; hence, sustained plankton growth can be maintained for longer periods through saprophytic food chain. Organic manures like cow dung and poultry manure can be applied at a rate of 5-10 tonnes ha^{-1} before 15 days of stocking and half of its dose can be used in case of poultry manure. It can be also be applied in split basis as follows: basal/ initial dose as $2/3^{rd}$ amount and second dose after a week of stocking.

Inorganic fertilizers are concentrated forms of nutrients such as N in the form of urea or ammonium sulphate and P as single or triple super phosphate. It promotes the production of primary producers, *i.e.* phytoplankton and varies in case of zooplankton. Therefore, combined form of fertilization (*i.e.* organic manure and inorganic fertilizers) will indorse quick plankton growth and sustain for longer duration in ponds.

(v) Aquatic Weeds Control

Fish cultures in large earthen ponds are usually infested with submerged, emergent floating and marginal weeds. The problems caused by the above weeds in fish ponds includes: a) compete for nutrients with phytoplankton, b) preventing light penetration, c) causing oxygen super saturation (in day time) and deficit (in night time), d) hindering free movement and reducing the space for cultured fish, e) inhabiting aquatic insects and predatory fish and finally, f) causing problems during harvesting by obstructing netting operations. Thus, the solution for controlling abundance aquatic weeds is by employing manual, mechanical, chemical and biological methods.

(vi) Predatory and Weed Fish Eradication

In the nursery and rearing ponds, the survival rate of fry is severely affected by predatory fishes. In addition, the weed fishes will compete for food, space and oxygen in ponds; thus reduces the stocked fish production. Eradication of predatory and weed fishes are very essential, as these species will breed prior to the onset of carp breeding and infest fry and fingerlings. The common predatory fish and weed fishes encountered in the culture ponds are listed in Table 11.1)

Table 11.1: List of Predatory Fish and Weed Fishes Commonly Found in Fish Ponds

Name of the Predatory Fishes	*Name of the Weed Fishes*
Murrels (Snakeheads)	*Puntius sp.*
Wallago attu	*Barbas* sp.
Clarias batrachus	*Danio* sp.
Heteropneustis fossilis	*Aplocheilus* sp.
Ompak sp.	*Anabas* sp.

The most effective way to control is to dewater the pond and then sun drying; some piscicides can be used in ponds which cannot be dewatered. Plant origin piscicides such as derris root powder (*rotenone* - contact poison) and mahua oilcake (*saponin* - causes RBC lysis, kills fish, frogs, snakes and turtles) are active ingredients and can be used at a rate of 4 to 20 ppm (mg l^{-1}) and 250 ppm respectively. Similarly, other piscicides such as tea seed cake @ 60 ppm, tamarind seed husk @ 50-100 ppm and jaggery @ 1 per cent can be used. Pesticides (chlorinate hydrocarbons and organophosphates) are also be used in eradication of weed and predatory fishes in culture ponds.

C. Stocking and Post-stocking Management

(i) Stocking

The nursery, rearing and grow-out/stocking ponds will be stocked with spawn, fry and fingerlings, respectively in the early morning or in late evenings or nights. In general, monoculture of fish is followed in nursery and rearing ponds, while polyculture in grow-out ponds. The stocking species needs to get acclimatized to prevent stress due to abrupt changes in water quality which results in poorer survival.

The stocking density to be for carp spawn in earthen nursery ponds is 300 to 500 numbers m^{-2} of pond area, where higher level of 1000-2000 numbers m^{-2} in cement tanks; while, in polyculture, the rearing ponds are stocked at a rate of 20 to 30 numbers m^{-2}; and grow-out ponds at 103000 numbers ha^{-1}.

(ii) Supplementary Feeding

When culture ponds stocked with higher stocking density of fishes, it certainly lowers the primary/natural food production provided through fertilization and manuring. Therefore, supplementary feeding is done with a mixture of rice bran/ wheat bran and oil cakes (groundnut cake or mustard cake or cotton seed cake) at 1:1 ratio. For improving the feed quality, fish meal, soya meal/flour and vitamin-mineral mix are added as feed ingredients and made into a ball shaped dough, then broadcasted into ponds. The feeding rate of fish vary according to its size and biomass estimated after periodic sampling. For instance, spawn fed @ 8-10 per cent of the biomass, fry @ 6-8 per cent; fingerlings @ 5 per cent of body weight and finally, decreased to 2-3 per cent biomass till harvest.

Cattle Husbandry

Cattle may be raised either for beef or dairy production. Usually, cow manure is used for fish farming due to its abundant availability. It is estimated that 8 tonnes of dung and 5000 litres of urine can be obtained annually from a cow of about 400 kg body weight (Kumar *et al.,* 2012). Fish have enough time to consume the edible portion of dung since the manure particles sink at the rate of 2-6 cm $minute^{-1}$. Since the cow manure is decomposed by micro-organisms in the rumen, the BOD is lower when compared to other livestock manures. The organic wastes (faeces and urine) added into the culture ponds will be highly beneficial for filter-feeding and omnivorous fishes such as silver carp, catla *etc.*

1. Breeds of Cattle

The cattle variety must be selected based on the purpose either for milk or for beef. Some of the varieties are given below:

- **Milch:** Sindhi, Tharparkar, Sahiwal, Hariana, Gir, Holstein Friesian, Black and White, Simmental, Jersey, and
- **Beef:** Shorthorn, Simmental, Hereford, Charolai, Jersey and Qincuan draft.

2. Housing of Cattle

The type of cattle housing depends on herd size, climatic conditions, labour availability, amount of capital investment, degree of mechanization and owner preference. The cow shed should be constructed in such a way that it gets sufficient sunlight. The floor of the shed must be made of concrete and must have slope at an angle to drain the water. Either thatch or asbestos can be used to construct the roof of the shed. Provisions for floor space should be made for suckling calf, older calf and cow (Tables 11.2 and 11.3 and Figure 11.3). Some of the important factors that should be considered while building a housing facility are:

- It must provide comfort to the cows so that maximum milk production is obtained,
- It should minimize injuries such as swollen hocks, stepped on teats and lameness,
- It must minimize exposure to environmental stresses,
- The feeding system in the housing should allow sufficient food for all cows with maximum labour efficiency,
- The housing facility must be easy to clean and fulfill sanitary requirements and keep disease incidence to a minimum,
- Choose a milking system that minimizes labour while producing highest milk yields from the dairy cows, and
- Choose a waste management system that will provide maximum economic return.

Table 11.2: Floor Space Requirements for Cattle

Standing space	1.2 × 1.2 m^2 per animal
Manger	Length : 0.6 m per animal
	Width : 0.6 m per animal
	Depth : 30 cm
Gutter	Width : 20-30 cm
	Depth: 2.5 cm with provision of gradient towards main drain.

Table 11.3: Pen Size Requirement for Cattle

Pen size	*Cover Area (m^2 $calf^{-1}$)*	*Open Area (m^2 $calf^{-1}$)*
Suckling calf	1 × 1	2 × 1
Older calf	2 × 1	2 × 2

The choice of a disposal system is based on:

- ☆ Size and type of operation,
- ☆ Availability of land and cropping practices,
- ☆ Population density around the location,
- ☆ Climatic conditions and
- ☆ Type of housing arrangement.

Figure 11.3: Housing/Shed of Cattle.

3. Types of Housing in Cattle

(a) Station or Tie Stall System

It is a one- or two- story construction. In a two-story barn, cows are confined on the first floor and feed materials mainly roughage are stored on the second floor. Cows are confined to individual stalls by a station or some type of restraining device. A paved area is included where the cattle are released for exercise. Cows usually spend 20 to 100 per cent of the time in stalls depending on pasture availability and season. Milking is done in the stalls with portable can milkers, a pipeline milking system and milk house, or in a milking parlour. The advantages of this type of housing are, more efficient feeding of individual cows, greater attention for individual cows, better display of cows, and greater environmental control.

(b) Free Stall Housing

It consists of four distinct parts known as the feeding area, stall area, milking parlour and exercise area. Cows are allowed free movement between feeding and stall areas. They are pastured up to 80 per cent of the time depending on climate and land availability. Milking is performed twice a day in the parlour. The housing can be made with cheaper pole-type construction or by more costly construction methods where greater environmental control is needed. The advantages of this system are less injuries, cleaner and happier cows, less usage of bedding materials, more efficient use of labour and lower space requirements per cow.

(c) Loose Housing

It is the oldest system for dairy cattle management. A loafing area or barn is provided which is heavily bedded for a manure pack to form. It is generally open to the environment with a paved exercise lot connecting the feeding and resting areas. The advantages are lower labour and building costs due to the adaptability of more mechanized feeding technology and lower construction costs of pole-type facilities.

(d) Partially Housed or Non-Protected Cow Yard System

It is used where minimal environment protection is needed. It employs an open or partially sheltered lot where cattle are fenced. A sun shade is built on the lot for partial protection of the cows from solar radiation and rain. Cattle are fed in feed bunks directly on the lots which are paved and semi-paved. Milking in this system is done to a milking parlour twice a day.

(e) Pastures

Pasturing is a good way to get full use out of forage materials on fields that are not suitable for harvesting. Milking cows must be returned to the milk house or milking parlour twice a day. If the herd is of considerable size or the fields are distant, the labour needed would be prohibitive. Pasturing is not ideal for calves which need greater attention and bulls that are likely to knock down fences. Considerable savings are realized through lower feed costs. If forage materials are not of high quality, additional concentrate feeding must be given to increase milk production. Shade should be provided by employing an open shelter or wooded area for protection from heat and solar radiation, as each of these have a deleterious effect on the cows' milk yield. Pastures are classified as following according to use:

- **Permanent –** always in grass because the land is unsuitable for crops;
- **Rotational –** fallow area used for grazing that is part of a crop rotational pattern;
- **Temporary –** area grown with a particular forage material that is used at times when other pastures are unproductive (*i.e.* early spring or late fall); and
- **Winter pasture –** areas available for winter grazing.

(i) Range and Pasture Production

Most or all of the feed requirement is provided by existing vegetation. Stocking rates are most generally selected to maintain a sustained yield of forage and are dependent upon local conditions including soil type, terrain and climates. Larger stocking rates are utilized for irrigated pasture or for pastures used in rotational grazing.

(ii) Feedlot Production

Cattle are fed and finished on unroofed feedlots in which the feed is all grown outside the animal pens. Beef cattle feedlots are designed and constructed to facilitate rapid and efficient gain requiring minimal input of labour. Usually it is constructed without shelter, however, construction can also include partial shelter for protection from either summer or winter extremes. Lot surfaces can be made of either completely earthen, or partially or totally impervious surface. Feedlot design also includes provisions for wind breaks, summer shade, mounding, cattle working areas, cattle receiving and shipping facilities, feed storage and processing, and provision for water in each pen. Surface shaping provides drainage to minimize mud. It is a common practice to increase space per animal during wet periods to minimize mud problems and to decrease space during dry periods as a dust control aid. Feedlot requirements per animal are based upon the nature of the surface, amount of shelter, local climate and management preferences of the operator (Table 11.4).

Table 11.4: Feeding of Cattle

Green fodder	30 to 40 kg day^{-1} cow^{-1}
Straw	3 to 4 kg day^{-1} cow^{-1}
Concentrated feed (Wheat bran + Rice bran + GNOC + Mineral mixture + Salt)	2 kg day^{-1} cow^{-1}

4. Health Management of Cattle

Hygiene maintenance in the cow shed is vital. The preventive measures must be adopted against coccidiosis, parasitic infection, *etc.* Immunization like FMD vaccine, anthrax spore vaccine, haemorrhagic septicaemia vaccine and black quarter vaccine should be done at proper time.

5. Requirement of Manure

For a fish pond of 1 hectare area, about 10 to 15 tonnes of dung and urine per year is required to maintain the pond fertility. This quantity can be generally obtained from 2 cattle. Kumar *et al.* (2012) mentioned that 5 to 6 cows will be sufficient to fertilize 1 hectare water body. In order to get better results, dung and urine can be mixed thoroughly and diluted with water. This mixture can be evenly broadcasted in the pond. Dung in the form of biogas slurry can also be used @ 80 to 120 kg of fresh slurry ha^{-1} pond day^{-1}. Application of cow manure should be suspended in case if the water colour turns from brown to dark green and if there is persistent cloudy weather.

6. Harvest and Production

Among livestock excreta, cow excreta are the most abundant, locally available and reliable manure. For instance, a cow weighing 450 kg can annually produce 12000 to 13000 kg of faeces and 8500 to 9000 litres of urine. If 0.024 kg of fresh cow manure is applied to 1 m^3 of water every day, the inorganic N and P content in the water sample will be 0.897 and 0.024 ppm, respectively. Also, these levels are very close to the inorganic N (0.97-2.06 ppm) levels estimated in high-yield fish ponds and so, the N/P ratio will be 36.9 (Table 11.5).

Table 11.5: Composition of Cow and Pig Excreta

	Moisture (Per cent)	*Organic Matter (Per cent)*	*N (ppm)*	*P_2O_5 (ppm)*	*K_2O (ppm)*
Milk cow dung	85	11.4	0.36	0.32	0.20
Cow dung	80-85	14.6	0.30-0.45	0.15-0.25	0.05-0.15
Cow urine	92-95	2.3	0.60-1.20	Traces	1.30-1.40

Source: FFRC Agricultural Handbook.

The susceptibility of cow manure helps the culture species to obtain more feed. In addition, the stacking process helps in reduction of oxygen consumption rate and thus, avoids formation of harmful gases. This can be elucidated as the BOD of cow manure is lower than (32 per cent) that of other livestock manures and estimated to be 20.6 g for 1 kg in 5 days.

Advantages

In general, farming of aquatic organisms provides high quality animal protein for human consumption worldwide. Indeed, the fish-cattle integrated system plays a significant role for a farmer by creating an additional income apart from fish production in terms of milk, meat, organic manure, *etc.* Fish cultured ponds can be controlled by the farmers through selection of fish species to be raised, application of cattle manure and thus, improving water management. Use of marginal land for cattle raising and fodder production along dyke slopes will be beneficial for fish farmers through effective utilisation of land devoted to fish farming, hence maximising the production and yield. IFS utilises the waste/by-products produced from livestock and agriculture for fish production, for example, 40 to 50 kg of organic manure from cow can produce 1 kg of fish. The silt obtained from the pond after fish harvest can be used as fertiliser for fodder crops and in turn, it is used as cattle or fish feed (Figure 11.4). Thus, reutilizing of waste is done efficiently in this cattle-fish integrated farming system.

Conclusion

Fish-cattle integrated farming provides a higher source of income to the farmer having a small land holding through little additional expenditure. In India, it can be expected that this type of farming will be increasing in the near future in all suitable agro-climatic regions; as this farming system is very sustainable in nature by following eco-friendly measures and ensuring higher returns than mono-fish culture as well as sustained levels of fish and other bio-resources production. In addition, it offers a believable solution to farmers in terms of, having a greater potential for augmenting higher production and yield, promoting employment to the livelihood, growth of rural economy, and finally improving socio-economic status of weaker community.

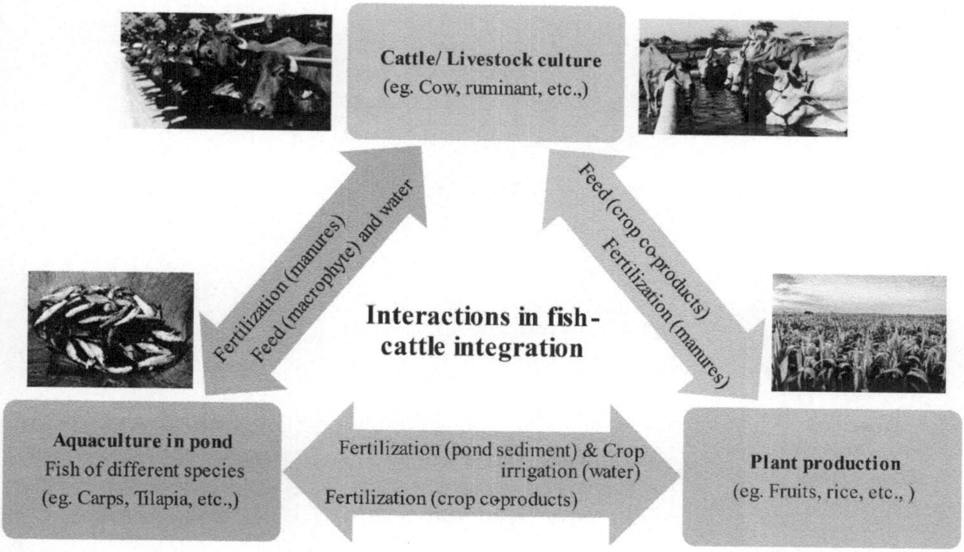

Figure 11.4: Interactions within Fish-Cattle Integrated Farming System.

REFERENCES

Brown, E. (2012). World fish farming: cultivation and economics. Springer Science and Business Media. Avi Publishing Co. (USA). pp. 263-426.

Coche, A.G. (1967). Fish culture in rice fields a world-wide synthesis. *Hydrobiologia*. **30**(1): 1-44.

DoF. (2020). Fisheries Profile of India. *In:* Department of Fisheries, Ministry of Fisheries, Animal Husbandry and Dairying, Government of India.

Kumar, K. and Ayyapan, S. (1998). Current practises in integrated aquaculture. *In*: Proceedings of Integrated aquaculture in Eastern India, Working Paper 5. Central Institute of Freshwater Aquaculture (Indian Council of Agricultural Research), Bhubaneswar. 25p.

Trivedi, S. P. (2020). Integrated Fish Farming (IFF) and Their Management. *In:* Extended Summaries prepared by Prof. Sunil P. Trivedi, Department of Zoology, University of Lucknow, Uttar Pradesh.

Rana, K., Perotti, M., Montanuro, S. and Inmink, A. (1998). Aquaculture production in China. *In*: FAO aquaculture newsletter. Dec. 1998. **20**. pp. 9-10.

Wiliman, R., Halwatt, M. and Barg, U. (1998). Integrating fisheries and agriculture to enhance fish production and food security. *In*: FAO Aquaculture Newsletter. Dec. 1998. **20**. pp.3-9.

Chapter 12

Fish-Mushroom Integration with Special Reference to Mushroom Compost Management

S.S. Rathore[1]*, N.K. Suyani[1] and M.I.R. Khan[2]

[1]College of Fisheries, Karnataka Veterinary, Animal and Fisheries Sciences University, Mangalore – 575002, Karnataka

[2]College of Fisheries, Central Agricultural University, Lembucherra, Agartala – 799210, Tripura

Corresponding author: sanjay.rathore941@gmail.com

ABSTRACT

Mushrooms are both a food and a medication source. Mushrooms have been added to the diet in large numbers, and they are easy to farm on carbon-rich agricultural leftovers. Mushroom farming not only produces food but also aids in the management of agricultural waste. In addition, mushroom compost or spent mushroom substrate is utilised as a biofertilizer and can be used in a variety of ways as part of integrated farm management. In many studies, mushrooms and mushroom compost were found to be beneficial components in raising farmer revenue and lowering farm input costs. The potential integration of mushrooms and mushroom composts in various components of the integrated farming system and farm management has been discussed and summarized in this paper.

Keywords: *Fish, Mushroom, Integration, Mushroom compost management.*

Introduction

The rising population at an alarming rate has resulted in the reduction of arable land, and there is no possibility of increasing additional area for cultivation, *i.e.*, there is hardly any scope for horizontal expansion of land for food, feed and fibre production, therefore, to feed the people with proper nutrient and quality food the vertical expansion farming by integrating appropriate farming component would be the alternate way to challenge the needs of food (Manjunatha *et al.*, 2014). The farming system and its management is the key and become thumb rule to check the food insecurities of the present population of the world and the numbers of the population in upcoming years. The changing climatic condition now become serious threats and constraints for agriculture production, in response to the climatic conditions and the changes, technological intervention with the effective farming system should be adopted, the integrated farming system and management would become the scientific approach to maximize the production by integration of different components of the agriculture in a specific space and place, accounting the periods. One of the components of agriculture is mushroom cultivation. In Mushroom cultivation, the Agric residue, plant waste, and different biomass are used as a substrate for edible mushroom cultivation, growing edible mushroom is an eco-friendly activity as it utilizes the waste from agriculture, horticulture, poultry, brewery, *etc.* (Ahlawat and Sagar 2007). The spent left after harvesting mushroom crop create a nuisance in the environment but several studies have shown that these spent (called spent mushroom compost: SMC) has potential use in the different farming system as soil conditioners to grow crops, vegetable, and grains to yield good production, the spent mushroom compost (SMC) or spent mushroom substrate (SMS) can be used in various soil organic mixtures (Lopez-Castro 2008) to produce grain in organic methods and could achieve sustainability in any farming system and could be the best soil amendments. Whereas, the mushroom crop is a cash crop that can be cultivated indoors under the limited area and less labour, as well as, it is women-friendly cultivation serving an additional income to the farmers with low input.

Major Types of Farming System

Agricultural practices were started since human civilization, in due course of time the ways of farming gradually started varying, and as time passes new methods and techniques were developed in the agricultural sector to meet the demands of people, therefore we find wide heterogeneity in structure and function of framing system depending upon the market opportunity, agro-climatic conditions, demands of the population of the region, resource endowment, infrastructure, economic development of the farming communities, *etc.* However, based on the farming system adopted by the farmers from time to time and region to region, let us highlight some of the important and visible forms of the farming system prevalent in different regions of the world.

a) **Integrated Farming System:** The integrated farming system may be defined as the linking together of two or more normally separate farming systems which become subsystems of a whole farming system. Two major features of an integrated farming system are: (i) waste or by-product utilization in which the wastes of one subsystem become an input to a second subsystem; and (ii) improved space utilization in which the two sub-systems essentially occupy part or all of the space required for an individual sub-system (Behera *et al.*, 2013). The integrated farming system may also be defined as the biological integration of farming system which integrates natural resources and regulation mechanism into farming activities to achieve maximum replacement of off-farm inputs (Choudhary *et al.*, 2019); secure sustainable production of high-quality food and other products through ecologically preferred technologies; sustained farm income; eliminate or reduce sources of present environment pollution generated by agriculture, and sustains the multiple functions of agriculture.

Figure 12.1: Mushroom-cum-Fish Farming in Diverse Aquaculture (Barik *et al.*, 2017).

b) **Specialized Farming System:** The specialized farming system is set for the production of a single cropping system or sequential farming and is the only source of income is said to be a specialized farming system, there are lots of variation if we compare specialized farming with the integrated framing system in respect to their structure, function, and management.

c) **Conventional Subsistence Farming System:** This system of farming is where the farmers usually follow single cropping or multiple crops that farmers manage for the subsistence of the farm family. There is no much emphasis given on resource management, optimization of resources in the farming system to enhance farm income.

d) **Indigenous Farming Systems:** Indigenous technical knowledge (ITK) is the knowledge that people in a given community have developed overtimes, and continue to develop. Indigenous farming is not just farming practiced by the indigenous people but it is the product of indigenous cultures that have always been deeply connected to their particular place. In the words of one of the indigenous farmer, Vena A-dae Romero, "indigenous people are as much part of the land as the land is part of us, we cultivate the land while the land cultivates us, this relationship that has supported my people since time immemorial is remembered daily when we place our fingers in the dirt, pull the weeds out from our fields, or plant our seeds with water, prayer, and hope, cook the food which we grow, and ingest the world with each bite of food we eat" (Pace, 2015).

e) **Organic farming system:** Organic agriculture is a production system that supports the health of the soil, ecosystem, and people," according to IFOAM (International Federation of Organic Agriculture Movement). Rather than using harmful inputs, organic farming focuses on ecological processes, biodiversity, and cycles that adapt to local conditions, and the main goal of organic farming is to build a sustainable farming system in harmony with nature (Barik 2017).

The Merit of Integrated Farm Management

As we discussed the concept of integrated farm management that it is a system of farming in which various types of agricultural components included or enterprises are taking place simultaneously. It helps farmers to boost their economy using resources sustainably. The merit of IFM depends on the components included in the farming system, however, the basic merit of all type of integrated farming system (IFS) are mentioned below:

- ☆ Enhance productivity,
- ☆ Profitability,
- ☆ Adoption of new technology,
- ☆ Environmental safety,
- ☆ Promotion of agro-industry,
- ☆ Increased input efficiency,
- ☆ Minimization of input cost,

- Fodder security for livestock,
- Recycling,
- Continuous income throughout the year,
- Energy saving,
- Helps in disease management,
- Can be applied to smallholder, medium and large farms,
- Supportive biodiversity conservation,
- Intensification of nutrient cycling,
- Diversification of crop varieties, and
- Achievement of sustainable agriculture.

The advantage and merit of IFM depend on the type of components included, every model develops for the effective production need to analyze which is depends on the resource availability, climatic condition, present level of resource utilization, the economics of the proposed model of integrated farming, and the farmer's skills (Rana 2015).

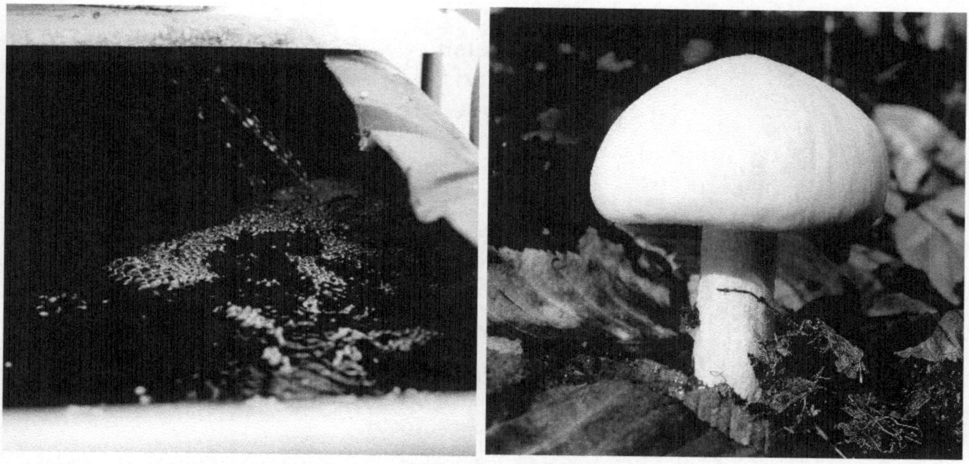

Figure 12.2: Mushroom-cum-Fish Farming with Response to Aquaponics (*Source*: Behera *et al.*, 2013).

Why Integrated Farm Management?

All over the world farmers do hard works but they do not make money, especially marginal farmer because there is very little left after they pay for all inputs (seed, livestock breeds, fertilizers, pesticides, labor, energy, feeds, *etc.*) but the emergence of the IFS enable the people of concerned to develop a feasible framework to improve the small size farmers to a larger one (CARDI, 2010). Integrated farming may ensure to feed the world and fulfil the multiple demands.

Mushrooms and Mushroom Composts in Integrated Farm Management

Making farmers self-sufficient by ensuring a balanced diet for family members, improving the standard of living by maximising total net returns and providing more employment, recycling crop residue, optimising resource use, minimising risk and uncertainty, and maintaining environmental harmony are among the goals (Patel and Dutta, 2004). In the current situation, crop production is declining day by day despite increased chemical input requirements; higher chemical input requirements resulted in poor soil quality, and excessive use of pesticides and other agri-chemicals has now reached a tipping point in the food chain, posing a threat to human health and the environment (Barik, 2017). With around 83 million people added to the world's population every year, the present population of 7.6 billion is estimated to reach 8.6 billion in 2030 and 9.8 billion in 2050. This rising trend in population size is expected to continue (UN, 2017). As a result, the necessity for IFM to combat hunger is critical. Some of the points that the IFM can control are listed below:

☆ Demand for food,

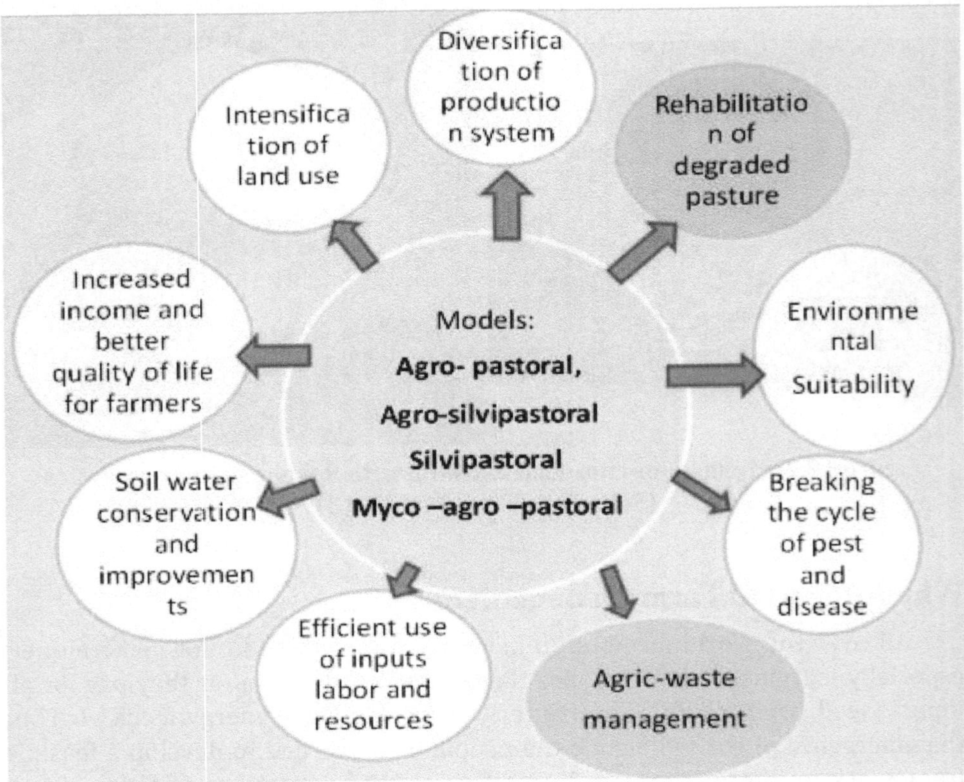

Figure 12.3: Immediate Goals and Results of Integrated System Application in Agro-ecosystems (*Source*: Balbino *et al.*, 2014).

- Changing climate,
- Unproductive land,
- Sustainability, and
- Helpful for the small landholder farmer.

Mushroom Production as a Part of Integrated Farming

Mushroom is a fruiting body of fungi, most of the cultivated mushrooms belong to the phylum *Basidiomycota,* although some Ascomycota such as members from the genera *Morchella* or *Tube*r has also been successfully cultivated and commercially exploited (Liu *et al.,* 2017, Rubini *et al.,* 2014). Mushrooms need substrate to grow, a mushroom substrate is a just a substance where mushroom mycelium will grow. Mushroom cultivation is landless farming using various substrates like paddy straw, wheat straw, sawdust, and other agricultural residues that are rich in carbon and nitrogen sources. There are generally two methods or formulas used for substrate preparation that employed in mushroom cultivation which is optimized depending on the species, these are composted materials and non-composted materials. Both are derived from agricultural residues such as cereals straw, plant fibre/husk, manure, or sawdust (Carrasco *et al.,* 2018).

Edible mushrooms need substrate either composted or non-composted if it is cultivated under the control condition for commercial production. The agric residue generated from the cropland after crop harvesting is generally burned in the crop field causing environmental problems. Therefore, this residue alternatively managed by using residues as a substrate for mushroom cultivations. In the Table 12.1, the range of agric residue required for most common mushrooms has been listed, which shows the utility of agric residue. Growing mushroom on agric residue generates additional income for the farmers (Sharma *et al.,* 2017). The total mushroom production in India is approximately 0.13 million tonnes, out of the total mushroom produced, white button mushroom share is 73 per cent followed by oyster mushroom (16 per cent), paddy straw mushroom (7 per cent) and milky mushroom (3 per cent) and Indian mushroom industry generated revenue of ₹ 7282.26 lacs by exporting 1054 quintals of white button mushroom in canned and frozen form. In India, approximately 500-550 million tonnes (MT) of crop residue is generated on-farm and off-farm annually from its production of 110 MT of wheat, 122 MT of rice, 71 MT of maize, 26 MT of millets, 141 MT of sugarcane, 8 MT of fibre crops (Jute, Mesta, Cotton) and 28 MT of pulses (Devi *et al.,* 2017). Agric waste generated from agriculture can be beneficially utilized in a various agro-based application and other industrial application, however, the cost of collection and transportations is higher than the revenue generated by beneficial use of such agric residues (Bhuvaneshwari *et al.,* 2019), but it could be utilized beneficially if mushroom farm established on the site of the farm, therefore, mushroom would be good components to be integrated into the agricultural enterprise.

Mushroom-Fish System

Mushrooms are fleshy fungus that are valued and appreciated for their meaty bite and flavour. *Agaricus bisporus* (white button), *Volvoriella* sp. (paddy straw), *Pleurotus* sp. (oyster), and *Calocybe indica* (milky) are four famous mushroom species in India. Mushroom growing necessitates a high level of humidity, which an aquaculture environment provides.

Table 12.1: Substrate Used in Most Common Edible Mushrooms

Mushroom	*Substrate*		*Remarks*
	Composted	*Non-Composted*	
Pleurotus	–	Wheat straw, sawdust, paddy straw, soybean residue, groundnut skull, maize cob, bagasse, cornstalks, waste cotton, stalks and leaves ofbanana, rice bran, *etc.*	In a matter of a week, these fungi convert 100 kg Agaric residue into 50 to 70 kg fresh mushroom
Tremella	–	Wood log and sawdust, cottonseed hulls	–
Flammulina	–	Hardwood logs, Sawdust and Rice bran	Fruits only at low temperature
Volvariella	Composting substrate depends on the material used in the composting, Wheat straw, paddy straw and cottonseed hulls are generally composted	Paddy straw	It grows at relatively high temperature
Lentinus	Composted sawdust use in bag logs methods	Wood logs	Highly medicinal tonic
Agaricus	Composting material is subject to substrate availability, the most common material used is wheat straw, paddy straw, cow dung, buffalo dung, horse manure, chick manure, *etc.*	Wood logs	Most popular edible mushroom in the world
Ganoderma		Wood log and sawdust bag log	Highly medicinal use for medicine, Mushroom of Immortal

Source: Ab Rhaman *et al.,* 2022.

Paddy straw, which has become rich in protein, organic material, and other important matter as a result of mushroom cultivation, is used for cow feeding. It has been discovered that feeding this straw to cattle increases milk production. Through the detritus food chain, cattle excrement is recycled in fish pond fertilisation, increasing pond production. Spent mushroom substrate (SMS) is a mushroom compost that includes 1.9, 0.4, and 2.4 per cent (fresh) nitrogen, phosphorus, and potassium, respectively, and can be utilized in aquaculture and agriculture (Ayyappan *et al.*, 2019).

Importance of Mushroom and its By-Product in Supplementation to Farmer Income

Mushrooms production and consumption are on the set of a rise in the world, whereas India responds positively to the growths. Mushroom cultivation activities can play an important role in supporting the local economy by contributing to subsistence food security, nutrition, and medicine; generating additional employment and income through local, regional, and national trade; and offering opportunities for processing enterprises (Marshall and Nair, 2009). Mushroom cultivation contributes to strengthening the livelihoods through reducing vulnerability to poverty and generating reliable sources of income. Mushroom cultivation does not require access to land, so it becomes an attractive activity for both the rural farmers and peri-urban dwellers. Mushroom cultivation might serve as a means of generating employment, particularly for rural women and youths to raise their social status (Hassan, 2012). Mushrooms are a good source of nutrition like protein, carbohydrate, minerals, vitamins, fat, ash and fiber. The percentage of nutritional value varies from mushroom to mushroom (Table 12.2).

Table 12.2: Nutritive Values of different Mushrooms (g per 100g dry weight)

Mushroom	*Carbo-hydrate*	*Fibre*	*Protein*	*Fat*	*Ash*	*Energy (kcal)*
Agaricus bisporous	46.17	20.9	33.48	3.1	5.7	499
Pleurotus sajor-caju	63.4	48.6	19.23	6.7	6.32	412
Lentinus edodes	47.6	28.8	32.93	3.73	5.2	387
Pleurotus ostreatus	57.6	8.7	30.4	2.2	9.8	265
Volvariella volvaceae	54.8	5.5	37.5	2.6	1.1	305
Calocybe indica	64.26	3.4	17.69	4.1	7.43	391
Flammulina veltipes	73.1	3.7	17.6	1.9	7.4	378
Auricularia auricula	82.8	19.8	4.2	8.3	4.7	351

Source: Stamets, 2005 (*A. bisporous, P. sajor-caju, Lentinula edodes*); FAO, 1972 (*Pleurotus ostreatus, V. volvaceae*); Doshi and Sharma, 1995 (*Calocybe indica*); Crison and Sand, 1978 (*Flammulina velutipes* and *Auricularia* spp.).

Mushrooms also contain a high amount of calcium, phosphorus, sodium, and potassium and low but available form of iron and magnesium which shows good health effects (Table 12.3). Mushroom with its regards to nutritional value and digestibility is gaining importance in today's human diet. It has also been examined that mushroom contains vitamins like thiamine, riboflavin, niacin, biotin, and vitamin-C (Chandra, 2006). Mushrooms have been utilized for food and medicine, mushrooms contain different primary and secondary metabolites, showing beneficial health effects, there have been numbers of research findings on the therapeutic application of mushrooms, metabolites like indole, and phenolic compounds, carbohydrates, fatty acids, proteins, free amino acids, sterols, carotenoids, enzymes, vitamins and elements with biological activity the presence of these compounds and elements conditions the nutrient and therapeutic activity of mushrooms *e.g.* immune-modulatory, anti-inflammatory, antioxidant, antiviral, antimicrobial and anti-genotoxic properties, anti-asthmatic, anti-diabetic activity, anti-obesity activity, prevent cardiovascular activity, anti-cancer activity, anti-aging, enhance bone health, Aphrodaic activity, ant-arthritis, *etc.* (Barbieri *et al.*, 2017; Breene, 1990; Khan *et al.*, 2013; Nahata, 2013; Sheikh *et al.*, 2015; Vyas *et al.*, 2016; Radwan *et al.*, 2012). The importance of the mushroom could be mentioned in adding supplementation of income through mushroom cultivation as the demands mushroom in pharmaceutical industry increase due to its neutraceutical application, moreover, mushrooms are being used for the production of mushroom products like mushroom pappad, mushroom pickle, mushroom biscuits, mushroom beverage, and mushroom in various value-added production as food health.

Table 12.3: Mineral Composition of Some of the Edible Mushroom (mg per 100g dry weight)

Mushroom	*Ca*	*P*	*Fe*	*Na*	*K*
Agaricus bisporous	23	1429	0.2	-	4762
Lentinus edodes	33	1348	15.2	837	3793
Pleurotus ostreatus	98	476	8.5	61	-
Volvariella volvaceae	71	677	17.1	374	3455

Source: Chandra (2006).

Mushrooms and Mushroom Composts in Integrated Farm Management

According to Wakchaure (2011) and Wakchaure *et al.* (2010), mushroom cultivation technologies have been developed to the point where a layman could easily start a mushroom farm using their unwanted agricultural equipment and earn extra money. There are a number of mushroom farms established in India, and the government is also providing loans to promote its cultivation.

Management of Agricultural Waste through Mushroom Cultivation

The 70 per cent of Indian depends directly on agriculture and 30 per cent of the country's GDP contributed by the agricultural products. The cultivation of crops is the primary base of survival, therefore it is obvious that a huge amount of agricultural residue generated, which is either dumped or burnt causing environmental pollutions, this generated issue to manage the stubbles of the crops to prevents its burning, there are numbers of strategies have been put forward through various organizations, it has been seen that the cultivation of mushroom using this stubble could be converted into an income source. Apart from mushroom cultivation, crop residues are commonly used as bedding material for animals, livestock feed, soil mulching, bio-gas generation, bio-manure/compost, thatching for rural homes, biomass energy production, fuel for domestic and industrial use, *etc.* for the management of the crop residue (MoA, 2014). However, a large portion of crop residue is burnt 'on-farm' primarily to clean the field for sowing the next crop, the problem of 'on-farm' burning of crop residues is intensifying in recent years due to a shortage of human labor, high cost of removing the crop residue from the field and mechanized harvesting of crops (IARI, 2012). As per available estimates, burning of crop residues is predominant in four states, namely, Haryana, Punjab, Uttar Pradesh and West Bengal (Bhuvaneshwari *et al.,* 2019). Since, mushrooms are heterotrophic, they need substrate rich in carbon, nitrogen, minerals, and vitamins to grow and the crop resides are found to be the best substrate, which is converted into edible mass out of waste, moreover, the substrate left after mushroom harvesting are a good source of soil conditioner and used as a source of biofertilizer. Crop residues management is vital for the long-term sustainability of Indian agriculture. Hence burning of residues must be avoided and should be used positively to improve the economic returns of farmers and reduce the environmental impacts. A potential and promising agro-residue management that can attract the farmer would be mushroom cultivation. This technique has various benefits such as the reduction in environmental pollution; generation of employment, particularly for rural women and youth; mode of short return agricultural business offering an immediate benefit to the community; enrichment in human diet and fitness with high nutrition and neutraceutical compounds; improvement of economic standards of the family (Figure 12.2) shows the utilization of agric residue and its management (Gowda and Manvi, 2019).

Mushroom in Animal Husbandry

Over the year persisting research on mushrooms potential in numerous health benefits, found to have a broad spectrum of action, and based on recent literature data the use of mushrooms is found to be justified in the feeding of farm animals, in poultry farm, and fisheries. Bederska-Lojewska *et al.* (2017) have reviewed the current status of knowledge concerning the use of edible mushroom as a feed additive with the dietary and health-promoting activities in the nutrition of broiler chickens and laying hens. In poultry farm, antibiotic is common to check the health and increase the production performance has led to human health hazards, therefore a natural source of the antibiotic substitute is being investigated, medicinal mushrooms are found to shows positive effects on immunity on the broiler and laying hens (Mahfuz and Piao, 2019). The waste of oyster mushroom left from the mushroom industry used as supplements and 1 per cent supplementation showed a positive response on the broiler chicken health (Fard *et al.,* 2014). In fisheries mushroom, supplemented food is currently culminating in recent literature and several research findings reported a positive impact of mushroom supplemented food in fish culture, Mohd-Din *et al.* (2012) have used mushroom supplementation as a prebiotic compound on red Tilapia fingerlings, they found weight gain in the fingerlings significant compare to control, replacing a 64-80 per cent proportion of dietary fish meal with fermented mushroom bran hydrolysate improve the growth, digestive enzyme activity, and antioxidant capacity of allogynogenetic crucian carp (fish) (Zhang *et al.,* 2017). Moreover, Dube and Dwivedi (1996) found that the oyster-mushroom waste supplemented water tank of fish *Cirrhinus mrigala* not only highly nutritive colonized detritus to the fish as direct feed, but also produced rich plankton in the treated water tank compare to the control. In Korea, dietary supplementation of microbially-fermented spent mushroom substrates evaluated for the weight gain, carcass characteristics, and economic efficiency of Hanwoo steers and have found to be positive effects that could replace a part of conventional roughage (Kim *et al.,* 2012). Inclusion of 5 per cent spent mushroom compost (SMC) in the diet of grower geese can be supplemented in the feed without adversely influencing growth performance or meat characteristics, the level of 5 per cent spent mushroom compost provides better meat sensory attributes for grower geese and promote good health performance (Chang *et al.,* 2016).

Mushroom Compost in Horticulture

SMC (spent mushroom compost) is the residual compost waste generated or left out after mushroom productions, this compost is readily available and its formulation generally consists of a combination of wheat straw, horse manure, ground chalk, *etc.* composted all together. SMC contains lots of salt and organic material along with enzymes and other nutrients that make the SMC suitable habitats for different microbes including bacteria and fungi, synergistically they are found to be beneficial in disease suppression and plant growth promotion, therefore, an integrated management approach could be adopted to check the severity of the pathogens. Disease suppressive properties of composts rely on several factors including microbial activity, microbial population dynamics, nutrient concentrations, and other associated chemical and physical factors (Patil *et al.,* 2018) it is seen that using compost as a soil amendment shows positive and significant effects on growth parameter of the crop production and plant promotion and induces microbiostasis, different study suggests that SMC could be used a source of bio-control agents against different plant diseases due to its abilities to produce organic salts and enzymes and have been reported that SMC has antagonistic fungi (*Trichoderma sp.*), and bacteria, SMC not only controls the disease but also reduces the use of fungicides and inorganic fertilizer, moreover it saves the money cost required for fertilization and conserve the soil, as SMC has great potential to remediate soil contamination. Spent mushroom substrate (SMS) and spent mushroom compost are two-term people often confused, SMS are those residues which are left after the mushroom harvest from the non-composted substrate and the SMS are those residues which left after the mushroom harvest from the composted substrate, therefore, both the residue found to have vast different on their nutrients contains available for the use as biofertilizer. Spent mushroom substrate (SMS) normally contains 1.9:0.4:2.4 per cent, N-P-K before weathering and 1.9:0.6:1.0, N-P-K after weathering for 8-16 months (Ahlawat and Sagar, 2007). Fresh spent mushroom contains Carbon: Nitrogen ratio of 13:1 on average (Fidanza *et al.,* 2010), the nutrients in the SMS and SMC depends on the mushroom species from which the spent material was derived, and material where mushrooms are grown, the characteristics of SMC and SMS highly effected by horticultural conditions, post]crop management, age of spent material, its storage time or post]crop composting. SMS and SMC have shared a common role in the soil as a soil conditioner like, improves soil structure, provides plant nutrients, increases plant nutrient availability, increases soil microbial populations, increases soil cation exchange capacity, increases plant root structure, increases soil aeration, improves soil water status, and reduces soil compaction, regulates pH of the soil. Therefore, SMC and SMS are suitable to use in horticulture, in (Table 12.4) the application of SMS and SMC in different cultivar has been tabulated. Mushroom spent substrate and Mushroom spent compost could be good as amendments in soil because of the following reasons:

- ☆ Low cost, consistently high-quality organic fertilizer,
- ☆ Reduce or eliminate the liming cost,
- ☆ Improves soil health,
- ☆ Improves crops health,
- ☆ Reduce the use of inorganic fertilizer,
- ☆ It is an organic approach,
- ☆ Reduce soil toxicity due to persisted pesticides and weedicides,
- ☆ Increases the soil microflora,
- ☆ Good bioremediator,
- ☆ Best in IFM, and
- ☆ Good in bio-control agent in plants disease management.

Table 12.4: Application of SMS and SMC in the different Horticultural Cultivars

Cultivars	*Spent substrate*		*References*
	SMC	*SMS*	
Amaranthus hybridus	+	-	Jonathan *et al.* (2012)
Bush Beans, Collards, Squash, Tomato	+	-	Stephens *et al.* (1989)
Salvia officinalis	+	-	Lopez-Castro (2008)
Capsicum, Tomato, Cauliflower, Pea, Potato, Ginger, Garlic, Wheat, Paddy, Maize and Apple	+	+	Sagar *et al.* (2009)
Tomato, Garlic, Onion, Brinjal, Cauliflower, Wheat, Capsicum, Maize, Pea, Spinach, Broccoli, Lettuce	+	-	Ahlawat and Sagar (2007)
Beetroot, Cucumber, Vineyards, Barley, Marigold, Green gram, Sweet potato	+	+	Rinker (2017)

Conclusion

Since the agriculture sector has been going through many challenges due to environmental threats and the increasing population. The increasing population and the challenges for food security have made concerns for a systematic approach to counteract food insecurity and increased the farmer's income. Though the application of the inorganic fertilizer and pesticides increase agriculture production but has worst effects on soils and human health, therefore the integrated farming system and the integrated farm management is promising methods to overcome the future threats in food security and agricultural land degradations. In this review, it has been reflected that how Mushroom and mushroom compost in IFM and IFS could provide progressive economic growth, employment opportunities, family nutritional requirements, livestock management, agricultural nutrients and optimal utilization of resources of the farming enterprises, *etc.* Furthermore, in this article, we have discussed the integrated farming system models existing in the country where mushrooms could be added as a component in any model that exists today. But there is very little documentation that has been made to reach the farmers. Hence step to be taken for the documentation of such kinds of farming models and to disseminate it to the needy farmers. The scientific community and research station has to initiate steps to evaluate and standardize the model so that the farmers can improve their standard of living and income. From this review, it is concluded that the integration of Mushroom and Mushroom compost could be a promising enterprise for the marginal and small farmers, particularly those who have fewer farm holdings.

REFERENCES

Ab Rhaman, S.M.S., Naher, L. and Siddiquee, S. (2022). Mushroom Quality Related with Various Substrates' Bioaccumulation and Translocation of Heavy Metals. *Journal of Fungi.* **8**(1): 42. (https://www.ncbi.nlm.nih.gov/pmc/articles/PMC8778862/)

Ahlawat, O.P. and Sagar, M.P. (2007). Management of spent mushroom substrate. *In: National Research Centre for Mushroom, Indian Council of Agricultural Research, Chambaghat, Solan, Himachal Pradesh.*

Balbino, L.C., Kichel, A.N., Bungenstab, D.J. and Almeida, R.G.D. (2014). Integrated systems: what they are, their advantages and limitations. *In: Integrated Crop-livestock-forestry Systems, a Brazilian Experience for Sustainable Farming.* pp. 11-18.

Barbieri, A., Quagliariello, V., Del Vecchio, V., Falco, M., Luciano, A., Amruthraj, N.J., Nasti, G., Ottaiano, A., Berretta, M., Laffaioli, R.V. and Arra, C. (2017). Anticancer and anti- inflammatory properties of *Ganoderma lucidum* extract effects on melanoma and triple-negative breast cancer treatment. *Nutrients.* **9**(3): 1-9.

Barik, A.K. (2017). Organic farming in India: Present status, challenges and technological breakthrough. The Third International Conference on Bio-Resource and Stress Management, November 8-11, 2017. pp. 84-93.

Bederska-Lojewska, D., Swi₹tkiewicz, S. and Muszynska, B. (2017). The use of Basidiomycota mushrooms in poultry nutrition – A review. *Animal Feed Science and Technology.* **230:** 59-69.

Behera, U.K., Dass, A., Rautaray, S.K., Choudhary, A.K. and Rana, D.S. (2013). Integrated farming system research in India: An overview. *Integrated Farming Systems for Enhancing Livelihood of Small and Marginal Farmers. Division of Agronomy, Indian Agricultural Research Institute, New Delhi.* pp. 40-78.

Bhuvaneshwari, S., Hettiarachchi, H. and Meegoda, J.N. (2019). Crop residue burning in India: Policy challenges and potential solutions. *International Journal of Environmental Research and Public Health.* **16**(5): 1-19.

Breene, W.M. (1990). Nutritional and medicinal value of specialty mushrooms. *Journal of Food Protection.* **53**(10): 883-894.

CARDI. (2010). A Manual on Integrated Farming Systems (IFS). *Caribbean Agricultural Research and Development Institute, Ministry of Economic Development, Belize.* pp. 1-58.

Carrasco, J., Zied, D.C., Pardo, J.E., Preston, G.M. and Pardo-Gimenez, A. (2018). Supplementation in mushroom crops and its impact on yield and quality. *AMB Express.* **8**(1): 1-9.

Chandra, S. and Samsher. (2006). Nutritional and medicinal aspects of edible mushrooms. *International Journal of Agricultural Science.* **2**(2): 647-651.

Chang, S.C., Lin, M.J., Chao, Y.P., Chiang, C.J., Jea, Y.S. and Lee, T.T. (2016). Effects of spent mushroom compost meal on growth performance and meat characteristics of grower geese. *Revista Brasileira de Zootecnia.* **45**(6): 281-287.

Choudhary, S.K., Kumar, R.A.H. and Gupta, S.K. (2019). Integrated farming system (IFS) is possible way out for double farmer's income. *Journal of Pharmacognosy and Phytochemistry.* **5:** 282-289.

Devi, S., Gupta, C., Jat, S.L. and Parmar, M.S. (2017). Crop residue recycling for economic and environmental sustainability: The case of India. *Open Agriculture.* **2**(1): 486-494.

Dube, K. and Dwivedi, A. (1996). Oyster mushroom waste as manure in fish culture: A preliminary study. *Journal of the Indian Fisheries Association.* **26:** 111-114.

Fard, S.H., Toghyani, M. and Tabeidian, S.A. (2014). Effect of oyster mushroom wastes on performance, immune responses and intestinal morphology of broiler chickens. *International Journal of Recycling of Organic Waste in Agriculture.* **3**(4): 141-146.

Fidanza, M.A., Sanford, D.L., Beyer, D.M. and Aurentz, D.J. (2010). Analysis of fresh mushroom compost. *HortTechnology.* **20**(2): 449-453.

Gowda, N.A.N. and Manvi, D. (2019). Agro-residues disinfection methods for mushroom cultivation. *Agricultural Reviews.* **40**(2): 93-103.

IARI. (2012). Crop residues management with conservation agriculture/ : Potential, constraints and policy needs. *Indian Agricultural Research Institute (IARI), Indian Council of Agricultural Research, New Delhi, India.* pp. 1-32.

Jonathan, S.G., Oyetunji, O.J. and Asemoloye, M.A. (2012). Supplementation of spent mushroom compost (SMC) of *Pleurotus ostreatus* (Jackuin Ex. Fr.) Kummer as a soil amendment for the growth of *Amaranthus hybridus* Lin. A Nigerian green vegetable. *BTAIJ.* **6**(12): 396-403.

Khan, M.A., Tania, M., Liu, R. and Rahman, M.M. (2013). *Hericium erinaceus*: An edible mushroom with medicinal values. *Journal of Complementary and Integrative Medicine.* **10**(1): 253-258.

Kim, Y.I., Lee, Y.H., Kim, K.H., Oh, Y.K., Moon, Y.H. and Kwak, W.S. (2012). Effects of supplementing microbially-fermented spent mushroom substrates on growth performance and carcass characteristics of Hanwoo steers (a field study). *Asian-Australasian Journal of Animal Sciences.* **25**(11): 1575-1581.

Liu, Q., Ma, H., Zhang, Y. and Dong, C. (2017). Artificial cultivation of true morels: Current state, issues and perspectives. *Critical Reviews in Biotechnology.* **38**(2): 259-271.

Lopez-Castro, R.I., Delmastro, S. and Curvetto, N.R. (2008). Spent oyster mushroom substrate in a mix with organic soil for plant pot cultivation. *Micologia Aplicada International.* **20**(1): 17-26.

Mahfuz, S. and Piao, X. (2019). Use of medicinal mushrooms in layer ration. *Animal.* **9:** 1-13.

Manjunatha, S.B., Shivmurthy, D., Satyareddi, S.A., Nagaraj, M.V. and Basavesha, K.N. (2014). Integrated farming system – An holistic approach/ : A review. *Journal of Analytical Atomic Spectrometry.* **3**(4): 30-38.

Marshall, E. and Nair N.G. (2009). Make money by growing mushroom. *Rural Infrastructure and Agro-Industries Division Food and Agriculture Organization of the United Nations, Rome, Italy.* pp. 1-64.

MoA. (2014). National policy for management of crop residues (NPMCR). Ministry of Agriculture, Department of Agriculture and Cooperation, Government of India, New Delhi. pp. 1-9.

Mohd-Din, A.R.J., Razak, S.A. and Sabaratnam, V. (2012). Effect of mushroom supplementation as a prebiotic compound in super worm based diet on growth performance of red tilapia fingerlings. *Sains Malaysiana.* **41**(10): 1197-1203.

Nahata, A. (2013). *Ganoderma lucidum*: A potent medicinal mushroom with numerous health benefits. *Pharmaceutica Analytica Acta.* **4**(10): 1-4.

Pace, K. (2015). Indigenous agriculture and sustainable food. *Sustainable Food Center, Austin.*

Patel, R.H. and Dutta, S. (2004). Integrated farming system approach for sustainable yield and economic efficiency – A review. *Agricultural Reviews.* **25**(3): 219-224.

Patil, M.G., Rathod, P.K. and Patil, V.D. (2018). Compost: A tool for managing soil borne plant pathogens. *International Journal of Current Microbiology and Applied Sciences.* **6:** 272-280.

Radwan, F.F., Perez, J.M. and Haque, A. (2012). Apoptotic and immune restoration effects of ganoderic acids define a new prospective for complementary treatment of cancer. *J. Clin. Cellular Immunology.* **1**(S3): 1-7.

Rana, S.S. (2015). Recent advances in integrated farming systems. *Department of Agronomy, College of Agriculture, CSK Himachal Pradesh Krishi Vishvavidyalaya, Palampur.* pp. 1-204.

Rinker, D.L. (2017). Spent mushroom substrate uses. *In:* Zied, D.C. and Pardo-Gimenez, A. (eds.). Edible and Medicinal Mushrooms: Technology and Applications. John Wiley and Sons Ltd., Hoboken, New Jersey. pp. 427-454.

Rubini, A., Riccioni, C., Belfori, B. and Paolocci, F. (2014). Impact of the competition between mating types on the cultivation of *Tuber melanosporum*: Romeo and Juliet and the matter of space and time. *Mycorrhiza.* **24**(1): 19-27.

Sagar, M.P., Ahlawat, O.P., Raj, D., Vijay, B. and Indurani, C. (2009). Indigenous technical knowledge about the use of spent mushroom substrate. *Indian Journal of Traditional Knowledge.* **8**(2): 242-248.

Sharma, V.P., Annepu, S.K., Gautam, Y., Singh, M. and Kamal, S. (2017). Status of mushroom production in India. *Mushroom Research.* **26**(2): 111-120.

Sheikh, I.A., Vyas, D., Lone, R. and Singh, V. (2015). *Ganoderma lucidum* and *Trametes hirsuta* as potent antioxidants in free radical systems in vitro. *World Journal of Pharmacy and Pharmaceutical Sciences.* **4**(5): 1695-1710.

Stephens, J.M., Henry, G.C., Castro, B.F. and Bennett, D.L. (1989). Mushroom compost as a soil amendment for vegetable gardens. *Proceedings of the Florida State Horticultural Society.* **102**: 108-111.

UN. (2017). World population projected to reach 9.8 billion in 2050, and 11.2 billion in 2100. *Department of Economic and Social Affairs, United Nations, New York.*

Vyas, D., Seikh, I.A. and Tiwari, G.K. (2016). Role of mushroom in maintaining mental health with special reference to anti-convulsant activity. *The International Journal of Indian Psychology.* **4**(1): 73-92.

Wakchaure, G.C. (2011). Mushrooms-value added products. *In:* Singh, M., Vijay, B., Kamal, S. and Wakchaure, G.C. (eds.). Mushrooms-cultivation, marketing and consumption. Directorate of Mushroom Research, Solan, Himachal Pradesh, India. pp. 233-238.

Wakchaure, G.C., Shirur, M., Manikandan, K. and Rana L. (2010). Development and evaluation of oyster mushroom value added products. *Mushroom Research.* **19**(1): 40-44.

Zhang, D., Zhang, Y., Liu, B., Jiang, Y., Zhou, Q., Wang, J. and Kuang, Q. (2017). Effect of replacing fish meal with fermented mushroom bran hydrolysate on the growth, digestive enzyme activity, and antioxidant capacity of allogynogenetic Crucian Carp (*Carassius auratus gibelio*). *Turkish Journal of Fisheries and Aquatic Sciences.* **17:** 1039-1048.

Chapter 13

Aquaculture-cum-Poultry Farming System for Augmenting Farmers' Income

A. Devivaraprasad Reddy[1]*, T. Vijaya Nirmala[1], CH. Balakrishna[2], J. Venkat Satish[1], E. KarunaSree[1], K. Venkata Subbaiah[1], V. Deepthi[1], B. Srinivasulu[3] and J.V. Prasad[4]

[1]*Krishi Vigyan Kendra, Dr. Y.S.R. Horticultural University, Venkataramannagudem, West Godavari District – 534112, Andhra Pradesh*
[2]*Krishi Vigyan Kendra, Acharya N.G. Ranga Agricultural University, Amadalavalsa, Srikakulam District – 532185, Andhra Pradesh*
[3]*Administrative Officer, Dr. Y.S.R. Horticultural University, Venkataramannagudem, West Godavari District – 534101, Andhra Pradesh*
[4]*ICAR-Agricultural Technology Application Research Institute (ATARI), Zone-10, Hyderabad – 500059, Telangana*
Corresponding author: prasad8reddy@gmail.com

More than 80 per cent farmers belong to of small and marginal and are in agricultural situations. Despite their tremendous efforts, they do not generate much money because there is little money left after paying all inputs. Marginal and small farmers at large are uneducated, financially disadvantaged and operate in resource-poor and risk-prone varied settings. Their small spread holdings are unsuitable for modern agricultural machinery and they are forced to work in a variety of risky and resource-poor environments. Aquaculture sector and poultry farming is the fast growing sector. The vast majority of farmers cultivate their fields just for farming, with no scientific or technical backup. A holistic all-around strategy is essential to make farming commercially profitable and ecologically sound and sustainable. This, in turn, is contingent on the adoption of enhanced technology and the prudent use of resources. Backyard poultry is an important part of rural people's lives since it provides them with additional cash and nutritional security. Farmers make a lot of money from aquaculture and poultry farming. By integrating diverse farm components like aquaculture coupled with poultry and recycling agricultural leftovers and by-products inside the farm, the method attempts to increase employment and revenue from small-holdings.

***Keywords**: Aquaculture, Poultry, Backyard poultry, Income generation.*

Introduction

In 2050, the world's population is predicted to exceed 9.8 billion people, putting even more strain on current food supplies. The aquaculture and animal husbandry industry can play a critical role in meeting the world's growing protein need. The potential for domesticated animals to thrive in India is limitless, and livestock is the backbone of the rural economy. Aquaculture, as one of the world's fastest-growing industries, presents enormous opportunities in both marine and inland fisheries in India, which has a wide coastline. The country has the second-largest proportion of the worldwide aquaculture industry, with the aquaculture sector accounting for around 68 per cent of the country's total fish production. In terms of employment, the industry employs around 28 million people in India. The fisheries sector earned US $ 5956.93 in exports in 2020-21 (MPEDA, 2021).

Small and marginal farmers are increasing day by day and it is estimated that more than 85 per cent was transformed having the per capita land availability of less than one hectare of land. It is difficult to have the farm resources/machineries *viz.,* animals, tractors, bore wells, *etc.* In addition to this, the inputs for these conditions are increasing and making farming an unprofitable and unsustainable enterprise.

The advancement of sustainable animal husbandry is exciting because it can improve all aspects of food safety and security. Animal ownership can flexibly encourage expanded food at the communal and public levels by increasing food output from animals and increasing yields through increased fertiliser use. Furthermore, animal ownership and creation represent a significant source of resources and earnings, enabling the use of food through self-creation or resource usage, thereby increasing food availability. Additionally, owning and raising domesticated animals improves nutrition and dietary variety by boosting the usage of animal-derived nutrients, such as essential vitamins and high amounts of protein, which supplement food consumption (Carletto *et al.,* 2015; Hoddinott, 2015).

Food safety necessitates the simple availability and accessibility of food to individuals, as well as proper nutritional evaluation for all members of the family unit, whether received or made by creation. Poultry is crucial for food security since it offers both food and a way to bring in income, which anybody may use to buy food (Sonaiya *et al.,* 1994). In this chapter, elucidates about the entrepreneurship development in the cultivation of the aquaculture and poultry bird rearing and making the farmers beneficial.

Poultry Farming

In India, backyard poultry rearing is still popular in tribal and rural communities, despite the presence of large commercial poultry operations. Backyard poultry is a type of low-input domestic fowl that has been kept by rural communities in most developed nations for generations (Akinola and Essien, 2011). One of the viable alternative methods for strengthening rural family incomes, producing additional cash, and supplementing protein intake in rural and tribal areas is free-range and semi-intensive backyard poultry farming. Backyard poultry raising is an important source of income for some of the world's poorest people. Under lawn poultry systems, women are generally accountable for the consideration of the flying creature, as well as the board (Deka *et al.,* 2014). Feasibly, it is the key commodity fully owned by rural people for additional income generation (Ramdas, 2009). Poultry rearing enabled the women to be financially active and socially active since her circle of acquaintances extended because of exchanging poultry and poultry goods. She should create an organisation that includes contact with a larger group of people who can help her in her time of need (Ayesha and Mahwish, 2015). Females keep backyard poultry in scavenging initiatives to complement and increase their incomes.

Poultry farming is a common activity in rural areas, with most rural homes having enough birds in their backyard. The variations are chosen according to the necessity and purpose, either meat or egg. Production and productivity of eggs and meat are influenced by many factors and some of them are diversity, feed, health, and environmental circumstances. The poultry shelter can be built on the pond dyke or on the dyke's side, with the excreta falling straight into the water or being collected using sheets. Permanent buildings or locally available materials can be used to build a home. When they are young, each bird wants around 1 square meter of space, and when they are older, they require approximately 2 square meters.

The day old chicks are to be kept in the brooding chamber with sufficient lighting and vaccination as per schedule. After four weeks old chicks can be transferred either to the cages or houses near to the pond. Due to bacterial action, the fall of the birds that fall on the litter eventually mixes with the litter material. When the litter depth gets lower, to retain adequate depth, more organic matter is added. Super phosphate or lime is used to keep the litter dry if it is wet. Every day, the litter is stirred to provide aeration. It becomes completely built-up litter in around 10-12 months, with a very high manurial value. Until chicks are put, the shed should be properly washed, disinfected and resting (Figure 1).

Poultry bird varieties *viz.* Vanaraja, Srinidhi, Rajasri or Kuroiler are the dual-purpose birds for meat and egg and Gramapriya, BV-380 is ideal for this integration of egg purpose birds. For one hectare of water distribution area, approximately 500 to 600 birds are expected. After vaccination against infectious diseases and providing other required prophylactic measures to protect against epidemics, about eight-week-old chicks are housed in poultry houses near the pond. Grower mash is supplied to the birds in the Intense rearing method between the ages of 9 to 20 weeks @ 50 to 70 g bird^{-1} day^{-1}. Layer mash feed can be fed to the birds above the age of 20 weeks @ 80 to 120 g bird^{-1} day^{-1}. Nirmala *et al.* (2017) said that under the free-range rearing programme, the birds were released to scavenge in the fields and ponds around them in order to meet their nutrition needs. The birds generally feed on kitchen waste, earthworms, grasshoppers, ants, green grass, leafy vegetables, seeds, *etc.* during scavenging. In addition to scavenging, all the participants gave their birds a handful of broken rice, bajra, sorghum, corn, *etc.* Both the birds are supplied with a sufficient supply of water at all times. The production of eggs starts at the age of 22 weeks and then slowly decreases.

Figure 13.1: Chick Brooding – Indoor Systems in Controlled Conditions.

The egg production ranges from 140 to 200 per year depending on the breed, feeding, health status and environment conditions where the birds are living. The birds are disposed of at the age of 18 months. Karunasree *et al.* (2017) announced that the average body weights of Vanaraja, Gramapriya, and Aseel chickens at 48 weeks of age were recorded as 3.22 kg, 2.50 kg, and 1.50 kg, respectively. This information showed that Gramapriya (178) produced more eggs than Vanaraja (160), and Aseel (62) bird. Fertilization with poultry manure results in the production in one year of 3000 to 4000 kg of fish, 60,000 to 100,000 eggs and more than 3,500 kg meat from a hectare of pond. The fish production of 10 tonnes per hectare might be attained by raising tilapia, common carp, and murrels with stocking densities of 20,000 fingerlings ha^{-1} and 4,000 chicks ha^{-1}.

One of the possible alternative systems for enhancing the livelihood of rural households is the backyard poultry production under the free range and semi-intensive system, which provides additional revenue and supplements the protein intake of rural and tribal people. Backyard chickens' meat and eggs are far more highly appreciated than those of "organic" hens and eggs, with prices that are 50 to 100 per cent more (Nirmala *et al.,* 2016).

Poultry and fish farming systems can be combined to cut the cost of fish farming's feeds and fertilisers. Raising chickens close to or over a fish pond hastens chicken excrement to simply fall into the fish pond and be recycled. In this approach, deep litter is used to rear chickens. About 3 per cent N, 2 per cent P and 2 per cent K are present in litter made from chicken droppings. In a fish pond, poultry waste serves as a good source of fertiliser that aids in the production of phytoplankton and zooplankton, which are used as fish food (Kochewad *et al.*, 2017).

Traditionally, desi breeds, whose annual egg production capacity is only 60 to 80, are utilised for backyard chicken raising, thus making the backyard poultry less economical (Figure 13.2). Therefore, the enhanced types, which resemble native chickens, are currently being widely introduced in the area in order to raise the productivity of backyard poultry production (Singh *et al.*, 2002).

Figure 13.2: Improved Backyard Poultry Birds Rearing in Open Grazing System.

Body weight is a direct reflection of growth and it influences the production and reproduction characteristics of birds. The average body weights at 48 weeks of age were recorded as 3.22 kg, 2.50 kg and 1.50 kg in cases of Vanaraja, Gramapriya and Aseel chicken, respectively. Egg production determines the success of a poultry enterprise. According to a comparative estimate of egg production, Gramapriya (178) produces more eggs than chickens Vanaraja (160) and Aseel (62) combined. This disparity in egg production may be caused by improved chicken types and the genetic makeup of desi chickens (Karunasree *et al.*, 2017).

Selection of Poultry Birds

Poultry birds can be selected based on the average body weight and egg production and the following varieties are selected as per mentioned in the Figure 13.3 and 13.4

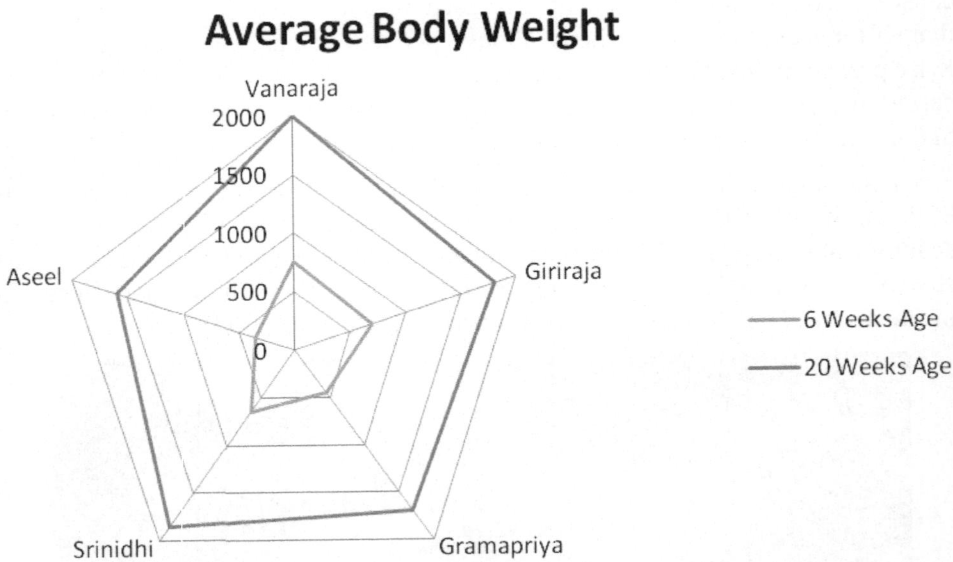

Figure 13.3: Average Body Weight of different Poultry Birds.

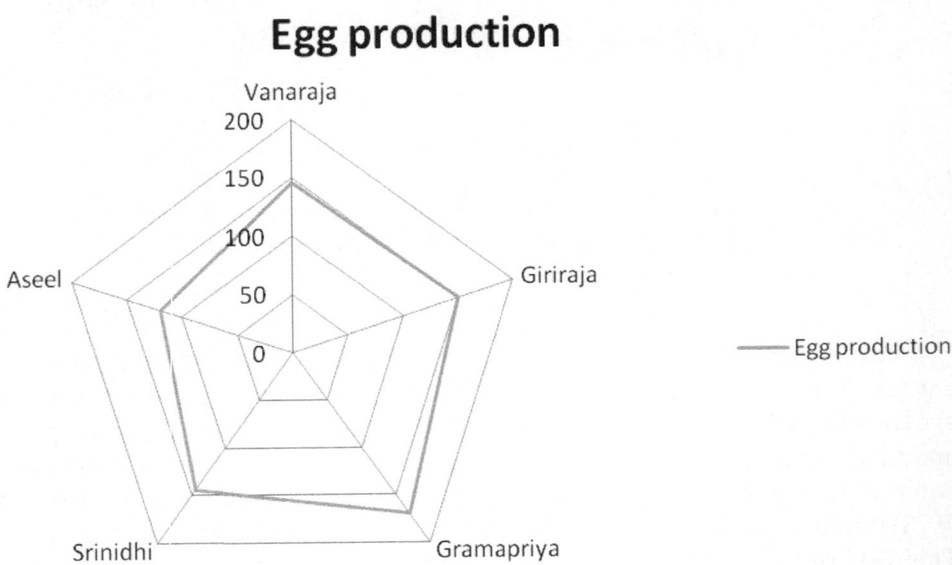

Figure 13.4: Egg Production Potential of different Poultry Birds.

Housing of Poultry Birds

Poultry house can be constructed over the pond or on the embankment of pond. The height of house should be 2.1 to 2.4 m at side and 3.0 to 3.3 m at the centre to provide slope on the either side. The roofing materials like asbestos sheets, or thatches may be used for construction. The width of poultry house should not exceed 9 m to have effective cross ventilation and length of the house may be as per the requirements. The droppings of the birds directly fall into the pond or collected from the floor and applied to the pond. Floor space requirement for each bird is 1 to 2 sq. ft. (Figures 13.5–13.7).

Figure 13.5: Poultry Cage with Improved Poultry Birds.

Figure 13.6: Poultry Cage Installed on the Pond Dyke.

Selection of Birds

Vanaraja, Giriraja, Srinidhi or Kuroiler are dual purpose, meat and egg birds, whereas Gramapriya, Rajasri and desi birds like Aseel, Kadaknath, and Denki are appropriate for this integration. For one-hectare water spread area, 500 to 600 birds are necessary. Around eight week old chicks are kept in a poultry house near the pond after being vaccinated against viral diseases and receiving other necessary prophylactic measures to prevent epidemics.

Figure 13.7: Floating Poultry Cage-cum-Aquaculture.

Feeding and Watering of Birds

Improved backyard poultry birds can efficiently utilize the natural food base available in scavenging. Feed the birds with different cereals available in rural areas (like bajra, ragi, jowar, korra, broken rice, rice bran) to sustain the production under free range condition. In case of intensive system of rearing provide grower mash @ 50 to 70 g bird^{-1} day^{-1} during the age of 9 to 20 weeks. Whereas, layer mash is provided to the birds above 20 weeks age @ 80 to 120 g bird^{-1} day^{-1}. An ample supply of water is made available to all the birds throughout the day.

Health Care of Birds

Marek's illness, RD/ND, IBD, and Fowl pox are all threats to these birds. Anti-stress drugs must be given on the day before vaccination to ensure a greater immune response. To avoid cannibalism, trace minerals and common salt should be given to the meal. Vaccination against Ranikhet Disease and Fowl Pox should be done at every six months (Figure 13.8).

Table 13.1: Vaccination Schedule of Poultry Birds

Age	*Name of the Vaccine*
First day	Marek's Disease (MD)
5th to7th day	Ranikhet Diseae (RD)/Newcastle disease (ND) – Lasota strain
14th day	Infectious Bursal Disease (IBD)
21st day	Fowl Pox
28th day	Ranikhet Diseae (RD)/Newcastle disease (ND) – Lasota strain
9th week	Ranikhet Diseae (RD)/Newcastle disease (ND) – R2B or RDK strain
12th week	Fowl pox

Figure 13.8: Vaccination of Poultry Chicks.

Aquaculture

Aquaculture sector is considered to be growth engines and this sector will be one of the options to address the food security under the human population growth scenario of 9.8 billion people on the globe in 2050. However this sector should continue its production increase in a sustainable manner (Blanchard *et al.,* 2017). In India, Aquaculture is performed in two ways *i.e.* freshwater aquaculture and brackish water aquaculture with capture and culture techniques of aquaculture. Carp fish production is the mainstay of aquaculture followed by shrimp, freshwater prawn, freshwater pearl culture, crab fattening, and ornamental fish farming *etc.* are practicing.

Composite fish culture is the age old practice in India and which is very much suitable for integration with other farming sectors. By using composite fish culture, available resources within the pond are utilized as well as pond productivity

(Reddy *et al.*, 2019). The production and productivity of carps can also be improved through polyculture. In order to efficiently manage pond productivity, perennial ponds with a size of 0.4 to 1 ha and a depth of 2 to 3 meters are preferred. For carrying the carp culture, rainwater harvesting systems are also a good alternative. Loamy soil with a pH between 6.5 and 7.5 is highly suitable. Based on the soil pH condition, apply quick lime or agriculture lime @ 200 to 1200 kg ha^{-1} on two to three intervals to correct the soil pH. Cow dung and gypsum ($CaSO_4$) are used to correct sodic soils. In addition, soil management and water quality parameters are crucial for the success of the pond culture. A light green or brown colour is preferred for culture. Aquatic weeds are undesirable to culture practices and can be controlled by mechanical removal or chemical or biological methods. Mahua oil cake (2500 to 3000 kg ha^{-1}) or Tea Seed Cake (750 to 1000 kg ha^{-1}) or bleaching powder (300 to 500 kg ha^{-1}) can be used to eliminate predatory and weed fishes. It is always recommended to dewater small ponds to eliminate trash and predatory fishes (Shyam and Kumar, 2001).

Composite culture involving a combination of 6 species, *viz.* catla (*Catla catla*), silver carp (*Hypophthalmichthys molitrix*), rohu (*Labeo rohita*), grass carp (*Ctenopharyngodon idella*), mrigal (*Cirrhinus mrigala*) and common carp (*Cyprinus carpio*). The surface feeder (catla and silver carp), column feeder (rohu and grass carp), and bottom feeder (mrigal and common carp) with a ratio of 3:4:3, respectively depending upon the water level maintained in the fish pond (Reddy *et al.*, 2019). Weeds control by biological means using herbivorous fishes like grass carp, *Puntius javanicus*, *Puntius punchellus* and Gorami. Stocking density (SD) is assessed by the criteria of species selected, productivity and size of the pond. Generally, the ponds are stocked @ 5000 to 7000 number of fingerlings (10-15 cm). In zero-point culture, the ponds are stocked with the advanced fingerlings (150 to 250 grams) @ 500-1000 numbers (Ramakrishna *et al.*, 2013). At the time of stocking, fish bath in either 3 per cent common salt or 3 to 5 per cent potassium permanganate for 15 to 30 seconds will reduce the bacterial infections, transferring from one pond to another pond.

Figure 13.9: Harvesting of Fish in Aquaculture-cum-Poultry Rearing System.

Poultry, goat, sheep droppings, cattle manure, vermicompost act as fertilisation for production of natural phytoplankton and zooplankton. The fishes feed on that plankton. Usually in integrated farming systems, additional supplementary feed is not required. However, the supplementary feeding like farm made feeds or pelleted feed are to be fed to produces high productivity and more returns and till the reared fish attains a weight of table size (>700 g) (Reddy *et al.*, 2019). Regular monitoring of water quality management during culture practices improves the success of culture. Subsequent accumulation of animal excreta and fish faeces affects the water quality and also leading to slow growth and diseases in fish. Regular sampling of fish at fixed intervals will know the health condition and growth of the fish can be monitored. High stocking density leads to occurrence of diseases. In this scenario, replacement of water with freshwater can solve the issues. Dissolved oxygen (DO) is the critical factor for survival of fish. If the factor increase in stocking density, phytoplankton, and aquatic weeds reduces the DO. Due to this, in the early morning hours before the sun rises, the fish come to the surface and tries to take the surface. During this time, aeration is required. Aeration in the pond can be done by using the paddle wheel aerator, motor for bring down to upside, using boat motor by disturbing, and using chemicals like calcium oxide in the name of oxygen tablets.

Figure 13.10: Selling of Fish in Aquaculture-cum-Poultry Rearing System.

Partial harvesting can be done after 6 months in case of zero-point culture and after 8 months in case of stunted fingerlings and after 10 months in case of fingerlings stocked in the pond. Cast nets or drag nets are periodically harvested depending on fish need. Since, fish attains its marketable size and harvesting of marketing size also reduces the stocking density and pressure in the pond. Release of species once again after harvesting, maintains its ecological balance (Figure 9 and 10). Catla had the fastest growth, followed by rohu, mrigal, and common carp. At the time of the final harvest, catla reached its maximum size of about 3 kg. This may be due to the consumption of plankton and feed. Fish production can be obtained @ 5 to 12 tonnes ha^{-1} year^{-1}. The fish production is dependent on the

Figure 13.11: Poultry-cum-Aquaculture Entrepreneur Rearing and Feeding the Improved Poultry Birds.

stocking density and supplementary feeding. The net income from the fish culture is about ₹ 120000 to 250000 depending of the management practices implemented by the farmers (Figure 13.11) (Reddy *et al.*, 2018; Reddy *et al.*, 2019).

Conclusion

Several researchers have proposed agricultural systems approach to research and development to satisfy the numerous aims of poverty reduction, food and nutrition security, competitiveness, and sustainability. A farming system is the consequence of complex interactions between a number of interdependent enterprises/components, in which a single farmer distributes specific quantities and quality of four production variables, namely, land, labour, money, and management, to which it has access. The farming system concept is a useful tool for managing natural and human resources in developing countries like India. It is a multidisciplinary whole-farm strategy that may be used to solve the difficulties faced by small and marginal farmers. Aquaculture coupled with poultry rearing has many advantages and can augment the farmer's income by choosing the right varieties/breeds/strains as per the market demand.

REFERENCES

Akinola, L.A.F. and Essien, A. (2011). Relevance of rural poultry production in developing countries with special reference to Africa. *World's Poultry Science Journal.* **67:** 697-705.

Ayesha, A. and Mahwish, Z. (2015). Poultry farming: An Anthropological take on women empowerment through food security at household level. *The Explorer Islamabad: Journal of Social Sciences.* **1**(6): 203-206.

Blanchard, J.L. Watson, R.A., Fulton, E.A., Cottrell, R.S., Nash, K.L., Buchholz, AB., Büchner, M., Carozza, D.A., Cheung, W.W.L., Elliott, J., Davidson, L.N.K., Dulvy, N.K., Dunne, J.P., Eddy, T.D., Galbraith, E., Lotze, H.K., Maury, O., Müller, C., Tittensor, D.P. and S. Jennings. (2017). Linked sustainability challenges and trade-offs among fisheries, aquaculture and agriculture. *Nature Ecology and Evolution*. **1:** 1240-1249.

Carletto, G., Ruel, M., Winters, P., Zezza, A. (2016). Farm-Level Pathways to Improved Nutritional Status: Introduction to the Special Issue. *The Journal of Development Studies*. **51**(8): 945-957.

Deka, R.J., Zakir, A.M.M. and Kayastha, R.B. (2014). Improvement of rural livelihood through rearing of Chara-Chemballi ducks in Assam. *World's Poultry Science Journal*. **70**(2): 397-404.

Hoddinott, J., Headey, D. and Dereje, M. (2015). Cows, missing milk markets and nutrition in rural Ethiopia. *The Journal of Development Studies*. **51:** 958-975.

Karunasree, E., Vijaya Nirmala, T. and Reddy, R.V.S.K. (2017). Comparative performance of different varieties of chicken under backyard system of rearing in tribal and rural areas of West Godavari district, Andhra Pradesh. *International Journal of Science, Environment and Technology*. **6**(2): 1100-1104.

Kochewad, S.A., Meena, L.R., Devender K., Sanjeev K., Meena, L.K., Singh, S.P. and Khushyal S. (2017). Livestock-based integrated farming systems for livelihood improvement of small and marginal farmers. *South Asian Journal Food Technology Environment*. **3**(1): 526-532.

Mandal, M.K., Khandnekar, N. and Khandnekar, P. (2006). Backyard poultry farming in Bareilly district of Uttar Pradesh, India: An Analysis. *Livestock Research for Rural Development*. Volume 18, Article #101. Retrieved September 29, 2022, from http://www.lrrd.org/lrrd18/7/mand18101.htm

MPEDA. (2021). Marine Products Exports. (Source: https://mpeda.gov.in/?page_id=438; accessed on 25.03.2022)

Nirmala, T.V., Prasad, S., Subrahmanyeswari, B. and Kumari, K.N. (2017). Status of Indigenous chicken rearing in Krishna district of Andhra Pradesh. *The Indian Journal of Veterinary Science and Biotechnology*. **12**(3): 52-55.

Ramakrishna, R., Shipton, T.A. and Hasan, M.R. (2013). Feeding and feed management of Indian major carps in Andhra Pradesh, India. FAO Fisheries and Aquaculture Technical Paper No. 578. Rome, FAO. 90 pp.

Ramdas, S.R. (2009). Reclaiming endangered livelihoods: Untold stories of indigenous women and backyard poultry. *World's Poultry Science Journal*. **65**(2): 241-250.

Reddy, A.D., Nirmala, T.V., Karunasree, E., Subbaiah, K.V., Raju, G.S., Deepthi, V., Reddy, R.V.S.K., Srinivasulu, B., Prasad, J.V. and Prasad. Y.G. (2019). Diversification in agriculture through fisheries based integrated farming system for doubling farmers' income in agency areas of West Godavari district of Andhra Pradesh – A case study. *International Journal of Chemical Studies.* **SP6:** 930-932.

Reddy, R.V.S.K., Karunasree, E., Reddy, A.D., Deepthi, V., Nirmala, T.V., Raju G.S., Subbaiah KV and Prasad, J.V. (2018). An evaluation study on viable integrated farming system (IFS) model in Godavari delta of Andhra Pradesh. *Journal of Pharmacognocy and Phytochemistry.* **SPI**: 2361-2368.

Shyam, R. and Kumar, K. (2001). Fish-Livestock Integrated Farming. Central Institute of Freshwater Aquaculture, Bhubaneswar, Odisha. 68 pp.

Singh, R.V., Saxena, V.K. and Sharma, D. (2002). Technological developments in the poultry sub-sector; *In*: Technology options for sustainable livestock production in India. Proceedings of the workshop on Documentation, Adoption and Impact of Livestock Technologies in India, pp. 99-103.

Sonaiya, E.B., Vernooy, R. and Kealey, K.M. (1994). Food Systems under Stress: Research/Action Issues and Needs in West Africa. *In*: Food Systems under Stress in Africa: African-Canadian Research Cooperation. Proceedings of a Workshop held in Ottawa, Ontario, Canada International Development Research Centre.

Chapter 14

Integrated Fish and Duck Farming System: A Potential Model for an Entrepreneurial Endeavour

Banlam Jingshai Marbaniang[1]* and Gouranga Biswas[2]

[1]Research Scholar, ICAR-Central Institute of Fisheries Education, Versova, Andheri (West), Mumbai – 400061, Maharashtra
[2]Senior Scientist, ICAR-Central Institute of Fisheries Education (Kolkata Centre), 2, GN Block, Sector V, Bidhannagar, Kolkata – 700091, West Bengal
Corresponding author: banlamarbaniang@gmail.com

ABSTRACT

In this chapter, we will discuss various aspects of fish and duck farming as an integrated model in utilizing their maximum potential and achieve a better productivity and profitability as compared to traditional farming systems. Several works done has proven that an integration of these two units can result in an overall better performance of the farm. Benefit-cost ratio achieved from this system varies between 1.91 and 3.11, whereas it is 1.4 to 1.83 in traditional farming. It is highly advisable to farmers and entrepreneurs to take up this concept and bloom in the industry. The production cost of this system is highly reasonable as a lot of amount is saved in terms of feed for fishes, manuring cost etc., which are produced by the ducks as by-products and utilized by fishes as inputs for plankton production and feeding. Duck meat and egg sale covers the production cost and the fish sale is entirely a net profit for the farmer. More work needs to be done to tackle constraints such as disease outbreaks, availability of quality seeds, popularization of duck meat and eggs and most importantly the need of the hour is the work at extension level in adoption of concept and diffusion of technology.

Keywords: *Integrated fish-duck farming, Profitability, Benefit cost ratio, Mutual relationship.*

Introduction

Integrated fish farming is a method of raising fish in conjunction with other elements such as cattle, agriculture, and horticulture, with fish being the most critical component. The basic idea behind creating this type of system is to make the most of the limited feedstuffs available (Strasbourg, 2013) and establish a mutual interaction between the animals and crops. Both components contribute to a mutually beneficial result.

Pigs, ducks, and chicken, a widely used approach in Asia, have produced the highest yields in integrated fish farming thus far (Pandey, 2016). Numerous studies and analyses have shown that an integrated system can provide a farmer with the most outstanding value and profit rather than fish farming alone. Several factors contribute to a farmer's total profitability in integrated fish farming which includes lower feed cost (excreta from one component serves as feed for the other), a natural process for manuring the pond system for natural plankton generation, which lowers manure costs are a few to mention.

Figure 14.1: Integrated Fish and Duck Farming System at CIBA, Kakdwip (*Courtesy*: ICAR-CIBA, Kakdwip Research Centre).

In Asian and European cultures, raising ducks in fish ponds is a long-standing tradition. In India, states like Andhra Pradesh, Odisha, West Bengal, Bihar, Kerala, Tamil Nadu, Karnataka, and North-Eastern states like Assam, Tripura, Manipur, and Mizoram use this system extensively (Kumar and Mishra, 2011). Duck farming, among other integration systems, contributes to a greater benefit cost ratio (BCR) for the farmer because the overall result includes not only fish and duck meat, but also eggs in large quantities, which adds to the farm's entire output. The *'Indian Runner'* duck breed is the mostly used duck breed in India's fish and duck integration system (Strasbourg, 2013). According to Thakur *et al.* (2018), fish cum duck farming is the finest type of integrated livestock and poultry farming, taking into consideration the utilization of manure, pond area, number of outputs, input-output connection *etc.*

Ducks serve as a natural aeration system by paddling across the pond's surface, adding up oxygen to the pond's total dissolved oxygen concentration, which is then inhaled by the fish. Furthermore, the duck droppings naturally fall into the water column, enabling natural pond manuring, which eventually contributes to nutrients add-up for the development of natural fish food. In turn, the pond provides the duck with a diverse diet *viz.* insects, small fish, and other species, and, as a semi-enclosed biological system, the pond protects them from disease by providing a safe and disease-free environment (Strasbourg, 2013). Fish and duck integration can become very productive to fulfil demand and supply while also bringing considerable production from a unit area, which is especially important for small-scale farmers (Sapcota and Begum, 2022). In Northeast India, duck farming act as a primary source of income for 26.8 per cent of the population, while for the remaining farmers, it served as a secondary source of income in addition to agriculture (Zaman *et al.,* 2005).

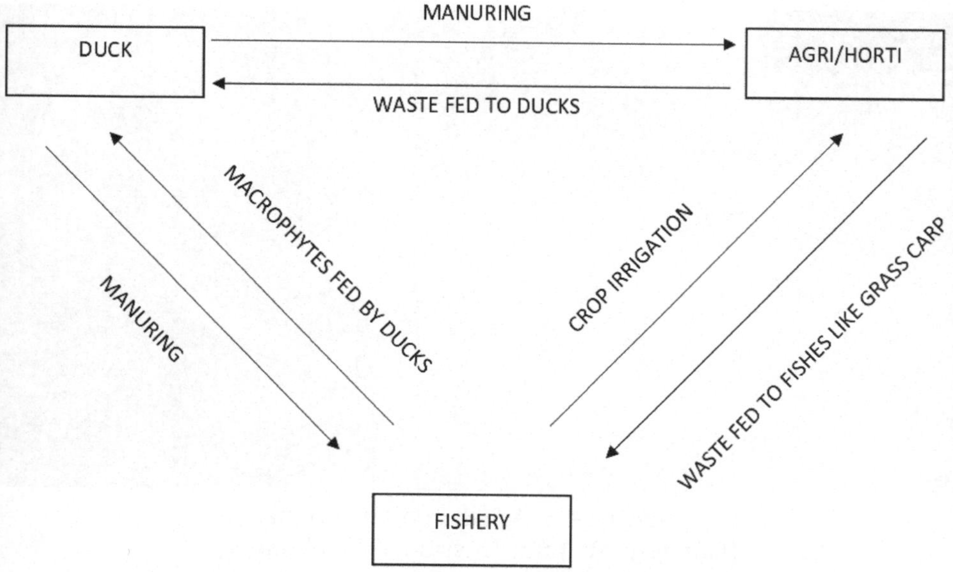

Figure 14.2: Triangular Model of Integrated Fish Farming for better Production in Minimal Investment.

Present Scenario

In India's 2.36 million hectares of tanks and ponds, culture-based fisheries predominate and account for the majority of total fish production. The current tank and pond production are 8.5 million MT. The Department of Fisheries has emphasised expanding the horizontal space under tanks and ponds to meet a target production of 13.5 million MT as a major contributor to production (GoI, DoF).

Annually, it is predicted that 28,390 million fingerlings and 202.3 lakh tonnes of feed will be required to reach the 13.5 million MT goal. However, if we transition from fish farming to an integrated fish and duck farming system, we can save a lot of money since the symbiotic relationship that exists between the ducks and fishes will come into play and eventually enhance the farm's economics.

The overall number of chickens has climbed by 16.81 per cent, bringing the total number of chickens to 851.81 million in 2019. Kumar (2021) has mentioned that, duck meat output climbed from 0.026 million tonnes to 0.15 million tonnes in 20 years, a 577 per cent rise in rate of growth. According to a recent assessment on the world's top fifteen duck meat producers, eight Asian countries are in the top fifteen (FAO, 2010). According to demographic dynamics and the distribution of the duck population, most of them are found in the Eastern, North Eastern, and Southern states of the region (Patil *et al.,* 2020). One of the most important ecologically balanced property technologies is integrated fish and duck farming methods, which can be found in a variety of forms, primarily in East and Southeast Asian countries (Majumdar, 2018).

Figure 14.3: A Fish Pond in Assam
(*Courtesy*: Mr. Jakir Hussain, CIFE-Mumbai).

Integrated Fish and Duck Farming

I. Duck Farming

A. Types of Duck Farming System

1. **Extensive:** A simple and traditional farming method in which a few locally accessible low-cost ducks are farmed primarily for family use (Soltan, 2016) or as a side-line source of revenue (Churchill *et al.,* 2022). There

is no additional feed provided; only kitchen trash is permitted, and the ducks are free to roam about the pond and surrounding regions in search of food. The link between the duck and fish subsystems is negligible in this strategy.

Figure 14.4: Duck Raising for Family Use in Harapur, West Bengal (*Courtesy*: Banlam J. Marbaniang, COF-CAU, Tripura).

2. **Semi-intensive:** In this approach, egg-laying ducks are raised and fed a sufficient amount of high-quality feed with the necessary nutritional contents to aid in egg laying. A higher stocking density per unit area of pond is maintained, which has a good influence on the pond's manuring mechanism. The biggest limitation of this strategy is the expensive cost of raising egg-laying ducks, which is difficult to offset through egg sales (Soltan, 2016).
3. **Intensive:** Ducks are raised at higher densities in confined settings. The duck house is built along the dyke or in the pond itself, where the uneaten feed and excreta fall directly into the pond, where the fish devour it and manuring occurs. Infrastructure, extra feed, and health management costs are all greater with this strategy. It also necessitates good and tight management procedures.

B. Models of Integration

1. **Mass Rearing in Open Water:** Ducks are allowed to wander freely in open waters such as lakes and reservoirs during the day and are sheltered in enclosures built into the banks at night. The grazing method is another name for this form of rearing. In this type of arrangement, 1000 ducks are raised on average. This strategy can also be used as an integrated management model for the development of major water bodies (Bao-tong and Hua-zhu, 1984).

2. **Rearing Ducks in Centralized Enclosures near the Pond:** Duck pens are built along the pond's edge where enough area of concrete dry and wet runs is provided in this method of rearing. Every day, the duck dung and uneaten feed are dumped into the wet run. The stocking density is kept at 4 ducks per square metre (Strasbourg, 2013). This system has the advantage of centralised fish-cum-duck farming management (Bao-tong and Hua-zhu, 1984).
3. **Raising Ducks in Fish Pond:** The ducks are raised in pens built above the water's surface in this method of upbringing. In this technique, uneaten feed and duck excrement fall straight into the pond water and are evenly dispersed throughout the water column, resulting in a manuring effect and improving the pond's primary productivity. This type of model has a higher pen investment cost and is only used in intensive farming systems when duck stocking density is kept at 4.5 individuals per m^2 (Bao-tong and Hua-zhu, 1984).

C. Breeds of Ducks

Ducks of exotic and local kinds are available for production in India. According to Sankaralingam and Mahanta (2022), India has 24 native duck breeds, each with 34 variants. Pekin, Muscovy, Khaki Campbell and Indian Runner are examples of exotic duck breeds. In Asia, the runner and mule are commonly utilised for meat and/or egg production and are readily available. Khaki Campbell ducks are suitable for fish cum duck farming (Biswas, 2015).

Table 14.1: Different Breeds of Ducks

Sl.No.	*Variety/Breed*	*Origin*
1.	Brown Tsaiya	Taiwan Province of China
2.	Indian Runner	Malaysia
3.	Khaki Campbell	England
4.	Pekin	China
5.	Muscovy	America
6.	Rouen	France
7.	Aylesbury	England
8.	Itik Java	Malaysia
9.	Mule	Taiwan
10.	Nageshwari	Assam, India
11.	Kuttanad (Chara and Chambeli)	Kerala, India
12.	Arni	Tamil Nadu, India
13.	Pati	Assam, India
14.	Maithili	Bihar, India
15.	Sylhet mete	Eastern India

The world's top egg-producing breeds include Brown Tsaiya from Taiwan Province of China, Indian Runner from Malaysia, and Khaki Campbell from England (Jalaludeen and Churchill, 2022). The Indian Runner, Khaki Campbell, Pekin, and Muscovy are the breeds of ducks that are most prevalent in rural areas (Sonaiya and Swan 2004). With a high development rate and efficient feed converters, one of the best duck breeds for producing eggs and meat is Khaki Campbell (Patil *et al.,* 2020). Rouen, Muscovy, Pekin, and Aylesbury are some of the most popular meat breeds.

Two further extraordinary germplasms that have the capacity to produce up to 330 eggs in a single laying year are Indian Runner from the East Indies and Brown Tsaiya from Taiwan. Aylesbury for meat, Campbells for eggs, and Orpington for dual purposes are to England's credit (Su, 2022). The *'Desi'* type duck was found in large numbers in India's Lakhimpur and Dhemaj districts, whilst the Nageshwari variant was only found in small numbers in Assam's Cachar and Karimganj districts (Patil *et al.,* 2020). Nageshwari, Sylhet mete, and Kuttanad ducks are among the most commonly reared indigenous kinds in India (Sapcota and Begum, 2022). Khaki Campbells are extensively utilised, according to Das *et al.* (2003), because they perform better.

D. Pre-requisites, Infrastructure and Facilities Required

For proper farm administration and animal comfort, an integrated fish and duck farming system requires numerous facilities and components. One of the most critical elements of a fish farm is a well-equipped duck house that provides shade from the sun, protection from severe rains, and most importantly, a safe haven for the birds at night. The house should have a height of 4-5 feet (Soltan, 2016). A duck house can be built along the dike or above the water surface, depending on the type of system. Because construction costs are lower and management is easier in the tropics, most duck houses are erected on the pond dyke rather than on the pond surface (Soltan, 2016). Components within the duck house include a feeder, bedding, proper ventilation system for good air circulation inside, curtains on the walls to prevent wind-driven rain during stormy weather, and a bridge leading to the duck house for feeding and collecting eggs, as well as another leading to the water for ducks to ascend and descend to the pond. Meat ducks can be reared on an earth/ cemented floor, but egg ducks need dry hay or grass to avoid shattering of their eggs. Egg production can be aided by supplementing natural light with artificial light for up to 15 hours, so a lighting facility is also necessary in the house. Seasonal ponds that hold water for 8 to 9 months are perfect for this type of practise (Sapcota and Begum, 2022). Transportation access, a feed storage area, and other equipment are all necessary in addition to the above-mentioned buildings and infrastructures.

E. Ideal Housing

The duck house in an integrated fish and duck farming system should be made of locally available materials such as bamboo, thatches and cane (Sapcota and Begum, 2022). If the duck house is built just above the pond's surface, the bedding should be set such that the duck excreta and uneaten feed fall straight into the fish

Table 14.2: Duck House on the Dyke *vs.* Duck House above Pond Surface

Duck House on the Dyke	*Duck House Above Pond Surface*
Cost effective	Incurs higher investment
Out of the total excreta output, 50-60 per cent falls in the pond.	Manure falls directly on the pond hence completely utilized
Incurs cost and time of manure collection, storing and transporting	All these costs are eliminated
May give rise to environmental pollution	Potential pollution problem eliminated
Uneaten feed not properly utilized	15 per cent uneaten granulated feed effectively consumed by the fishes
Inefficient recycling of uneaten feed and manure	Efficient recycling of uneaten feed and manure
Ducks descending and ascending through the dyke creates erosion ultimately causing its destruction	No effect to dyke
Ducks descending and ascending through the dyke creates turbidity and siltation since they add soil to the water and this ultimately effect light penetration and photosynthesis is disturbed.	No such problem

Figure 14.5: Duck House Erected above Pond Water.
In frame: Mr. Prasanna Rabha, Assam
(*Courtesy*: Mr. Jakir Hussain, CIFE, Mumbai).

pond, acting as a supplement for fish production. Excreta and uneaten feed should be manually dumped into the pond water if the house is built near the shore. The housing system can be built using floats or securely erected with stilts slightly above the pond level (Sapcota and Begum, 2022). The pond embankments are partially enclosed with net to create a wet run. As discussed in the preceding section, there are a number of other crucial considerations to be taken into account.

F. Preparation of Fish Pond

The actions or activities required to prepare a pond for the introduction of fishes are referred to as pond preparation. Because the fish will spend 100 per cent of their time and thrive in it until harvest, water chemistry is one of the most important factors affecting the pond's production and well-being. In terms of water chemistry, various water parameters must be thoroughly monitored on a regular basis and kept at optimal levels to support fish development and survival, prior to the introduction of fishes. pH, dissolved oxygen, alkalinity, hardness, ammonia, transparency are a few key water properties to mention that is to be maintained optimally.

Table 14.3: Ideal Level of Water Parameters (Bhatnagar and Devi, 2013, Mohanty *et al.*, 2014)

Sl.No.	*Parameter*	*Optimum Level*
1	pH	7.5-8.5
2	Dissolved oxygen	> 4 ppm
3	Alkalinity	75-200 ppm
4	Hardness	75-150 ppm
5	Temperature	25-32°C
6	Total Ammonia Nitrogen	< 1 ppm
7	Nitrate	< 5 ppm
8	Nitrite	< 0.01 ppm
9	Secchi disk transparency	25-45 cm
10	Biological Oxygen Demand	< 10 ppm
11	Plankton density	3000-4500 no. litre^{-1}

Note: Values given with respect to carp culture.

According to Sapcota and Begum (2022), the following steps should be performed during preparation of a pond:

a) Application of lime @ 1200 kg ha^{-1} to adjust pH as basal dose,

b) Application of 5000 kg ha^{-1} cow dung for natural food production as basal dose,

c) Elimination of aquatic plants to facilitate photosynthesis and oxygen circulation by allowing sunlight to penetrate, and

d) Use of Mahua (*Bassia latifolia*) oil cake to control predatory and weed fishes that prey on cultured species and compete for nutrients, space, food, and oxygen.

These methods will effectively prepare the pond for the introduction of fish and the start of farming activities. Furthermore, all of the above characteristics should be evaluated on a frequent basis in order to maintain a proper growth environment for the fish and avoid any stress.

Figure 14.6: (a) Liming of Pond, (b) Collection of Cow Dung for Manuring (*Courtesy*: Banlam J. Marbaniang, COF-CAU, Tripura).

G. Raising of Ducklings

Ducklings should be purchased from government farms or certified providers to ensure that they are healthy and disease-free (Biswas, 2015). Ducklings are delicate creatures that require special attention and upkeep in all areas, including temperature, heat, space, feeding, and drinking. Day-old ducklings require a carefully controlled environment for the first 2-3 weeks of their lives (Soltan, 2016), after which they can be moved to pens near ponds.

H. Housing of Ducklings

Ducklings up to 15 days old are kept in bamboo basket enclosures, with heat given by electric lamps for the first 1-2 weeks. In the comprehensive method, no artificial heat is provided, but "Closed tents" are built to preserve the heat of the brooding shed. (Patil *et al.*, 2020). To absorb excess moisture from duckling excreta, the litter/bedding should be 3 cm thick or thicker (Gajendran and Veeramani, 2022).

I. Brooding Period of Ducklings

Ducklings can be successfully brooded in any brooder house and under any type of brooder used for chickens. In fact, ducklings are generally easier to brood than chickens. The layer ducklings should be brooded for 3 to 4 weeks, whereas, broiler ducklings should be brooded for 2 to 3 weeks. Although Muscovy ducks can be used for brooding ducklings, they should not be wasted for this purpose – they are more valuable for egg production or incubation. Discourage broodiness by placing broody ducks in broody coops, usually for 3-5 days, making sure they have continuous access to feed and water.

Figure 14.7: Floating Duck House Unit at CIBA, Kakdwip (*Courtesy*: ICAR-CIBA, Kakdwip Research Centre).

J. Temperature Maintenance

The temperature needs to be 32°C for the first week, and then lowered at a range of 3°C every week until a temperature of 24°C is maintained in the fourth week. After that, the ducklings can be transferred outside to a predator-proof house. By the time the ducklings are 3-5 weeks old, weather-dependent, they can spend warm, sunny days outside, carefully supervised and protected from predators.

K. Feeding

Ducklings are fed broken rice, crushed snails, and cooked rice until they are 15 days old, after which they are fed paddy grains, kitchen garbage, cooked arum root, cooked rice, and vegetables, in addition to grazing throughout the pond area (Patil *et al.,* 2020).

L. General Considerations for Raising Ducklings (Soltan, 2016)

1. Stocking density = 50-55 ducklings m^{-2},
2. Screen floor is laid using a mesh of 1-5 cm having a gauge of gauge (to facilitate falling of excreta and uneaten feed into the pond),
3. Feed = Starter pelleted feed in demand feeders,

Figure 14.8: (a) Cooking of Rice to Feed Ducklings; (b) Grinding of Feed Ingredient for Feeding [In frame: Mr. Prasanna Rabha, Assam; *Courtesy*: Mr. Jakir Hussain, CIFE, Mumbai].

4. Water = clean taped water in troughs of depth 5 to 7.5 cm (Patil *et al.*, 2020), allowing only the beak inside to prevent the ducklings from getting wet,
5. Air temperature = 30 to 32°C,
6. The ducklings are allowed into shallow splashing pools after 7 days (Patil *et al.*, 2020) or 3-4 days (Soltan, 2016) for acclimatization.

M. Raising of Ducks

1. **Stocking Density and Floor Space:** About 250-300 ducks are required to fertilize one hectare pond that produces 45 to 55 kg dung per duck per year (Sapcota and Begum, 2022) or 10,000 to 15,000 kg manure per year (Biswas, 2015). The stocking size should be reduced accordingly when raising meat ducks, because they produce a lot of waste (Sapcota and Begum, 2022). According to Patil *et al.* (2020), an intensive raising system requires 4 square feet of floor space per bird, whereas a semi-intensive system requires 3 square feet of night shelter and 10 to 12 square feet of outdoor run space per bird. It is sufficient to provide 6-7 cm of feeding space per bird (Patil *et al.*, 2020). Three laying birds can be contained in a box with dimensions of 30 cm × 30 cm × 45 cm. The number of ducks to be stocked and their ratio, as well as the breeds of fish being raised in the pond, depends on the climate (Sapcota and Begum, 2022).
2. **Feed and Feeding Management:** Ducks feed primarily on post-harvested paddy grains, as well as explore flooded areas in search of fish and insects. The ducks are fed locally available ingredients by farmers in

Kerala, Andhra Pradesh, and Tamil Nadu. A feed composed of 15 per cent broken rice, 30 per cent rice bran and 20 per cent concentrated feed is fed at the rate of 150 g duck^{-1} day^{-1} (Soltan, 2016), or 100 to 120 g duck^{-1} day^{-1} (Biswas, 2015) is sufficient if they get enough food scavenging outside. Furthermore, food consumed by ducks through grazing contains a reasonable amount of protein, reducing the expense of protein that would otherwise have been consumed through supplementary feed (Biswas, 2015). Alternatively, depending on the availability, affordability, and quality of feed components, duck feed can be made by combining broken rice, rice bran, finger millets, sesame cake, minerals, and vitamins (Sapcota and Begum, 2022). According to Patil *et al.* (2020), supplemental feeding of indigenous ducks has been observed to boost egg production. Supplemental feeding is essential throughout the winter and early summer seasons, when natural feed is in short supply. Ducks prefer moist mash over dry mash because it is easier to swallow. Approximately 10 per cent of the feed is washed away and used as fish food in the pond. Ducks eat a variety of foods found in and around the pond, including plants (especially duck weed), insects, molluscs and small fish. Ducklings are fed 4 to 5 times per day at first, then twice per day after they reach adulthood. Adult ducks can eat from a 10 cm feeder (Soltan, 2016). If pond feeding isn't an option, keep a drinker near the feeder. Using crumbled or pelleted feeds results in less feed waste and higher feed efficiency (Gajendran and Veeramani, 2022).

Table 14.4: Nutrient Requirement of Ducks (*Source*: Patil *et al.*, 2020)

Nutrient	*Starter (0-8 weeks)*	*Grower (9-20 weeks)*	*Layer*	*Breeder*
ME (kcal kg^{-1})	2750	2750	2650	2650
CP (per cent)	22	16	18	15
Lysine (per cent)	0.7	0.65	0.75	0.6
Methionine (per cent)	0.4	0.3	0.29	0.27
Calcium (per cent)	0.65	0.6	2.5	2.75
Phosphorous (per cent)	0.4	0.3	0.45	0.3
Vitamin A (IU)	2500	2500	6000	4000
Vitamin D_3 (IU)	400	400	1000	900
Vitamin E (mg)	10	10	20	10
Vitamin K (mg)	0.5	0.5	2	0.5
Riboflavin (ppm)	4	4	5	4
Vitamin B_5 (ppm)	11	11	15	11
Niacin (ppm)	55	55	55	55
Pyridoxine (ppm)	2.5	2.5	6	3

Table 14.5: Feed Ingredients for Preparing 10 kg Adult Feed (*Source*: Patil *et al.*, 2020)

Ingredients	*Quantity (kg)*	*Ingredients*	*Quantity (kg)*
Maize*	4	Salt	0.250
Rice Polish	3	Vitamin mix	0.250
Mustard Oil Cake	1	Antibiotics	0.075
Mineral mix	0.350	Coccidiostate	0.075

* Inclusion of maize in duck feed may be avoided as it is rich in magnesium which will favour mould growth in moist environment (Biswas, 2015).

N. Egg Laying and Breeding Management

Other than chicken eggs, Asia produces nearly 95 per cent of the world's duck eggs, and the majority of Asian nations have their own indigenous breeds for producing eggs. (Liu and Chuchill, 2022). Good genetics and management are required to maximise duck egg production.

Table 14.6: Egg-Laying Performance (Gajendran and Veeramani, 2022)

Parameter	*Performance Result*
Age at first egg	23 weeks
Age at peak of lay	32 weeks
Average duration of lay	42 weeks
Total no. of eggs per female	230
Hatchability of fertile eggs (per cent)	85-87
No. of ducklings per female housed	165-175
Average fertility (per cent)	94-96
Average egg weight (g)	86

Ducklings of one day old are brought and reared for one month before being stocked in a duck house. After 4-5 months, they begin laying eggs (Soltan, 2016). According to Patil *et al.* (2020), for good hatchability and fertility, a sex ratio of 1:6 (M: F) for intensive systems and 1:15-20 for extensive systems is desired, and according to Soltan (2016), a sex ratio of 1:5 (male to female) is maintained for fertilised egg production and 1:10-15 is recommended for commercial egg production. In general, egg-type ducks have higher fertility and hatchability than meat-type ducks (Liu and Churchill, 2022). Furthermore, according to Patil *et al.* (2020), in a rural duck rearing system, farmers often maintain a ratio of 1:20-25 and attain a fertility of 70-80 per cent. For better egg production, ad libitum feeding should be done with high-quality duck layer mash. (Gajendran and Veeramani, 2022). Common ducks are about 120 days old when they lay their first egg, while Muscovy ducks are 261 days old (Liu and Churchill, 2022). Up until the age of 40 weeks, ducks can lay about 140 eggs, and up until the age of 52 weeks, 200 eggs, according to sources, whereas Muscovy ducks can only produce 46 eggs in their first year of life. The average egg weight of native duck breeds is roughly 65 grams (Liu and Churchill, 2022). Restricted feeding at the age of 18-19 weeks is recommended for greater egg production during the laying stage, and lighting of 14-16 hours is required for optimal egg production (Gajendran and Veeramani, 2022).

O. Management Practices

Feeders and drinkers should be cleaned and dried on a daily basis to avoid disease-causing contamination (Soltan, 2016). To avoid mould formation, which could lead to the production of dangerous aflatoxin, duck feed should be stored in a cold, dry area (Biswas, 2015). In the event of an algal bloom, the duck house bedding is covered with plastic sheets to avoid further pond manuring, which will only exacerbate blooming (Sapcota and Begum, 2022). If the duck house is near the water and manuring is done by hand, the collected dung should be kept carefully for future use. In order to maintain a steady supply of excrement, it is desirable to keep various age groups of ducks together (Sapcota and Begum, 2022). A detachable fence could be used to divide different age groups. Drinkers should be kept at equal distances from one other, covering each section of the duck house and fresh, cool, potable water should always be provided to the birds. Spilling of water should be minimised. Litter management practises must be rigorously adhered to. Proper records on the number of ducklings received, feed intake, body weight, mortality, culling, medicines and vaccines administered, and other factors must be kept in order to analyse production efficiency and failures (Gajendran and Veeramani, 2022). Ducks grown on thick bedding systems have improved growth performance and feed conversion ratio (Sari *et al.*, 2013). The essential factors in better litter floor management include adding fresh litter material on a regular basis, replacing dirty or damp litter with fresh new litter, and adding disinfectants like super phosphate (Gajendran and Veeramani, 2022).

P. Health Management

Duck meat production for commercial purposes is a developing industry all over the world. Biosecurity and disease control are growing increasingly critical as production methods become more intensive (Chen *et al.*, 2022). Infections that affect ducks are comparable to those that affect chickens, but the former is more resistant to common diseases than the latter (Soltan, 2016), as well as turkey (Patil *et al.*, 2020), and the course of disease may differ among these animals (Biswas, 2015). Duck disease management necessitates early detection and effective treatment, as well as the elimination of any potential pathogens through strong biosecurity measures (Chen *et al.*, 2022). Botulism, duck plaque, duck cholera, and hepatitis are among the most prevalent diseases in ducks (Patil *et al.*, 2020). Disease outbreaks on farms are largely caused by inherited traits and weaknesses, unsanitary conditions, poor management techniques, damp litter, feed, and water, close contact, contaminated equipment, attendants and visitors, air, external parasites, free-flying birds, rats and flies, and other factors. (Patil *et al.*, 2020 and Biswas, 2015). In their study, Patil *et al.* (2020) found that duck cholera caused increased mortality in ducklings, with mortality rates of 10-15 per cent for ducklings and <10 per cent for adults. Internal parasites such as tape worms, flukes, and round worms are not a problem for ducks (Patil *et al.*, 2020).

Prevention and Control of Diseases (Biswas, 2015, Patil et al., 2020 and Chen et al., 2022)

a) Before introducing disease-free stock to farm premises, ensure that it has been quarantined,

b) Maintain proper biosecurity procedures,

c) Provide a stress-free environment, such as proper housing management and ventilation, among other things,

d) Separate the age groups of ducks,

e) Always advance from the youngest to the oldest ducks when checking them,

f) Immediately isolate ill ducks,

g) It is necessary to maintain good hygiene,

h) Clean feed, water, and floor space should all be supplied,

i) Rodents and birds should not be allowed access to the house,

j) Vaccine regimens should be adhered to on a regular basis,

k) Dead birds must be disposed-off correctly and promptly,

l) At the entrance, $KMNO_4$ solution should be utilised as footbaths,

m) It is necessary to utilise appropriate bedding material,

n) Bedding should be maintained dry and turned on a regular basis,

o) Parasites should be treated with anathematic medications,

p) The feed offered should include all of the necessary components to boost immunity and supplement with all of the vitamins and minerals needed on a regular basis,

q) Use of coccidiostats (anti-protozoal) and deworming on a regular basis, and

r) Keep accurate records.

Guidelines to Tackle a Disease Outbreak in Ducks (Biswas, 2015)

a) Restrict duck movement (sale/purchase),

b) Adhere to stringent hygiene guidelines, and

c) Seek advice from a veterinarian.

Q. Guidelines for Vaccination (Biswas, 2015)

a) Purchase vaccines from reputable sources and manufacturers,

b) Keep vaccines refrigerated until ready to use,

c) Administer the correct dose,

d) The vaccine must be used in less than 3 to 4 hours after dilution,

e) The expiration date should be double-checked,

f) Needles and syringes that have been sterilised should be used, and

g) Vaccination should be done during the cooler hours of the day.

Table 14.7: Vaccination Schedule (Biswas, 2015)

Vaccine	*Mode of Delivery*	*Dose*	*Age of Ducks*
Duck Cholera	Subcutaneous	1 ml (ducklings)	3 to 4 weeks
		2 ml (adults)	1 month after last shot
Duck Plaque	Subcutaneous	1 ml (adults)	8 to 12 weeks

Table 14.8: Important Diseases of Ducks, their Diagnosis and Treatment (Biswas, 2015)

Disease	*Type*	*Signs*	*Treatment*
Duck plaque	Viral	Blood-filled lumina of the colon and gizzard, free blood in body cavities, and vascular injury accompanying tissue haemorrhages	Prevented only by vaccination @ 8-12 weeks of age.
Duck Viral Hepatitis	Viral	Primarily affects ducklings that are 2–3 weeks old. Signs include haemorrhagic lesions on liver, opisthotonos, lethargy, lost balance, paddle spasmodically	Attenuated virus vaccine.
Duck cholera	Bacterial	Causes at 4 weeks age. Common lesions include pericarditis, arthritis, petechial and echymotic haemorrhages under the skin (Pink skin), in visceral organs, over the serous surface and intestine (Haemorrhagic enteritis), enlarged liver and spleen. Other symptoms include loss of appetite, high body temperature, thirst, diarrhoea, and sudden death.	Duck cholera at the age of 4 weeks and 18 weeks. Use of sulpha drugs; 30-60 ml of Sulpha Quinoxaline in 5 litres of drinking water for 7 days, or 30 ml of Sulphamezathine (33.1 per cent) in 5 litres of drinking water.

Disease	Type	Signs	Treatment
Botulism	Bacterial	Food poisoning caused by ingestion of bacterium growing on dead plant matter.	Prevent ducks from scavenging on dead plant matter.
Aflatoxicosis	Fungal	Liver lesions (high toxin content) or long-term consequences such sluggishness, non-thriftiness, hepatitis, and postponed death (low conc.)	No particular treatment, maintain proper storage conditions.

According to Chen *et al.* (2022), duck virus enteritis, avian influenza, parvovirus infection, *Riemerella anatipestifer* infection, reovirus infection, duck circovirus infection, and duck egg drop syndrome are additional common diseases that affect ducks.

Table 14.9: Common Medications for Ducks (Patil *et al.*, 2020)

Medication Type	Dose	Application
Electral	10 g litre^{-1}	On arrival of duckling
Anti-stress (Stressroak)	20 g litre^{-1}	On arrival and once every week
Multi-vitamins (Vimeral)	5 ml litre^{-1}	5 day continuously every month
Amprolium (Anti-coccidiostate)	1 g litre^{-1}	3 weeks age for 5-7 days and repeat every 2 months
	2-5 g/l	Diseased birds
De-wormer	15 g per 100 birds	3 weeks age for 5-7 days and repeat every 2 months
Tetracycline	5 g litre^{-1}	For 5-7 days during diarrhoea

II. Fish Rearing

A. Selection of Fish Species and Stocking Density

Fish species should be chosen with care, taking into account a variety of factors such as growth rate, compatibility, disease resistance, and so on. When it comes to a fish and duck integrated system, the fish chosen should be able to handle low dissolved oxygen and high organic content to some extent (Biswas, 2015 and Soltan, 2016). Furthermore, according to Soltan (2016), stocking density should be maintained in relation to duck density by calculating the amount of dung produced by a specific number of ducks that will be sufficient to fertilise the pond as needed. Depending on pond conditions and seed availability fingerlings of about 10 cm should only be stocked in this system, as little fishes may be preyed upon by ducks. However, according to Biswas (2015), fingerlings between 5 and 10 cm can be supplied, but ducks should be introduced when the fingerlings reach a size of 10 cm. A stocking density of 10,000 fingerlings per hectare (Patil *et al.*, 2020) or 600 fingerlings per 1000 m^2 can be maintained (Biswas, 2015). The species ratio should be maintained based on market demand and availability. According to Kumar and Mishra (2011), the notion of multi-species and diverse species practise should be promoted with regionally important combinations of medium and minor carps.

Figure 14.10: (a) Transportation of Seeds; (b) Length-Weight Data Collection; (c) Stocking of Seeds (*Courtesy*: Banlam J. Marbaniang, COF-CAU, Tripura).

Fish should be supplied in March and harvested in October-November in areas with harsh winters; however in areas with mild winters, fish can be placed in June-September and reared for a year until harvest (Sapcota and Begum, 2022).

Table 14.10: Recommended Guideline for Species Combination (Biswas, 2015)

Tropic Niche	*Fish Species (Six Species Combination)*	*Stocking Ratio (Per cent)*	*Stocking Ratio (per cent) (Only IMC/with Common Carp)*
Surface feeder	*Catla catla*	20	40/30
	Hypophthalmichthys molitrix	20	-
Column feeder	*Labeo rohita*	20	30/30
	Ctenopharyngodon idella	10	-
Bottom feeder	*Cyprinus carpio*	15	30/20
	Cirrhinus mrigala	15	20
Expected yield	-	500 per bigha*	-

Note: *1 bigha = 0.253 hectare.

B. Feed and Feeding Management of Fishes

The quality and quantity of feed, as well as the type and manner in which it is distributed to the fish, determine the best feeding management practises (Jeney and Woynarovich, 2011). Fishes can only be fed during the summer season at 0.5

per cent of body weight with mustard de-oiled cake and polished rice at a 1:1 ratio once daily in the morning hours (Pandey *et al.*, 2016). Duck manure is an excellent source of nutrients for the creation of natural fish food. Furthermore, while fish such as common carp can eat duck excrement directly, herbivore fish such as grass carp, java puthi and others must be fed grasses such as Napier grass, para grass, banana leaves, or chopped green cattle fodder (Biswas, 2015). Regular monitoring of plankton density and growth rate of fish should be carried out to ensure the availability of sufficient food and positive growth of fish.

Figure 14.11: Feeding of Fishes
(In frame: Prasanna Rabha, Assam; Courtesy: Jakir Hussain, CIFE, Mumbai and Banlam J. Marbaniang, COF-CAU, Tripura).

C. Utilization of Duck Manure

According to Kalita *et al.* (2014), duck excrement is the primary source of nutrients for the pond, which is used to produce natural fish food *viz.* phytoplankton (*Chlorella, Spirogyra, Ulothrix etc.*) and zooplankton (*Daphnia, Moina, Cyclops etc.*). As previously stated, fish such as common carp eat duck excrement directly. If the duck house is built over the pond surface, duck manure falls straight off the duck house through floor gaps, however, if the duck house is built along the dyke, manual application is required.

In the event of over fertilization (which can be detected by measuring plankton density or watching pond colour change), duck manure should be stopped immediately to prevent the establishment of intense algal blooms, and freshwater should be added and aeration should be done if possible (Sapcota and Begum, 2022). According to Kalita *et al.* (2015), the maximum plankton density was calculated to be 6253 unit litre^{-1} after duck droppings were introduced, compared to 1592 unit litre^{-1} before duck droppings were introduced.

D. Management Practices

Because the entire productivity and profitability of the fish farm is dependent on the environment in which the fish are kept, pond water body management is the backbone of the fish farm. Fish should not be stressed in any manner because

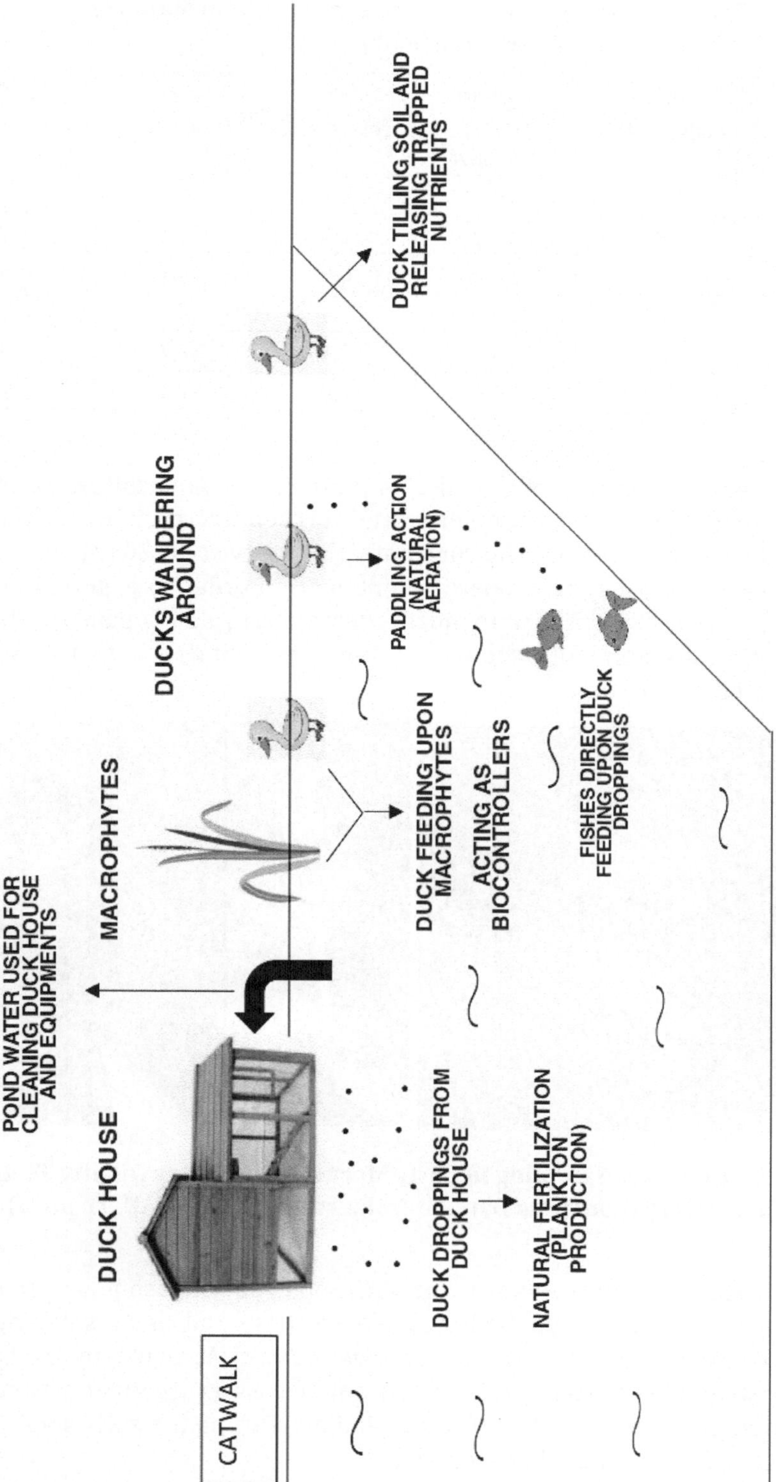

Figure 14.12: Integrated Fish-cum-Duck Farming Concept.

Table 14.11: Composition of different Animal Manures (Kumar and Mishra, 2011)

Manure Type	*Moisture (per cent)*	*Organic Matter (per cent)*	*N (per cent)*	*P_2O_5 (per cent)*	*K_2O (per cent)*
Cow	80-85	14.0	0.3	0.2	0.1
Pig	85	15.0	0.6	0.1	1.4
Duck	81	26.2	0.9	0.4	0.6
Chicken	78	25.5	1.4	0.8	0.6
Goat	10	-	2.7	1.7	1.9
Rabbit	10	37.0	2.0	1.3	1.2

this would reduce their rate of growth and productivity. Aquaculture ponds are living dynamic systems that go through a variety of chemical reactions and physical changes, causing them to fluctuate constantly (Mohanty *et al.,* 2014). As a result, various parameters must be properly maintained in order to ensure a healthy environment for the fishes. The water quality parameters (pH, DO, plankton density, hardness, alkalinity, ammonia, and so on) should be kept as constant as feasible throughout the culture time.

Figure 14.13(a): Water Sampling in Early Morning; (b) Water Quality Testing in the Laboratory (*Courtesy*: Banlam J. Marbaniang, COF-CAU, Tripura).

When visible, aquatic plants growing on the pond's surface should be removed because they obstruct sunlight, limiting photosynthesis and, as a result, lowering the dissolved oxygen content (Sapcota and Begum, 2022). At regular intervals, each parameter should be thoroughly examined, and necessary treatments should be implemented in the event of significant variations. During the early stages of an algal bloom, manuring should be stopped.

Figure 14.14: A Fish Farm in Assam
(Owner: Mr. Prasanna Rabha; *Courtesy*: Mr. Jakir Hussain, CIFE, Mumbai).

E. Health Management

Fish health is a critical determinant of the farm's overall success, and it is influenced by a variety of factors such as water quality, nutrition and feeding, and the presence of other species that may act as carriers.

F. Management of Environment

a) Removal of aquatic weeds,

b) Liming,

c) Fertilization, and

d) Water quality management.

G. Harvesting of Fishes

Depending on demand, fish might be harvested partially at intervals or totally towards the end of the season when the fish have reached table size. According to Sapcota and Begum (2022), duck eggs are collected every morning, and 1.5 to 2 year old ducks are sold for meat in the neighbouring markets. Fish harvesting is a time-consuming task that necessitates the use of equipment and labour. The drag net is the most common pond harvesting equipment in India, and fish are transported from the pond to the market in hundis or sold to fish sellers on the spot after harvesting.

Live fish is more expensive in most regions and is well-liked by a wide spectrum of customers. A total of 3500 to 4000 kg of fish, 17,000 to 18,000 eggs, and 450 to 600 kg of meat are expected to be produced with this technique (Biswas, 2015).

Table 14.12: Disease of Fishes, its Prevention and Treatment (Mishra, 2010)

Disease Type	*Signs/Symptoms*	*Prevention*	*Treatment*
Bacterial	☆ Lehargic, floats on the water's surface, occasionally has its head protrude out, has reddish sores, and loses scales.	☆ Disease prone ponds can be given lime (CaO) treatment @ 50 kg/ha during post monsoon which can be followed by bleaching powder treatment @ 0.5 ppm after one week.	☆ Lime, CaO @ 100 kg/ha during initial stages of lesions and bleaching application @1 ppm after 1 week ☆ Gammaxene treatment @ 1 ppm ☆ Benzalkonium chloride @ 10 mg $litre^{-1}$ for 5-10 min and 0.5-1.0 mg $litre^{-1}$ for prolonged immersion ☆ Oxytetracycline @ 75mg kg^{-1}, Amoxycilin @ 80 mg kg^{-1}, Furazolidone @ 75 mg kg^{-1}, Co-trimazine @ 30 mg kg^{-1}, Oxolinic @ acid 10 mg kg^{-1} (via feed).
Fungal	☆ Skin lesions, raised white "cotton wool" appearance in case of saprolegniosis (most imp. Fungal disease in fishes).	☆ To disinfect viscid eggs, submerge them in a 7 ppm malachite green solution for 10-15 minutes over the course of two days. ☆ Sprinkle 10-15 kg of malachite green solution at 70-100 ppm onto the hatching shelf twice daily, in the morning and evening, until the fry hatch out. ☆ Soak egg nests in 1-3 per cent NaCl solution for 20 minutes or 3-5 per cent formalin solution for 2-3 minutes.	☆ Malachite green solution or methylene blue solution, at concentrations of 0.15 to 0.2 ppm and 2 to 3 ppm, respectively, should be applied to the contaminated pond. ☆ Repeat the sprinkle with the same dosage at 3 to 4 days after the initial application if the desired result is not seen.

Disease Type	*Signs/Symptoms*	*Prevention*	*Treatment*
Invasive disease (Trichodinasis and Trichodinelliasis)	☆ Slides on the skin and gill, primarily enter the skin of young fish and feed on its tissue cells, destroying the delicate fish skin, causing the gill tissue to decay, and exposing the cartilage.	☆ Disinfect fingerlings with $CuSO_4$ solution @ 8 ppm for 20-30 minutes or NaCl solution @ 1 to 2 per cent for 2 to 10 minutes. ☆ Treat the pond with $CuSO_4$ and Fe_2SO_4 @ 0.7 ppm (ratio 5:2) ☆ Trichodina on young eel gills can be eliminated at 30 ppm Formalin solution, but it also degrades water quality and impairs appetite.	
Monogenean trematodes (Skin fluke and gill fluke)	☆ Attach with a hook to the body's surface, fins, and gills.	☆ To eradicate parasites created from recently hatched eggs, repeated chemical treatments are required. Good husbandry, as well as cleaning and sterilising the space, aid in getting rid of the eggs.	
Learneasis (Crustacean parasite)	☆ Attach to the surface of the skin	☆ Before stocking, bathe the Lernaea-infected fingerling in $KMnO_4$ solution @10 to 20 ppm for 1.5 to 2 hours and make a dramatic change in water quality by adding 400 kg of fermented cow dung or pig manure.	
Argulus (crustacean parasite)	☆ Located behind the fins or on the fish's head.	☆ Organophosphorus treatments used repeatedly are successful in eliminating juvenile stages. All stages, including eggs, will be killed by draining and drying. ☆ Regularly removing flora that has eggs attached will assist to lower the parasite population.	

Table 14.13: General Treatment of Parasitic Diseases (Mishra, 2010)

Parasite	*Chemical*	*Treatment Procedure*
Myxobacteria, Costia, Trichdina, White spot, Gyrodactylus	Chloramine T	Bath treatment at the rate of 50 to 200 mg per 100 litres
Ectoparasites	Copper sulphate	Long term bath treatment @ 0.1 ppm
	Formalin	160-250 ppm for 1 hour, For prolonged treatment, 2ml per 100 litres for 12-24 hours.
	Sodium chloride	Bath treatment @ 1.0 to 1.5 kg per 100 litres for 20 minutes, Dip treatment @ 2 to 3 kg 100 litres for few seconds, For prolonged treatment, use 500 g per 100 litres water.
Sealice	Dichlorvos	Bath treatment at 1 ppm for up to 60 min
Argulus	Trichlorphon	0.1 to 2 ppm for long term exposure
Protozoa and Monogenean	Potassium permanganate	Bath treatment @ 500 mg per 100 litres water (1 hour) or dip treatment @ 100g per 100 litres water for 30 seconds.
Ichthyophthirius	Leteux-Meyer Mixture (3.68 g malachite green in 1 lit formalin)	2.5 ml per 100 litres for 1 hour or 1.5 ml per 100 litres (prolonged treatment) which is to be repeated 3 times at 3 to 4 days interval.
General therapeutic measure	Lime Application	Application @ 100 kg ha^{-1}

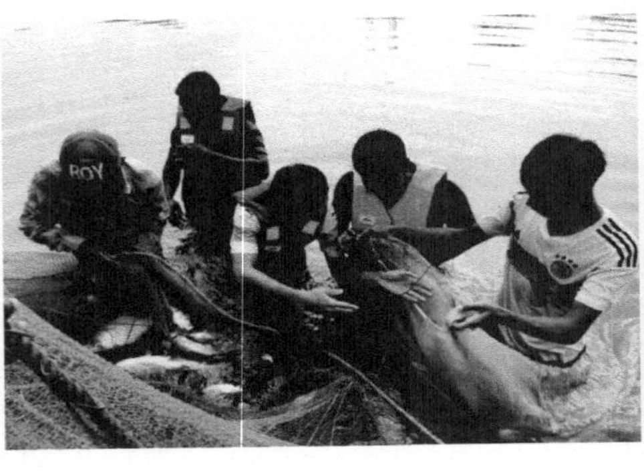

Figure 14.15: (a) Harvesting of Fishes; (b) Sale of Fishes (*Courtesy*: Mr. Banlam J. Marbaniang, COF-CAU, Tripura).

Table 14.14: Growth of Fishes in Integration with Duck Farming (Kalita *et al.*, 2015)

Fish	*Initial Weight (g)*	*Final Weight (g)*
C. catla	18-20	2200 – 2400
L. rohita	13-16	1100 – 1300
C. mrigala	10-12	900 – 1000
H. molitrix	15-17	1400 – 1500
C. carpio	13-15	1400 – 1500
C. idella	15-17	900 – 1000

H. Economics of Fish-cum-Duck Farming

Several studies have shown that integrated fish and duck farming provide farmers with a higher profit ratio than fish farming alone. According to Majhi (2018), a 1 bigha (1333.33 m^2 or 0.134 ha) area of pond yielded an average net profit of ₹ 20,015 and ₹ 29,290 in traditional pisciculture and integrated fish and duck farming, respectively. The benefit-cost ratio obtained from integrated fish-duck farming is 3.11 (Thakur *et al.*, 2018), 2.27 (Pandey *et al.*, 2016), 1.91 (Kumar *et al.*, 2020), 2.26 (Saikia *et al.*, 2020), but it is 1.40 (Thakur *et al.*, 2018) and 1.83 (Saikia *et al.*, 2020) in traditional farming.

Table 14.14: Economics of Fish and Duck Farming Model (Thakur *et al.*, 2018)

Sl.No.	*Particulars*	*Normal Pond (0.5 ha)*	*Integrated Pond (0.4 ha)*
1.	Fixed Cost	14400	11520
2.	Variable Cost	7500	30800
3.	Total Cost (₹)	21900	42320
4.	Total Cost per Hectare (₹)	43800	105800
5.	Returns Sale of fish (₹ 100 kg^{-1})	52600	79200
6.	Sale of egg, (₹ 4 egg^{-1})	-	8494
7.	Duck meat (₹ 250 kg^{-1})	-	42660
8.	Ducklings (₹ 80 $duckling^{-1}$)	-	43684
9.	Gross Return (₹)	52600	174038
10.	Net return (₹)	30700	131718
11.	Net return per Hectare (₹)	61400	329295
12.	BCR	1.4	3.11

Advantages of Fish-Duck Integration

a) Ducks use their paddling activity to aerate the pond water as they move around it,

b) Duck droppings provide nutrients for the formation of natural fish food,

c) Ducks' uneaten food serves as fish food,

d) Fish ponds provide a great environment for ducks, increasing their survival by 3.5 per cent (Majhi, 2018),

e) Housing ducks does not necessitate additional land,

f) Ducks eat the creatures that live in the pond, which provides additional nourishment while lowering protein requirements by 2 to 3 per cent (Majhi, 2018),

g) Through the selling of duck eggs, meat, and fish, this strategy provides a high profit with low expenditure,

h) Ducks prey on fingerling predators and aid in their growth (Sapcota and Begum *et al.*, 2022),

i) Ducks eat aquatic plants and hence act as bio-controllers of plant growth,

j) When ducks dig for food in the bottom soil, they release critical trapped nutrients into the water column, resulting in a good effect in terms of natural food production (Sapcota and Begum *et al.*, 2022),

k) Duck raising can make full use of the water surface of ponds,

l) Duck droppings and leftover feed from each duck can boost fish production to 37.5 kg ha^{-1} (Majhi, 2018),

m) Fish production costs are lowered by 60 per cent as compared to other methods because no additional fertilisers or supplementary feed are required (Biswas, 2015),

n) Income from eggs and meat covers the cost of duck feed, and profit from fish farming becomes the farmer's net profit (Kumar and Mishra, 2011), and

o) Aids in the reduction of greenhouse gas emissions (Sapcota and Begum, 2022).

Constraints of Fish-Duck Integration

a) Expensive duck feed,

b) Expensive veterinary medicines,

c) The remote distance between veterinary hospitals and centres,

d) High veterinary consultation fees,

e) Lack of raw materials for duck home construction,

f) A lack of high-quality feed,

g) Diseased ducks, particularly those infected with Salmonella, represent a threat to fish due to their droppings (El-nabarawy *et al.*, 2020),

h) Severe winter make it uneconomic to provide enough warmth, resulting in increased mortality and stalled growth,

i) A high mortality rate during the first three to four weeks of duckling rearing as a result of a lack of technical expertise (Gangwar *et al.*, 2013), and

j) Due to the poor adoption rate of this type of system and the low acceptance of eggs and duck meat, a marketing barrier is finally developed (Biswas, 2015).

Case Study

An interview was carried out by Mr. Jakir Hussain (M.F.Sc. 2021-2023, ICAR-CIFE, Mumbai) *via* phone call with one of the progressive farmers named Shri. Prasanna Rabha, aged 41 years old from Kothalbari, Assam. He has had an experience of over 10 years in Integrated fish farming business which he is practicing in his own land. He is carrying out his fish cum duck farming practice in an area of 2 hectares with a stocking ratio of 1 duck per 15 fish and he is maintaining a stocking density of 800 ducks to 15,000 fishes (400 ducks ha^{-1} and 7500 fishes ha^{-1}). The variety of ducks he has selected is Indian Runner, White Pecking and Khaki Campbell which has been integrated with carps *viz.* Catla, Rohu, Mrigal, Grass carp, Silver carp and Common carp. The duck house is constructed just above the pond water and the ducks are fed with rice bran, cooked rice and formulated feed purchased from Godrej company. He follows a culture period of 8 months through which he achieves a fish production of 6.5 tonnes (3.25 tonnes ha^{-1}), 700 kg duck meat (350 kg ha^{-1}) and 12,200 eggs (200 eggs day^{-1} for 2 months). The fixed cost of the farm stands at ₹ 825000.00 which covers pond and duck house construction whereas, the operational cost stands at ₹ 701000.00 which includes duck and fish seed cost, feed cost, medicines and vaccination charges, wages, electricity and other miscellaneous requirements. The output is sold @ ₹ 150 kg^{-1}, ₹ 400 kg^{-1} and ₹ 5 $piece^{-1}$ for fish, duck meat and egg, respectively. The gross income from fish, meat and egg sale is ₹ 975000.00, ₹ 280000.00 and ₹ 60000.00, respectively and the net income (excluding fixed cost) is ₹ 614000.00. According to his experience, the major constraint faced is the high cost of feed and unavailability of quality seed. For carrying out his farming practice he gets technical assistance and support from Krishi Vigyan Kendra, Udalguri. Finally, his message to all beginners is to take up fish and duck farming business since production of food is always profitable.

Future Perspective

Numerous studies and works done on this area of aqua and livestock farming have proven to be more profitable for farmers, indicating that fish and duck farming has a bright future ahead of it. It can pave the way for entrepreneurs to excel in this field and build a stable and compounding income for themselves, as well as a medium for women to start their own venture, promoting women empowerment (Churchill and Jalaludeen, 2022). The need of the hour is to popularise this system among people, work on technological advancements, and disseminate technology to farmers, women, and entrepreneurs so that a good market structure and customer base can be created, as this has been the major constraint (Biswas, 2015) preventing this type of farming from flourishing within the industry.

Conclusion

This chapter has covered maximum aspects of Integrated Fish and Duck Farming which has proven that this system is a profitable one for farmers and entrepreneurs to start a venture and set their future. More work needs to be done at different levels like, popularization, adoption and diffusion, technological interventions, better management practices and so on and so forth but nevertheless, this system has positively shown several evidences based on different works and trials done and moreover many farmers are having a better livelihood by practicing such a farming system.

REFERENCES

Bao-tong, H. and Hua-zhu, Y. (1984) Integrated Management of Fish-cum-Duck Forming and its Economic Efficiency and Revenue. Network of Aquaculture Centres in Asia, Bangkok, Thailand. NACA/WP/84/14.

Bhatnagar, A. and Devi, P. (2013). Water quality guidelines for the management of pond fish culture. *International Journal of Environmental Science.* **3:** 1980-2009.

Biswas, A. (2015). Fish-Duck Integrated Farming. Accessed January 30, 2022 Portal:

C.H. Su. (2022). Breeds of Domestic Ducks. *In:* Duck Production and Management Strategies (Jalaludeen, A.; Churchil, R. R. and Baéza, E. *Eds.*) pp. 57-96 Springer Nature Singapore Pte Ltd.

Chen, Y.P., Yu, C.F. and Shih, Y.H. (2022) Duck Diseases and Disease Management. *In:* Duck Production and Management Strategies (Jalaludeen, A.; Churchil, R. R. and Baéza, E. Eds.) pp. 549-580 Springer Nature Singapore Pte Ltd.

Churchil, R.R. and Jalaludeen, A. (2022). Duck Farming: Opportunities, Constraints and Policy Recommendations. *In:* Duck Production and Management Strategies (Jalaludeen, A.; Churchil, R. R. and Baéza, E. Eds.) pp. 617-657 Springer Nature Singapore Pte Ltd.

Das, G.B., Haque, M.E., Ali, M.L., Chandra, G.C. and Das, C. (2003). Performance of Indian Runner, Zending and Khaki Campbell in integrated duck-cum-fish farming system. *Pakistan Journal of Biological Sciences.* **6**(3): 198-201.

El-nabarawy, A.M., Shakal, M.A., Hegazy, A.M. and Batikh, M.M. (2020). Comparative Clinicopathological Study of Salmonellosis in Integrated Fish-Duck Farming. *Journal of World's Poultry Research.* **10**(2S): 184-194.

FAO. (2010). Asia dominates world waterfowl production, Food and Agriculture Organization.

Gajendran, K. and Veeramani, P. (2022). Intensive Duck Rearing. *In:* Duck Production and Management Strategies (Jalaludeen, A.; Churchil, R. R. and Baéza, E. *Eds.*) pp. 265-302, Springer Nature Singapore Pte Ltd.

Gangwara, L.S., Saran, S. and Kumar, S. (2013). Integrated Poultry-Fish Farming Systems for Sustainable Rural Livelihood Security in Kumaon Hills of Uttarakhand. *Agricultural Economics Research Review*. **Vol. 26 (Conference Number):** 181-188.

Jalaludeen, A. and Churchil, R.R. (2022). Duck Production: An Overview. *In:* Duck Production and Management Strategies (Jalaludeen, A.; Churchil, R. R. and Baéza, E. Eds.) pp. 1-56, Springer Nature Singapore Pte Ltd.

Jeney, Z. and Woynarovich, A. (2011). Carp Culture. *In:* Better Management Practices for Carp Production in Central and Eastern Europe, the Caucasus And Central Asia (Woynarovich, A.; Bueno, P.B.; Altan, Ö.; Jeney, Zs.; Reantaso, M.; Xinhua, Y. and Van Anrooy, R. *Eds.*) pp. 23-53, FAO Fisheries and Aquaculture Technical Paper No. 566, Ankara. FAO, 2011.

Kalita, H.K., Talukdar, B., Baishya, R.A., Basumatary, S., Das, J., Dutta, A., Barman, R.C., and Sarma, D. (2015). Role of duck droppings on pond productivity through fish-integrated farming system in agro climatic condition, Assam. *Nature Environment and Pollution Technology*. **14:** 541-543.

Kumar, A. (2021). Duck Farming: A Profitable Source of Income to Farmers. Just Agriculture Multi-disciplinary e-newsletter. Vol.2 Issue-2, Oct. 2021.

Kumar, K. and Mishra, B. (2011). Integrated Fish Farming. *In:* Ayyappan, S., Moza U., Gopalakrishnan A., Meenakumari B., Jena, J.K. and Pandey, A.K. (eds.) Handbook of Fisheries and Aquaculture, 2nd Edn., ICAR New Delhi, DKMA, pp. 439-448.

Kumar, R., Monobrullah, M., Samal, S.K. and Sen, A.R. (2020). Fish-cum-duck integrated farming in floodplain ecosystems: technology assessment and refinement. *Food and Scientific Reports*, ISSN 2582-5437.

Liu, M.H.C. and Churchil, R.R. (2022). Duck Genetics and Breeding. *In:* Duck Production and Management Strategies (Jalaludeen, A., Churchil, R.R. and Baéza, E. *Eds.*), pp. 97-156, Springer Nature Singapore Pvt. Ltd.

Majhi, A. (2018). Role of Integrated Duck cum Fish Farming and Its Economic Efficiency: A Study in Purulia District, West Bengal. *International Journal of Information Research and Review*. **05(**5): 5443-5450.

Majumdar, P., Pegu, C., Behera, B.K. and Das, B.K. (2018). Future Aspects of Integrated Fish Farming. *Acta Scientific Agriculture*. 2(12): 45-47.

Mishra, S. (2010). Fish Disease Management in Integrated Farming System. *Webmed Central Parasitology*. **1**(9): WMC00663.

Mohanty, R.K., Kumar, A., Mishra, A., Panda, D. and Patil, D. (2014). Water Budgeting and Management: Enhancing Aquacultural Water Productivity. Bulletin No. 63. Directorate of Water Management, Indian Council of Agricultural Research, Chandrasekharpur, Bhubaneshwar, India, 62 p.

Pandey, A., Dhawan, A., Srivastava, A. and Singh, A. (2016). Integrated fish-cum-duck farming: productivity enhancement through technological interventions. *Indian Journal of Ecology*. **43**: 894-889.

Pandey, N. and Malik, D.S. (2016). Integrated Fish Farming. pp. 190, Daya Publishing House, New Delhi.

Patil, S., Khune, V.N. and Karunamay, S. (2020). Duck farming: A Potential to Reduce Poverty in Rural Households in Indian Communities – A review. *The Pharma Innovation Journal*. **9**(SP-4): 130-135.

Saikia, P., Nath, K.D., Kalita, D.N. and Hussain, S.M. (2020). Integrated Fish-cum-Duck farming system: A Tool for Increasing Farmer's Income. *Journal of Krishi Vigyan*. **(Special Issue):** 162-167. DOI: 10.5958/2349-4433.2020.00104.X

Sankaralingam, S. and Mahanta, J.D. (2022). Nomadic (Transhumant) Duck Farming Practices. *In:* Duck Production and Management Strategies (Jalaludeen, A., Churchil, R.R. and Baéza, E. *Eds.*) pp. 187-246, Springer Nature Singapore Pvt. Ltd.

Sapcota, D. and Begum, K. (2022). Integrated Duck Farming. *In:* Duck Production and Management Strategies (Jalaludeen, A., Churchil, R.R. and Baéza, E. *Eds.*) pp. 247-264, Springer Nature Singapore Pvt. Ltd.

Sari, M., Önk, K., Isik, S., Tilki, M. and Tufan, T. (2013). Effects of housing system, slaughter age, and sex on slaughter and carcass traits of native Turkish ducks. *Turkish Journal of Veterinary and Animal Sciences*. **37:** 694-700.

Soltan, M. (2016). Aquaculture systems. Retrieved February 02, 2022 from https://www.researchgate.net/publication/308928410_Aquaculture_systems

Sonaiya, E.B. and Swan, S.E.J. (2004). Small-scale Poultry Production Technical Guide. pp. 119, *FAO Animal Production and Health Manual*, FAO, UN, Rome.

Strasbourg, L.K. (2013). Fish Feeding in Integrated Fish Farming. pp. 272, Random Exports, New Delhi.

Thakur, T.K., Singh, P.K., Verma, L.R. and Ayam, G.P. (2018). Effect of integrated fish cum duck farming system on productivity of fish ponds. *Asian Journal of Animal Sciences*. **13**(1): 16-20.

Zaman, G., Goswami, R.N., Aziz, A., Nahardeka, N., Roy, T.C. and Mahanta, J.D. (2005). Farming system of Nageswari ducks in North-Eastern India (Assam). *World's Poultry Science Journal*. **61:** 687-693.

Chapter 15

Common Practiced Integrated Fish Farming Systems in the Northern States of India

Rajinder Kaur[1]*, P.S. Tanwar[1] and Tarang Kumar Shah[2]

[1]Krishi Vigyan Kendra, Barnala, Guru Angad Dev Veterinary and Animal Sciences University, Ludhiana – 141004, Punjab
[2]Department of Fisheries, Dr. G.C. Negi College of Veterinary and Animal Sciences, C.S.K. Himachal Pradesh Agriculture University, Palampur – 176061, Himachal Pradesh
Corresponding author: rajindergoraya7@gmail.com

Introduction

Integrated farming may be defined as a sequential linkage between two or more farming activities. When the fish becomes a major commodity of this system it is known as integrated fish farming (IFF). Thus, the practice of combining fish culture with agriculture or livestock for full utilization of resources and increased production is commonly known as Integrated Fish Farming. Integrated fish farming is based on the concept that '*there is no waste*', and '*waste is only a misplaced resource which can become a valuable material for another product*' (FAO, 1977). In integrated farming system, closely related and inter dependent farming systems are practiced together which ensures the optimum utilisation of the available natural resources resulting in reduction of input cost and ultimately increasing the net profit. The integration of different farming systems makes it possible for low-income farmers to profitably engage in the integrated enterprise for higher productivity and profitability.

IFS has introduced a change in the farming technique for maximum production in the cropping pattern, producing a variety of produce from same land and takes care of optimal utilization of the available resources. The farm wastes are better recycled for productive purposes in the integrated farming system. IFS basically approaches for the integration of various agricultural enterprises like trees, crops, livestock, fish *etc.* to achieve sustainability, food, nutrition, and livelihood security (Bhatt *et al.*, 2011).

The integrated fish farming system produces high yield with low input with the fish receiving limited supplementary feed. Animal waste and undigested and spilt food particles are directly consumed by the fish and some portion of waste acts as nutrients and organic substrates for many microorganisms which in turn consumed directly by fish or by invertebrate fish food organisms.

Types of Integrated Fish Farming

Basically, the integrated fish farming systems are grouped under two broad categories –

a) **Agri-based Fish Farming:** Fish cum agriculture involves rearing of fish with horticulture or agriculture include crops, vegetables, fruits *etc.* It is basically based on producing plant and animal protein in the same area.

b) **Livestock-based Fish Farming:** The fish-cum live-stock farming is realised as innovation for recycling of organic wastes as well as production of high-class protein at low cost.

Most Suitable Integrated Fish Farming Models

IFS is basically based on utilising the waste as resource for enhancing production. The inter-relation and inter-dependency of fisheries, livestock and agriculture has originated the concept of IFS. Livestock waste directly as manure or slurry can be bring into utilisation for enhancing the plankton production which directly reduces the fish feed cost. In fish farming, fish feed costs about 60-70 per cent of the total input cost which can be easily reduced in integrated farming system. Cattle dung can be used as input for biogas production which may further be utilised as fuel. Any two-tier, three-tier or four-tier integrated farming system can be adopted depending on the area, environmental conditions, topography, soil and water quality, demand of local consumers, financial status of the farmer *etc.* Different types of integrated fish farming models are:

Fish cum Horticulture/Agriculture Fish cum Dairy Farming Fish cum Poultry Farming Fish cum Pig Farming Fish cum Goat Farming Fish cum Duck Farming	Two Tier Integrated System
Fish – Livestock (Dairy/Poultry/Pig/ Goat) – Agriculture	Three Tier Integrated System
Fish-Poultry-Pig-Agriculture Fish-Livestock-Agriculture-Apiculture	Four Tier Integrated System

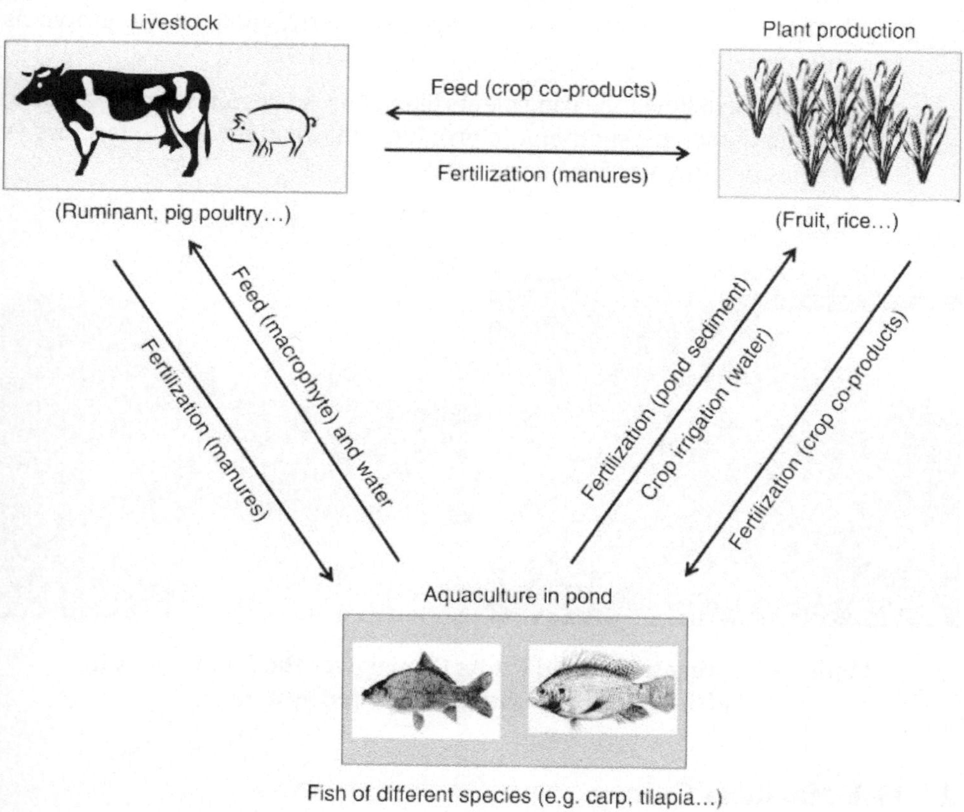

Figure 15.1: Simplified Flow Diagram of Interaction within Integrated Fish Farming System.

Agri-based Integrated Fish Farming

1. Fish-Horticulture Integrated System

In integrated fish farming system, the most common one is pond-dyke system where fish is cultured in ponds and dykes are being utilized for horticulture. Crops of fruits, vegetables and flowers can be cultivated on inner and outer dykes and adjoining areas of the fishpond. In this integration system, pond water is being used for irrigation purpose and pond silt, which is a high-quality manure can be used for crops, vegetables and fruit bearing plants. The success of the system depends on the selection of plants. They should be of dwarf type, less shady, evergreen, seasonal and highly remunerative.

- **Seasonal vegetables** like spinach, radish, carrot, cabbage, turnip, bottle gourd, cucumber *etc.* can be easily grown over the pond dykes.
- Dwarf variety fruit bearing plants like amla, banana, papaya, orange, lemon are also suitable while ginger, turmeric, chilli can be grown as intercrops.
- Plantation of flower bearing plants like tuberose, rose, jasmine, gladiolus, marigold and chrysanthemum provide additional income to farmers as well as beautify the farm area.

Figure 15.2: Vegetables and Fruits Grown over the Pond Dykes in Fish-cum-Horticulture Integrated System.

2. Fish-Makhana Culture

In this integrated farming model, along with pisciculture, Makhana (*Euryale ferox*) is grown in the pond. It is an aquatic plant with gigantic floating leaves which grows in standing shallow water with a depth of 0.2 to 2.0 m. Its leaves are thorny, oval or round in shape with 0.2 to 1.0 m diameter and are of green color on the upper side and purple red on the lower side. It is a common practice in Bihar to grow air breathing fish species with makhana plants (Verma *et al.*, 2008) and the suitable ones are *Heteropneustes fossilis* (singhi), *Clarius batracus* (magur) and *Anabas testudenius* (climbing perch).

Figure 15.3: Fish-cum-Makhana Integration.

Makhan crop plant pieces decompose to form organic matter and as a result, release nutrients into the pond which directly boost the plankton population. Organic detritus act as a suitable substratum for the growth of zooplankton, insect larvae, nematodes, and gastropods, as well as providing food for bottom-dwelling fishes (mrigal and common carp). Fish help to keep makhana bugs under control and their feaces serve as organic manure for the makhana crop. Giant makhana leaves form a barrier between water surface and sunlight penetration. Hence, result in the reducion of dissolved oxygen content in the covered regions of the pond (Kumar *et al.*, 2012).

Culture of Fish and Makhana

Fish culture in makhana ponds can be divided into different categories. After makhana, there comes a secondary crop of fish. During the makhana harvesting season, the fish are continuously cultured and transferred to specially constructed ponds or canals. After the makhana crop is harvested, some airbreathing fishes are released (Verma *et al.*, 2008). Such as (magur, singhi, kawai, gorai, mud eel) trash fishes and the only crustacean (*Macrobrachium lammeri*) which generally get access and enter through paddy fields into ponds as wild fish are also being harvested by farmers as additional animal crop without introducing seed of these fishes (Kumar *et al.*, 2012; Singh *et al.*, 2017).

Cultivation of Makhana

Cultivation of makhana starts with the clearing of pond or other water body following disseminating the seeds, thinning and filling gaps, safeguarding the plant, harvesting it, and eventually collecting the seeds are all part of the makhana cultivation process (Dey *et al.*, 2020). Water bodies that are currently under cultivation do not require broadcasting since saplings grow from the seeds that are left behind. Experienced producers sweep the entire water body's floor, creating heaps of sunken seeds that are scooped out using the appropriate equipment. Smaller and lighter makhana seeds are collected with small and fine nets (Verma *et al.*, 2008).

Makhana Cycle in Fish Ponds

- ☆ Perennial ponds with a thick layer of very nutritious muddy bottom work best.
- ☆ Sowing of seed generally takes place between the month of October and November. In ponds that are sown annually, 90 -100 kilograms seed/ha is necessary for the first time planting, whereas just 35 kg seed/ha is required for the first time sowing.
- ☆ In new ponds, seeds are distributed on the pond's surface which sink to the bottom and germinate in the mud. During February and March month, the geminated seeds sprout.
- ☆ After germination, thinning operations are carried out and further, saplings are transplanted throughout the available water area, with a gap of roughly 1 m between two plants.
- ☆ During the April and May months, the massive, thorny and elliptical leaves float on the surface of the whole water body.
- ☆ In June, the fruits begin to appear followed by bursting of mature fruit which scatter seeds across the pond's bottom.
- ☆ During this time, farmers break the leaves into pieces and discard them for decay.
- ☆ During the months of September and October, the scattered seeds at the bottom are manually collected. Seeds are gathered and stacked in several heaps (Kumar *et al.*, 2012; Dey *et al.*, 2020).

2. Fish-Singhara Integrated Farming

Singhara is an aquatic fruit and known as water chestnut in English and Singhara in Urdu. It is a good source of various elements like calcium, iron, potassium and zinc and can be grown positively in freshwater ponds, marshes, seasonal and perennial lakes. Fruits start to appear in the month of September and continue till January. Plant grows best at water depth of 0.5-1.0m requiring bright sunlight and high organic content.

Deep ponds are either utilised exclusively for fish cultivation or at the most the outer peripheral portions with shallow water are used for singhara cultivation. Such ponds could be put under integrated aquaculture in the form of concurrent singhara-cum-fish cultivation. Suitable fish species are airbreathing magur, singhi, kawai *etc.*

Singhara Culture

Singhara can be cultvated using two methods *i.e.,* firstly using natural seedings from previous crops and secondly by preparation and transplanting of seedlings.

- ☆ While selecting fruits for seeds, spines are cut with sharp knife and the selected seeds will be stored further only after curing.
- ☆ The seeding material should be kept in large earthen pots which are filled with freshwater and left undisturbed for two to three days.
- ☆ Water is changed on daily basis for a week.
- ☆ The process is continued until the hard, thick outer skin of the fruit rots and the loose coating of the seed is detached from the fruits and the thin, stony, inner coat becomes visible.
- ☆ The curing of seed material is done at room temperature and is completed in about 35 to 40 days.
- ☆ Seeds so cured can easily be stored in the same earthen pot without water and covered with a moist cloth or gunny bag to provide high humidity and low temperature.
- ☆ These containers are kept in a cool, shady place and can be stored for up to 3 to 4 months, without affecting seed viability.

Plantation

- ☆ Before broadcasting, the seeds are coated with a layer of soil on the opposite face of the germinated portion in order to add extra weight on the non-germinated face and to assure that after broadcasting, in the manner of a shuttle-cock, the seeds will settle at the bottom with the germinated face up and the coated face down.
- ☆ They also can be manually sown.
- ☆ The stem starts emerging and gradually spreads out.
- ☆ During the months of June-July, seedlings are lifted from the nursery pond and transplanted into larger ponds, ditches, or reservoirs.
- ☆ For transplantation, the uprooted stems are cut into several smaller pieces.

Rearing of Plants

- ☆ Lateral shoots commonly known as suckers can be detached from the main mother seed nut for transplanting.

- ☆ Single seeded water nuts can develop 20-30 and sometimes even up to 50 such lateral suckers.
- ☆ Each of these laterally developed suckers may very well be able to send out 5-10 further shoots after transplanting.
- ☆ From sowing to later such formations takes about 40-50 days.
- ☆ Shoots also arise form the nodes, forming roots and new plants.

Livestock-based Integrated Fish Farming

1. Fish-Pig Integrated System

This system of integration is very common in China, Taiwan, Vietnam, Thailand, Malaysia and Hungary. In India, it is also getting popular and exotic breeds such as White Yorkshire, Landrace and Hampshire are reared in pig sheds near the fishpond or over the pond dykes. In India, the main market of pig is north-eastern states. Utilization of swine dung for augmenting fish production has developed into an industry in China, many South-East Asian and East European countries. Pigs are fed majorly on kitchen waste, aquatic plants and crop wastes. Maize, groundnut, wheat-bran, fishmeal, mineral mixes are provided as concentrate feed-mixture.

In this farming system, fish utilize the left-over feed spilled by pigs and their excreta, which is nutrient enriched and acts as a substitute to pond fertilizer and supplementary fish feed. Pig dung as manure for fishpond has certain advantages over other livestock waste. Pigs have a limited ability to consume roughages and as such, the pig dung does not contain higher amount of cellulose, hemicellulose, and lignin *etc.* which might resist the process of decomposition and cover the soil at the bottom of the pond (Bhagawati *et al.,* 2020). Further, fresh pig dung has about 70 per cent digestible food for fish which is an additional benefit in reducing the feed cost in aquaculture (Pillay, 1980). No additional land is required for piggery operation as the shed can be constructed utilizing the pond dyke land. Wastewater after washing the pig-sites and pigs can directly enter the pond which results in increasing the plankton production in pond and reduces the wastage of water. It results in higher production of animal protein per unit area and ensures high profit through less investment.

Pig Shed

Pig house with adequate accommodation and all the requirements are essential for the rearing of pigs. The pigs are raised under two systems *i.e.,* the Open Air and Indoor Systems. A combination of both can be followed in fish cum pig farming system. A single row of pig pens facing the pond is constructed on the pond embankment. An enclosed run is attached to the pen towards the pond so that the pigs get enough air, sunlight, exercise and dunging space.

The feeding and drinking troughs are also built to keep the pens dry and clean. The gates should be of sufficient size to provide open run. The floor of the run is cemented and connected via the drainage canal to the pond whereas the height of the pig house should not exceed 1.5 m. A shutter is provided in the drainage canal to stop the flow of wastes to the pond. The drainage canal is provided with a diversion channel to a pit, where, the wastes can be collected and the slurry can be used whenever and wherever necessary. The pig house can be constructed with locally available materials and is advisable to provide 1 - 1.5 m^2 space for each pig.

Earlier, the structure was made of wood and provided with a lattice type of floor which permits the excreta and uneaten food to fall directly into the pond water. Modern practices are to avoid direct washing of the wastes into the pond. The urine and dung of pigs are first allowed to the oxidation tanks (digestion chambers) where sedimentation and fermentation of the manure take place. The supernatant liquid, at regular intervals, are then discharged into the fish-ponds. The sludge that remains is utilised as fertilisers in agriculture. Alternatively, the pig manure may be kept in a heap on the pond embankment for later use.

Selection of Pigs

In India, four types of pigs are available such as wild pigs, domesticated pigs or indigenous pigs, exotic pigs and upgraded stock of exotic pigs. The Indian varieties are small sized with a slow growth rate and produce small litters. Its meat is of inferior quality. Exotic upgraded stock of pigs such as large - White Yorkshire, Middle - White Yorkshire, Berkshire, Hampshire and Hand Race are most suitable for raising with fish culture. These are well known for their quick growth and prolific breeding. They attain slaughter maturity size of 60 to 70 kg within six months.

Figure 15.4: Pig Shed Constructed along the Pond Dyke.

2. Fish-Duck Integrated System

The combination of duck and fish farming is seen presently as a means of reducing the cost of feed for ducks, at one hand, and on the other, the excreta of ducks act as an inexpensive way of fertilising ponds, which results in production of fish food organisms. In aquaculture, ducks are known as 'biological aerators' as the movement of ducks in the pond aerate the pond water. Ducks are habituated to consume juvenile frogs, tadpoles and dragonfly *etc.* and there by make a safe environment for fish (Saikia *et al.,* 2020).

Duck droppings provide essential nutrients go directly into the pond as the droppings are good sources of carbon, nitrogen and phosphorus, which in turn stimulate the growth of natural food organisms (Kumar, 2021). The fish-cum-duck integration system provides meat, eggs in addition to fish. It generates additional products and income to the farmer. Ducks are allowed to enter fish-ponds once the size of stocked fish reach 4-5 inches as the ducks can consume smaller size fish.

Duck House

It is preferable to build the duck house just on embankments so that they can be protected during the night and egg-laying and have quick access to the pond as needed. The opening in the bottom should be approximately 5 cm^2. Each duckling needs 1 to 1.5 square feet of area to be comfortable and for adequate ventilation, the house's height should be around 2.5 m (Lal *et al.,* 2022). A proper fencing of the pond is needed in fish cum duck farming to protect the ducks from dogs' attack.

Selection of Duck Species

Healthy ducks should be selected for duck cum fish farming for better growth and survival of ducks. Due to excellent disease resistance, rapid growth, and great egg-laying capability, Khaki Campbell and Indian runner are good duck varieties for farming practices. Khaki Campbell is one of the best egg and meat producing breed in ducks having fast growth rate and efficient feed converters. Duck breeds commonly used for meat production, egg production or for both purposes are easily available in India (Kumar, 2021).

Figure 15.5: Duck House over the Pond Dyke and Ducks Aerating the Pond Water.

3. Fish-Poultry-Horticulture Integrated System

It is the most practiced three-tier integrated system in which poultry shed can be constructed over the fishpond and pond water can be utilised for irrigating the crop. Suitable species for the integrated fish farming system are Rhode Island Red,

Leghorn and Babcock. The droppings of chicks rich in nitrogen and phosphorus would fertilise fishpond. Poultry house, when constructed above the water level using bamboo poles would fertilise fishponds directly. Fish can directly accept poultry droppings as feed whereas the uneaten ones can manure the fishpond. This directly reduces the input cost including fish feed, pond manure.

Broiler production provides good and immediate returns to farmers (Singh *et al.*, 2014). Procurement of quality chicks, housing, brooding, feeding, and disease management are important for this type of system.

Rearing Chicken in Shed

In fish-poultry integration, birds housed under intensive system are considered best. Birds are kept in confinement with no access to outside. In shed method, unnecessary roaming can be avoided energy loss and supply of good quality feed on need basis will yield better and higher income. Two types of shed rearing is practiced –

i) **Cage System:** In this method, on top of the fish-pond birds shed can be lined and arrange to fall the country chicken excreta into the pond and fish will accept it as feed. If poultry sheds are constructed over pond dyke and cultivated in cage system means every cage place the plate to collect the poultry droppings. This was used for fish feed.

ii) **Deep Mulching Method:** In this method, chicken shed can be constructed over pond dyke or nearby place. Over the cemented floor, rice husk, groundnut shell, or coconut fibre can be used as mulches and maintained for the height up to a feet from the floor. The droppings can be collected regularly and after decomposition it can be used as manure in fish-pond. Poultry manure contains 25.5 per cent organics, 1.63 per cent nitrogen, 1.54 per cent phosphorus and 0.83 per cent potassium.

Figure 15.6: Poultry Shed Constructed over the Fishpond; Seasonal Vegetables grown over the Pond Dyke and the Field Crops Irrigated using Pond Water.

4. Fish-Cattle-Horticulture Integrated System

In this system of integration where the cattle dung, oxidized in biogas plants is transferred to fishpond through small channels to act as manure. It is advantageous to construct cowsheds near the ponds as the cow dung acts as fertilizer for propagation of natural fish food in water like planktons. Cattle wastes and washings from the cattle sheds are conveyed through pipes into the ponds which acts as good fertiliser. Cattle wastes are generally collected in a pit for later use. Cattle shed can constructed near the find pond to utilise the nearby area. Cattle dung can be used as manure in fishpond as well as in horticulture crops. Cattles can take bath in the fish-pond and it will help in aeration of pond water.

Figure 157: Cattle Shed Constructed Near the Fishpond; Seasonal Vegetables grown over the Pond Dyke and the Field Crops are Irrigated using Pond Water.

5. Fish-Goat-Agriculture Integrated System

Goat is known as the poor man's cow and are reared by small and marginal farmers for their domestic needs as well as additional income generating unit. Wherever convenient, they can also include a fishpond, whose size can vary according to the availability of land.

Goats are very fond of leguminous fodders. Some of the common green roughages like by goat are cowpea, cabbage and cauliflower leaves, soyabean. These can easily be cultivated on the pond dyke without putting many efforts into it. Leaves of trees such as ber, neem, tamarind and peepal are also like by goats. For the irrigation, pond water can be utilised which not only irrigate the vegetables and other plants but also fertilises them.

Goat manure can be used in fishpond in two ways *i.e.*, firstly, it can directly be added to the fishpond through the outlet pipe from the goat shed. Secondly, the faeces can be collected in an earthen pit from where the slurry will be added to the fishpond depending on the pond requirement. The amount can be decided by evaluating the pond watercolour. Integration results in the higher fish production, other output products like meat, milk *etc.* form the same unit area.

Housing Management

Adequate housing provides conditions for good health and comfort to the animals, high reproduction and more efficient management. A low cost housing can be constructed near the pond embankment using locally available material like bamboo, wood thatch *etc.* Floor space requirement per goat is 10 sqft. The house should be oriented in such a way that the greatest amount of sunshine and air enter the house.

Space Requirement

- **Floor space –** 3 feet above the ground, and
- **Floor space requirement per goat:** 10 square feet per goat.

Species Selection

In Punjab and other northern states, most reared species of goat is Beetal. This breed is large in size, black coloured and is a good dairy type. Average daily milk yield is 2 kg. The twins and triplets are very common in this breed.

Table 15.1: Stocking Density of Animals in Integrated Fish Farming System

Sl.No.	*Crop*	*Stocking Density (per hectare)*
1.	Fish	10,000 (fingerlings)
2.	Cattle	2-3
3.	Pig	20-30
4.	Ducks	200-300
5.	Poultry Birds	400-500

Figure 15.8: Four Tier (Fish-Goat-Horticulture-Agriculture) Integrated Model.

Advantages of Integrated Fish Farming Systems

The scope of integration of a fish farm is considerably wide and the major ones are:

- ✰ The topmost advantage of integrated farming system is that it reduces and promotes the sustainable utilisation of water.
- ✰ Integrated fish farming systems utilise the waste of livestock and agriculture by-products for fish production. About 40 to 50 kg of organic manure can produce 1 kg of fish.
- ✰ Pond silt can be used as fertiliser for fodder crops which in turn can be used to raise live-stock and poultry or as fish feed. Thus, a recycling of waste is done in integrated fish farming system.
- ✰ Integrated fish farming systems produce a variety of products per unit area with less input cost.
- ✰ IFS is environmentally friendly as it reduces the waste load.
- ✰ IFS supports the optimum utilisation of water body, the water surface, the land, and the pond silt to increase food production for human consumption.

REFERENCES

Ayyappan, S. (2011). Handbook of Fisheries and Aquaculture. Indian Council of Agricultural Research, New Delhi.

Behera, U. K., Panigrahi, P. and Sarangi, A. (2012). Multiple water use protocols in integrated farming system for enhancing productivity. *Water Resources Management.* **26:** 2605-2623.

Bhagawati K., Tamuli K.K. and Saharia, P.K. (2020). Integrated Fish cum Pig Farming System for Enhancing Farm Income in Assam. *Journal of Krishi Vigyan.* **(Special Issue):** 209-213.

Bhatt, B.P., Bujarbaruah, K.M., Vinod, K. and Karunakaran, M. (2011). Integrated fish farming for nutritional security in eastern Himalayas, India. *Journal of Applied Aquaculture.* **23**(2): 157-165.

Bhuiyal, M.S.U., Karim, M.M. and Hossain, S.M.A. (2014). Study on integrated system model development. *Journal of the Bangladesh Agricultural University.* **12**(2): 325-336.

Chan, G.L. (2006). Integrated farming system. What Does Integrated Farming System Do? http://www.actahort.org/books/655/655-36.html

Debnath, C., Sahoo, L., Yadav, G.S., Datta, M. and Ngachan, S.V. (2013). Integrated fish farming for sustainable livelihood of farmers of Tripura. *Indian Farming.* **62**(10): 18-21.

Dey, A., Patel, S., Rashid, M., Singh, H.P. and Singh, R. (2020). Income enhancement of low-land farmers through makhana-cum-fish cultivation. *In Conference*: 5th Uttar Pradesh Agricultural Science Congress.

Edwards, P. (1983). The future potential of integrated farming systems in Asia. *Proc. V WCAP*. **1:** 273-281.

FAO. (1977). China: Recycling of organic wastes in agriculture. FAO Soil Bull., 40 - Rome.

Kumar, A. (2021). Duck Farming: A Profitable Source of Income to Farmers. *Just Agriculture*. **2**(2): Oct, 2021.

Kumar, L., Gupta, V.K., Jha, B.K., Singh, I.S., Bhatt, B.P. and Singh, A.K. (2012). Status of Makhana (*Euryale ferox Salisb*) Cultivation in India. Tech. Bull. No. R-32/PAT-21. ICAR-RCER, Patna P.31.

Lal, J., Shatrupa, Tameshwar, Kashyap, N. and Debbarma, J. (2022). Integrated Fish Farming: Fish cum Duck Farming. https://krishijagran.com/featured/integrated-fish-farming-fish-cum-duck-farming/

Manjappa, N., Patil, R. and Pavadip. (2017). Potential use of village tanks and farm ponds for aquaculture in Karnataka, India - A case study. *International Journal of Research in Applied, Natural and Social Sciences*. **5**(10): 45-50.

Misra, V.K., Singh, N.D., Mishra, B.P. and Tripathi, A.N. (2016). Study on Integrated Horticulture cum Fish Farming in District West Kameng, Arunachal Pradesh. *In*: National Conference entitled "*Horticulture in North Eastern Region*", organized by College of Horticulture and Forestry, Central Agricultural University, Pasighat, Arunachal Pradesh 16th – 18th January 2016, Arunachal Pradesh. pp 225.

Nanda, P.K. and Bandopadhyay, U.K. (2011). Recent advances in integrated livestock cum fish farming in India. Training manual of short course on advances in production of livestock management practices. IVRI, ERS, Belgachia, Kolkata.

Nath, K.D., Rajbangshi, P., Borah, D. and Saharia, P.K. (2020). Doubling Farmers Income through Adoption of Integrated Farming System – A Case Study. *Journal of Krishi Vigyan*. **(Special Issue):** 59-62.

Othman, K. (2006). Integrated farming system and multifunctionality of agriculture in Malaysia. *Acta Horticulturae*. **655:** 291-296.

Pillay, T.V.R. (1980). Aquaculture: Principles and Practices (Fishing New Books). New York: Wiley/Blackwell.

Saikia, P., Nath, K.D., Kalita, D.N. and Hussain, S.M. (2020). Integrated Fish-cum-Duck farming system: A Tool for Increasing Farmer's Income. *Journal of Krishi Vigyan*. **(Special Issue):** 162-167.

Sharma, S.K., Nainwal, G.C., Kumar, P. and Ujjaniia, N.C. (2016). Integrated fish cum poultry cum horticulture farming system: An approach to uplift the socio-economic status of fish farmers. *Progressive Research – An International Journal*. **11**(1): 411-414.

Singh, I.S., Kumar, L., Bhatt, B.P., Thakur, A.K., Choudhary, A.K. and Kumar, A. (2017). Integrated Aquaculture with Foxnut – A Case Study from North Bihar, India. *International Journal of Current Microbiology and Applied Sciences*. **6**(10): 4906-4912.

Singh, N.D., Sarkar, A., Biswas, P., Pal, P. and Upadhyay, A.D. (2018). Pond Niche Based Integrated Farming: A Case Study in Tripura, India. *International Journal of Bio-resource and Stress Management*. **9**(3): 359-364.

Singh, S., Singh, P., Verma, N. and Kumar, D. Integrated Fish Farming - Rationale and Scope. Aquafind. http://aquafind.com/articles/Integrated_Fish_Farming.php

Singh, U.P., Bisht, H.C.S. and Pandey, N.N. (2014). Poultry cum fish culture as an effective tool for environmentally sustainable and economically viable technology for the mid hills. *International Journal of Advanced Research*. **2**(2): 854-860.

Taiganides, E.P. (1978). Principles and techniques of animal waste management and utilization. *FAO Soils Bull Bulletin*. **36: 3**41-362.

Talukdar, P.K. and Sontaki, B.S. (2005). Correlates of adoption of composite fish culture practices by Fish farmers of Assam, India. *The Journal of Agricultural Science*. **1**(1): 12-18.

Verma, A.M., Jha, V. and Ahmad, S.H. (2008). Fish-Makhana Integration – A Case Study of Sustainable Aquafarming system in North Bihar. *Journal of Indian Fisheries Association*. **35:** 87-96.

Chapter 16

Integrated Farming System: An Emerging Approach for Sustainable Development of the Farmers

***Kajal Sengupta*[1], *Prasun Chatterjee*[2]* *and Adyant Kumar*[3]**

[1]Professor and Head and [2]Post-Graduate Student,
Division of Agronomy, Faculty of Agriculture and Rural Development (FARD), Ramakrishna Mission Vivekananda Educational and Research Institute (RKMVERI) [Deemed to-be-University], Narendrapur Campus, Kolkata – 700103, West Bengal
[3]Assistant Professor cum Junior Scientist, Department of Agronomy, Tilka Manjhi Agriculture College (Birsa Agricultural University, Ranchi), Punsiya, Godda – 814 133, Jharkhand
**Corresponding author: prasunchat1998@gmail.com*

ABSTRACT

Integrated farming system (IFS) is a holistic approach of farming which aims at fulfilling the multiple demands viz., farm resilience, farmer livelihoods, food security, ecosystem services etc. IFS are characterized by temporal and spatial mixing of crops along with livestock, fishery and other allied activities on a single farm. It is assumed that farming systems having complex nature are more productive at a system level than the single enterprise farms. IFS provide the various needs of small and marginal farmers, who are the backbone of agriculture in India. IFS have the potential to improve farm profitability (265 per cent) and employment (143 per cent) compared to single enterprise farms. IFS also enhance the recycling of nutrients through composting, mulching, and residue incorporation and, also help in reducing the purchase and use of external inputs. The recycling of nutrients in turn helps to improving the soil quality and health through incorporation of organic matter, which results in improving soil organic carbon up to 0.75 to 0.82 per cent. And this increase in carbon sequestration in the integrated farm ultimately results in biomass production by trees, reduced consumption of fertilizers and pesticides, over and above, results in significant reduction in the greenhouse gas emission. Additionally, the IFS play a significant role in biodiversity conservation by adopting a diversified cropping system and integrating indigenous livestock breeds. It also supports in promoting the non-crop biodiversity through ecological management strategies and the use of organic management practices. This results in a linked system making it sustainable and climate-resilient. The main challenge associated with adoption of IFS is it requires skill, knowledge, resources, labour, and capital which are not always available with small and marginal farmers. There is a need for integrating productivity, profitability, and environmental sustainability variables in a single evaluation framework to effectively generate information toward enhancing adaptability of IFS.

***Keywords**: Integrated farming system (IFS), Components, Enterprises, Doubling income.*

Introduction

The advent of green revolution in India in late-1960s had made the country self-sufficient in food grain production (Behera and France, 2016). Now-a-days, the country has achieved the meritorious heights in food production by extensive cultivation of few crops. At the same time farming turned non-profitable for farmers and unsustainable to environment. Subsequent economic liberalization in early-1990s, the farmers gradually started focusing on a few enterprises due to several imposing factors including shrinking farm sizes, fluctuating commodity prices, livelihood diversification and shortage of labour during peak agriculture season, which had a severe impact on food and nutritional security of millions of poor farm households.

In India, about 70 per cent of rural population is engaged in agriculture and 80 per cent of population live, directly or indirectly on income derived from agriculture. There are about 115 million operational holdings in the country, out of which 80 per cent holdings are marginal and small farmers (Manjunatha *et al.*, 2014). The farmers used to maintain diverse enterprises (agriculture, horticulture, dairy, poultry, goatry, duckery, agroforestry, fodder, sericulture, apiculture *etc.*) since immemorial time for livelihood sustainability. Diverse enterprises complement each other for food, nutritional security, soil health, economic and environmental sustainability. Because of above-mentioned reasons and increase in nuclear families, gradually the maintenance of diverse enterprises on farm has lost its viability.

For the survival of mankind apart from food (cereal, pulses, oilseeds, milk, fruit, honey, meat *etc.*) require fibre, fuel, fodder and feed for milk production *etc.* which balance the environment and farm productivity. The agricultural experts are judging the options and strategies for achieving this enviable target. One of the options is to evaluate the potential of age-old integrated farming system (IFS) in enhancing income of farm families within the reasonable time period. Thus, IFS, as a multidisciplinary approach, helps in elevating the farm family income in short period and it also shows the way to resilience of the farm productivity. With the integration of more number of farm enterprises, high farm productivity can be achieved in addition to nutritional security, social, economic and environmental benefits (Mynavathi and Jayanthi, 2015). It plays a key role in increasing the farm production with less investment and appears to be very effective in solving the problems of small and marginal farmers by the integration of various farm enterprises, optimal utilization of resources, recycling of the crop residues and by-products within the farm and employing the family labour, thus, results in increasing the income and employment Sheikh *et al.* (2021).

Role of Integrated Farming System

- ☆ Higher productivity and enhanced farm income,
- ☆ Effective recycling of resources,
- ☆ Minimize environmental pollution, and
- ☆ Employment generation.

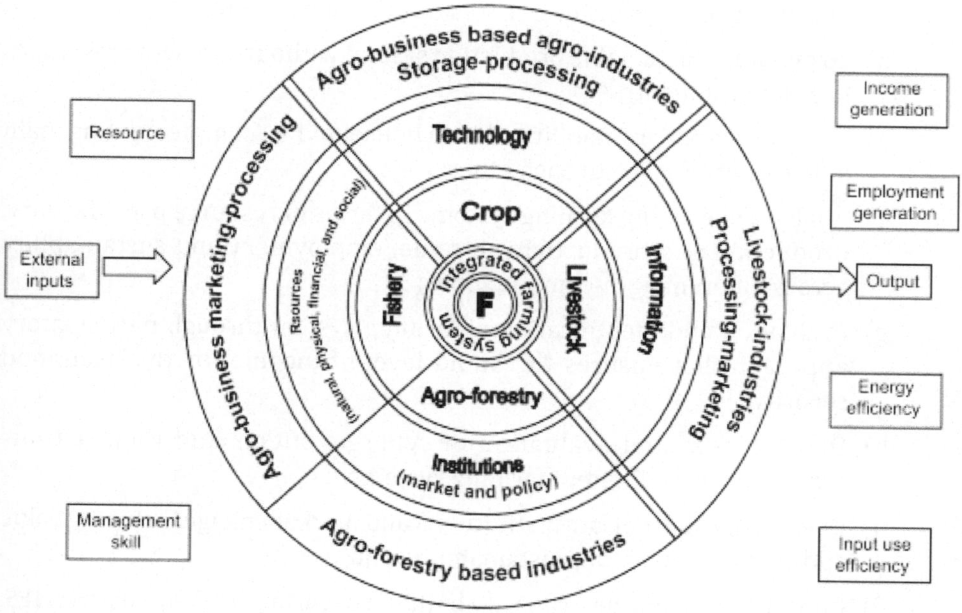

Figure 16.1: Representation of a Farming System Influenced by the Socio-economic, Political and Biophysical Environment and Managed by the Farmer (*Source*: Behera and France, 2016).

Apart from this, IFS plays a significant role in ensuring food and nutritional security and providing a balance food to the rural poor. Food security needs to be redefined as 'livelihood security for the household and all member within which ensures both physical and economic asses to balanced diet, safe drinking water, environmental sanitation, primary education and basic health care. The integration of components like dairy, poultry, goatry, fisheries, mushroom, *etc.* along with crop production not only meets the food demand but also caters to the need of protein, fat, vitamins and minerals required for good health. These components can ensure round the year availability of good quality food. When the living standard is improved then the cereals consumption is reduced and subsequently replaced by other products such as milk, eggs, fish, *etc.* which is rich in animal protein. Thus, it can be emphasized that integration of livestock, horticulture and agroforestry systems will lead to better diversity and quality of food.

Objectives of Integrated Farming Systems

a) To identify existing farming systems in specific area and assess their relative viability;

b) To formulate farming system models involving main and allied enterprises for different farming situations;

c) To ensure op-tional utilization and conservation of available re-sources and effective recycling of farm residues within system;

d) To maintain sustainable production system without damaging resources base environment;

e) To raise overall profitability of farm household by complementing main allied enterprises with each other;

f) Analysis of existing farming systems in terms of resource use efficiency, production and productivity, income, employment and sustainability across different agro-climatic zones;

g) Evaluation and identification of farming system through participatory approach that ensures threshold level of income for the livelihood security;

h) Development and evaluation of synergic effects and their actions associated with different farming systems;

i) Developing appropriate institutional and market linkage including value addition for enhancing sustainability; and

j) Imparting training and capacity building of various stakeholders on IFS.

Goal of Integrated Farming System

The goals of integrated farming systems are maximization of yield of all component enterprises to provide steady and stable in-come rejuvenation/ amelioration of system's productivity and achieve agro-ecological equilibrium; avoid build-up of insect-pests, diseases and weed population through natural cropping system management and keep them at low level of intensity; and reducing the use of chemicals (fertilizers and pesticides) to provide chemical-free healthy produce and environment to the society.

Components of Integrated Farming System

1. **Crop Production** – The crops and cropping system should be location-specific. It should generate enough food to meet the domestic and market purposes. It can be achieved by adopting intensive cropping and multiple cropping. Inclusion of pulse and oilseed crop will help to meet their requirements without having to purchase them. Forage crops can also be taken in sequence to provide fodder for the cattle.
2. **Dairy Farming** – The dairy animals not only produce milk, but also are good source of farmyard manure, which is good source of organic matter for improving soil fertility. The farm by-products in turn are gainfully utilized for feeding the animals. The per capita availability of milk is still about 220 g day^{-1} against the minimum requirement of 250 g day^{-1} as recommended by Indian Council of Medical Research. Thus, their integration will meet the milk demand of the household.
3. **Goat Rearing** – Goat rearing can be done in wide range of environments such as dry, hot, wet and cold, high mountains or low lying plains. It requires very low investment and maintenance costs. Goat meat is the highly preferred meat in the country and it is mainly reared for milk, meat, skin and hide.
4. **Sheep Rearing** – Sheep with its multi-facet utility for wool, meat, milk, skins and manure, form an important component of rural economy particularly in the arid, semi-arid and mountainous areas of the country. They are excellent gleaners and make use of much of the waste feed. They consume large quantities of roughage, converting a relatively cheap food into a good cash product. It provides a dependable source of income to the shepherds through sale of wool and animals. They play an important role in the livelihood of a large percentage of small and marginal farmers and landless labourers engaged in sheep rearing.
5. **Fishery** – Fish consumption contributes a significant amount of animal protein to the diets of people worldwide. Fish is *nature's super food,* an important source of proteins and healthy fats, a unique source of essential nutrients, such as long-chain *omega-3* fatty acids, iodine, vitamin D, and calcium. It is especially of great importance for pregnant and lactating women. The nutrients present in fish promote optimum brain development, regulating the immune system and building healthy bones. A composite fish culture with the stocking density of 5000-7500 fingerlings per hectare and supplementary feeding can boost the total biomass production as compared to traditional fish farming.
6. **Poultry** – Backyard poultry is an age old practice, but this enterprise is gaining momentum as a new agricultural industry. It can be effectively reared by constructing a shade over the pond, thus requiring no extra space for its management. The average global consumption is 120 eggs

person^{-1} year^{-1} and in India, it is only 32-33 eggs person^{-1} year^{-1}. As per the nutritional recommendation, the per capita consumption is estimated at 180 eggs year^{-1} and 9 kg meat year^{-1}. So, there is a gap between the demand and supply of eggs which can be bridged by the integration of poultry in farming systems. Poultry meat and eggs can provide high-quality protein and micronutrients in readily available forms which, even in small quantities will substantially increase the nutrient levels of traditional diets which are mainly based on staple crops such as rice and wheat.

7. **Duck Rearing –** Ducks account for about 7 per cent of the poultry population in India. Ducks are quite hardy, easily brooded and are resistant to common avian diseases. Among the duck breeds, Khaki Campbell is known to be the highest egg producer in the world. Their eggs are especially high in vitamin B_{12}, selenium and choline and contain 9.60 grams of fat as compared to 5 grams of fat in chicken eggs. Duck eggs are also higher in *omega-3* fatty acids (71.40 milligrams) as compared to 37 milligrams in chicken eggs.
8. **Pig Farming –** Among the livestock species, pig finds an important place as it being reared by socio-economically weaker sections of the society. Pig as compared to other livestock species has a great potential to contribute to faster economic return to the farmers, because of certain inherent traits like high fecundity, better-feed conversion efficiency, early maturity and short generation interval. It has immense potential to ensure nutritional and economic security for the weaker sections of the society. Management of pigs is very much essential for optimum production of pork.
9. **Mushroom Production –** Mushrooms are fairly rich in vitamin C and B-complex. The protein have 60-70 per cent digestibility and contains almost all essential amino acids. It is also a rich source of minerals like Ca, P, K and Cu. They contain less fat and CHO and are, therefore, considered good for diabetic and blood pressure patients. Mushroom consumption can improve the nutritional level of poor farmers. Mushroom cultivation can easily be done by utilizing straw and compost that are produced on-farm without depending on external sources.
10. **Bee-Keeping –** Honey bees play a very important role in pollinating various agricultural and horticultural crops and increase their yield and improve the quality of produce. It is being increasingly realized that bees could be less expensive input for promoting sustainable and eco-friendly agriculture and enhancing crop productivity. Beekeeping has quadruple benefits – providing self-employment to rural and forest based population; production of honey, pollen and beeswax, venom, royal jelly, *etc.*; providing employment to rural educated youths in collecting, processing and marketing of bee/beehive-products and the most important; and cross-

pollination of various agricultural and horticultural crops and improving their quality and increasing yields. Thus, beekeeping is a low investment and skill industry having the potential to offer direct employment to lakhs of people especially hill dwellers, tribal and unemployed youth and farmers. The sustainability of beekeeping is therefore vital to the country's economic well-being and development.

11. **Biogas Plant** – Biomass resources such as cattle dung, agriculture wastes and other organic wastes have been one of the main energy sources for the mankind since the dawn of civilization. There is a vast scope to convert these energy sources into biogas. Biogas production is a clean low carbon technology for efficient management and conversion of fermentable organic wastes into clean cheap and versatile fuel and bio/organic manure. It has the potential for leveraging sustainable livelihood development as well as tackling local and global land, air and water pollution.
12. **Agroforestry** – Agroforestry is a self-sustaining land-management system which combines production of agricultural crops with woody components such as trees, shrubs, bamboos, canes, palms *etc.*, as also with that of the livestock simultaneously or sequentially, on the same unit of land and time, to satisfy the ecological as well as socio-economic needs of the people. In this system, forest tree species are regenerated and managed for their ability to produce not only wood, but leaves and or fruit that are suitable for food and or fodder. It is managed to yield multiple products like fruits, flowers, leaves, barks, roots, gums, honey, medicines and other non-timber forest products *etc.* which may be eaten and or utilized for other purposes.

Choice of Enterprise in Integrated Farming System

Farming enterprises include crop, vegetables, fruits, livestock, poultry, fish, tree crops, plantation crops, sericulture, lac culture, apiculture *etc.* A combination of one or more enterprises with cropping, when carefully chosen, planned and executed, gives greater dividends than a single enterprise, especially for small and marginal farmers. Farm as a unit is to be considered and planned for effective integration of the enterprises to be combined with crop production activity. The basic points that are to be considered while choosing appropriate enterprise in Integrated farming system are:

a) Soil and climatic features of an area/locality,
b) Resource availability with the farmer,
c) Present level of utilization of resources,
d) Return/income from the existing farming system,
e) Economics of proposed integrated farming system,

f) Farmer's managerial skill, and

g) Social customs prevailing in the locality.

For any sustainable and economically viable farming system, *in-situ* conservation of rain water to the maximum possible extent (which automatically takes care of soil conservation) and build-up of soil fertility and productivity largely through the incorporation of organic manures but only supplemented by minimum extent of need based chemical fertilizers/amendments are the primary considerations. *In -situ* rain water harvesting and conserving can be achieved by appropriate bunding followed by land levelling and strengthening of the bunds by planting trees and grasses on the bunds. Regular maintenance of bunds and interbund management practices such as carrying out all cultural operations across the slope, compartment bunding, mulching, application of organic manures to improve the infiltration rate of soils *etc.* will help to conserve soil and water. Soil fertility build up and its regular maintenance can be achieved on sustainable basis only through the application of organic manures such as FYM, compost, vermicompost, crop residues and green manures. Building up of soil fertility through organics is very important since with such a practice, stability of increasing trend in crop yields can be achieved as compared to sudden rise and fall of crop yields when chemical fertilizers are used. Regular use of organics encourages multiplication of beneficial soil microbial population and the soil borne pathogens are controlled. Organics such as vermicompost also make the plant resistant to certain pests and diseases. Moreover, quality of the produce from such a crop will be superior as compared to the produce obtained when chemical fertilizers are used. In irrigated agriculture, after few years of irrigation, degradation of soil is very common. But, wherever green manuring is practiced as a regular feature, degradation of soil has not taken place. At present, it has been the experience of most of the farmers that the farming under the existing situation is not a profitable enterprise. The major reason for such a situation is the high cost of cultivation which is increasing year after year. Therefore, any effort to reduce the cost of cultivation by avoiding/minimizing the use of chemicals and middlemen in agriculture will definitely bring in stability of agricultural production and farming will be a profitable concern.

Crop management practices such as crop rotation, mixed cropping, INM and IPM need to be incorporated in all the farming situations. Crop rotation and mixed cropping have many advantages. Build-up of pests/diseases and depletion of plant nutrients from the soil can be avoided by following the simple practice of crop rotation. It will also help to build soil fertility when a legume crop is involved in the crop rotation. Mixed cropping when followed in an appropriate manner will increase total output under normal situations but will act as an insurance against failure of crops under abnormal conditions. When a legume is involved as one of the component crops, in mixed cropping, it will improve soil fertility. Mixed cropping is also advantageous in minimizing the spread of pests and diseases and is efficient in the utilization of soil moisture, nutrients and sunlight. Instead of depending solely on chemical fertilizers and amendments, a combination of organic manures, biofertilizers and chemical fertilizers is always better to maintain good soil health and at the same time to harvest higher crop yields. In case of plant protection, sole dependence on chemicals and their indiscriminate use has resulted in making the insect pests resistant to chemical pesticides, thereby increasing the dosage of pesticides for the control of pests year after year. Moreover specific pesticides are to be used for particular pest at its particular stage. Many farmers have very little knowledge of such specific plant protection measures which is again complicated by almost frequent changes in the plant protection schedules. Under such situations, IPM has definite advantages, but complete elimination of chemicals in a phased manner from the IPM schedule and developing plant protection methods with appropriate cultural practices, cropping systems and use of bioagents and plant products will be ideal.

Mixed farming involving allied enterprises which are complementary to agriculture, such as dairy, apiculture, sericulture, inland fisheries, sheep/goat rearing and poultry will provide employment and additional income to the farmer throughout the year. It will also act as an insurance against crop failures due to climatic vagaries and will help to build soil fertility when animal wastes are properly utilized for preparation of manures.

Integration of Enterprises in Integrated Farming System

In agriculture, crop husbandry is the main activity. The income obtained from cropping is hardly sufficient to sustain the farm family throughout the year. Assured regular cash flow is possible when cropping is combined with other enterprises. Judicious combination of enterprises, keeping in view of the environmental conditions of a locality will pay greater dividends. At the same time, it will also promote effective recycling of residues/wastes.

Livestock is the best complementary enterprise with cropping, especially during the adverse years. Installation of a biogas plant in crop-livestock system will make use of the wastes, at the same time provides valuable manure and gas for cooking and lighting. In a wetland farm, there are greater avenues for fishery, duck farming and buffalo rearing. Utilizing the rice straw, mushroom production can be started. Under irrigated conditions (garden lands), inclusion of sericulture, poultry and piggery along with arable crop production is an accepted practice. The poultry component in this system can make use of the grains produced in the farm as feed. Pigs are the unique components that can be reared with the wastes which are unfit for human consumption. In rainfed farming, sheep and goat rearing form an integral part of the landscape. Sericulture can be introduced in rainfed farming, provided the climatic conditions permit it. Agro- forestry (silviculture and silvi-horticulture) are the other activities which can be included under dryland conditions. In the integrated system, selection of enterprise should be on the cardinal principle that there should be minimal competition and maximum complementary effect among the enterprises.

Integrated farming system meets spread out demand for food, income and diverse requirements of food grains, vegetables, milk, egg, meat *etc.*, thereby improving the nutrition of small-scale farmers with limited resources. Integration of different agriculturally related enterprises with crops provides ways to recycle the products and by products of one component as input to another and reduce the cost of production and increase the total income of the farm.

Sustainability of Integrated Farming System Models

The following features should be integrated to make the integrated farming system models more sustainable:

a) The model should be self-input generating, seeking minimum requirement of external resources from the market,

b) The model should be able to generate year-round employment and perennial income in contrast to the seasonal nature of income,

c) Waste of one component should be wealth for another component, implying that complementarity should exist between/among the various components,

d) The model should be energy efficient, economically viable and socially acceptable,

e) Rationality should be maintained among economic, ecological and social dimensions of IFS models,

f) The model should be capable of sustaining the farm family's nutritional needs as recommended by the Indian Council of Medical Research,

g) Ecosystem services should be taken into consideration while designing the IFS Models,

h) The model should effectively reduce greenhouse gas emission and check soil and nutrient erosions, and

i) IFS model is more likely to sustain when it is built over traditional systems, where due importance is given to indigenous crops and bio-diversity.

Significance of Integrated Farming System Research

1. Food and Nutritional Security

Sustaining household food has been an issue of prime importance to majority of the farmers belonging to the categories of small and marginal holdings. Farmers under these categories are economically poor, work in diverse, location specific, risk-prone environment invite attention to develop technologies for inter-disciplinary enterprises as farming system research. It envisages development of technologies by integration of allied enterprises with existing crop components for harnessing the productivity and profitability of each enterprise by taking into account the farmer priorities, socio-economic conditions and resource available with them. Thus farming system is the appropriate combination of farm enterprises viz., crops, livestock, fishery, poultry, mushroom, bee keeping and the means available to the farms serve as valuable manures for recycling within the system. It further ensures interaction with the environment without dislocation of the ecological and socio economic balance on one hand and attempt to meet the national goals on the other. Integration of economically viable, socially accepted and environmentally safe enterprises, along with existing farming systems has been found enterprising in many of the states including Tamil Nadu, Maharashtra, Punjab and Uttar Pradesh. While, assessing the need of farming system approach in Indian context described that having attained food security for the nation, the future strategy would necessitate a change in priorities through diversification to encompass farm level horticulture, agro-forestry, animal production and fisheries *etc.* in to the subsistence level farming avocation. This calls for holistic configuration of different farm enterprises.

2. Income and Livelihood Development

Integrated Farming System (IFS) is the main source of livelihood of nearly 65 per cent of rural masses dependant on agriculture. It is holistic approach and an integration of different farm enterprises to which a farm family allocates its resources in order to efficiently utilize the existing enterprises for enhancing the productivity and profitability of the farm. The input cost in traditional farming was more or less constant in the subsequent years, while it decreased by 25-35 per cent in case of IFS models and thus the IFS model proves to be profitable in the present scenario of decreased landholding (Sheikh *et al.*, 2021). The integration of farm enterprises provides the better resource generation and livelihood base in terms of increased food production, higher net income, improved productivity and reduced income imbalance between the agricultural labourer and urban factory worker. Sustainable intensification and diversification of IFS in small farms can achieve the food and nutritional security, dynamic resource integration and overall enhanced outcomes of sustainable livelihoods at the community level through an integrated process-based approach (Gill *et al.*, 2009). Thousands of small and marginal family farmers in resource-poor regions in Asia and Africa have converted their farming to this sustainable farming system to diversify farm production, increase cash income and improve the quality and quantity of food produced and the exploitation of unutilized resources. It usually takes three to four years to establish a well-integrated farm with market linkages to ensure nutrition and the livelihood of a family.

3. Sustainable Development

The declining profitability of farming is a major concern especially in the arid and semi-arid regions. The IFS has great promise for improving the profitability by reducing the cost of production and/or enhancement of productivity by adopting sustainable management (Manjunath and Itnal, 2003). The IFS has the ability to reduce cost of production through recycling wastes as by-products of one enterprise

Table 16.1: Food and Nutritional Security Provided by Integrated Farming System

Farming System	*Location and Area of the Model*	*Cereals (kg)*	*Pulses (kg)*	*Oilseeds (kg)*	*Milk (liters)*	*Fruits (kg)*	*Vege-tables (kg)*	*Eggs (nos.)*	*Green Fodder (kg)*	*Dry Fodder (kg)*	*Mush-room (kg)*	*Fuel Wood (kg)*
Crops + Horticulture + Dairy + Fish + Boundary plantations	Modipuram and 0.70 ha	3040	282	363	4228	5560	2960	-	20940	4095	457	1194
Crops + Dairy + Horticulture + Goat + Vermicompost + Azolla + Boundary plantations + Kitchen gardening	Siraguppa and 1.00 ha	4959	289.9	44.88	1403	106	1740	-	5423	7281	-	114.2
Crops + Dairy + Horticulture + Fishery + Poultry + Duckery + Goat + Apiary + Vermicompost + Biogas	Jorhat and 1.00 ha	2368	100	363	3600	270	1758	450	12180	3000	60	220
Crops + Dairy + Horticulture (fruits + vegetables)	Ludhiana and 1.00 ha	3513	127	138	12635	487	5516	2500	19780	4966	38	-

Farming System	Location and Area of the Model	Cereals (kg)	Pulses (kg)	Oilseeds (kg)	Milk (liters)	Fruits (kg)	Vegetables (kg)	Eggs (nos.)	Green Fodder (kg)	Dry Fodder (kg)	Mushroom (kg)	Fuel Wood (kg)
Crops + Dairy + Goat + Duckery + Vermicompost + Boundary plantations	Sabour and 1.00 ha	5220	135	140	3472	475	2065	1125	27800	7140	148	6.75
Crops + Dairy + Horticulture + Goat + Poultry + Biocompost	Durgapur and 1.45 ha	1738	239	283	5902	162	1581	675	26560	3761	-	810

Source: Ravisankar *et al.* (2019).

Table 16.2: Net Income (USD) of Cropping System and Integrated Farming System

Sl. No.	Single Enterprise	Net Income (USD)	Integrated Farming System	Net Income (USD)	Per cent Improvement	References
1.	Cotton + Pigeon pea,Maize + Pigeon pea	486	Rice, Cotton, Pigeon pea, Buffalo, Cow, Bullock	947	95	Rao *et al.* (2017)
2.	Rice-Rice	288	Rice + Fish + Poultry	648	125	Channabasavanna and Biradar (2007)
3.	Cropping alone	483	Cropping + Fish + Goat	1748	262	Jayanthi *et al.* (2003)
4.	Rice	306	Rice + Azolla + Fish	424	38	Balusamy *et al.* (2003)
5.	Arable farming	321	Crop + Dairy (2 cows) + 15 Goats + 10 Poultry + 10 Duck + Fish	599	86	Tiwari *et al.* (1999)
6.	Rice-Wheat	1083	Rice-Wheat + Dairy + Fish + Piggery	1509	39	Gill *et al.* (2005)
7.	Rice-Maize	780	Rice-Maize + Fish + Duck + Goat	2211	184	Kumar *et al.* (2012)
8.	Agro-Pastoral system	194	Agro-Pastoral System + Livestock	540	178	Gill *et al.* (2009)
9.	Rice-Wheat	1427	Rice-Wheat + Dairy + Kitchen garden + Mushroom	5071	255	Walia *et al.* (2016)
10.	Conventional system (Rice-Maize)	1845	Integrated farming system	8122	340	Goverdhan *et al.* (2020)
11.	Crop alone	1400	Paddy-Groundnut + Maize + Vegetable + Fruit + Piggery + Poultry + Fishery	4847	246	Patra and Samal (2018)
12.	Horticulture system	769	Horticulture + Fish	1659	116	Laxmi *et al.* (2015)
13.	Arable farming	204	Horti-Pasture	569	179	Singh *et al.* (2017)
14.	Tuber crops	758	Tuber crops + Vegetables + Field crops + Livestock	5383	610	Shankar *et al.* (2018)

Table 16.3: Employment (Man days) of Cropping System and Integrated Farming System

Sl. No.	Single Enterprise	Employment (Man days)	Integrated farming system	Employment (Man days)	Per cent Improvement	References
1.	Cotton + Pigeon pea, Maize + Pigeon pea	649	Rice, Cotton, Pigeon pea, Buffalo, Cow, Bullock	808	24	Rao *et al.* (2017)
2.	Rice-Rice	437	Rice + Fish + Poultry	618	41	Channabasavanna and Biradar (2007)
3.	Cropping alone	369	Cropping + Fish + Goat	576	56	Jayanthi *et al.* (2003)
4.	Rice-Wheat	115	Rice- Wheat + Vegetable + Goat + Poultry + Mushroom	287	150	Kumar *et al.* (2012)
5.	Makhana	247	Makhana + Fish + Water chestnut	347	40	Kumar *et al.* (2012)
6.	Rice-Wheat	293	Rice-Wheat + Dairy + Kitchen garden + Mushroom	775	165	Walia *et al.* (2016)
7.	Tuber crops	199	Tuber crops + Vegetables + Field crops + Livestock	261	31	Shankar *et al.* (2018)

becomes input to other enterprises (Ravisankar *et al.*, 2007) and minimizing the need for external inputs (Ryschawy *et al.*, 2012). The concept of integrated system research takes various enterprises and resource inputs at the farm into consideration for planning production of crops, selecting cropping systems and combining various enterprises to develop integrated farming systems having sustainable agriculture production.

Table 16.4: Employment Opportunities Provided by Integrated Farming System

	Probable Areas	*Job Opportunities*
1.	**Livestock and fodder component**	Caring for animals, and
		Maintaining fodder block of forages, legumes and azolla unit.
2.	**Organic farming**	Production of organic inputs such as vermicompost, compost *etc.*
		Post-harvest farm to market supply chains, and
		Value addition to organic produce to increase the income.
3.	**Agro-ecotourism**	Linking tourism to traditional,
		Farming landscapes, and
		Developing herbal garden, biodiversity park, fish farm *etc.*
4.	**Management of resources**	Rising nurseries to supply planting material,
		Participatory seed production to fetch higher market price, and
		Planting diverse tree species and maintaining diverse economically important species.

Source: Paramesh *et al.* (2021).

Crop residues can be used for feeding to animal, while enhancing the agricultural productivity should be done through utilization of manure from livestock by intensifying nutrients that improve soil fertility as well as reducing the use of chemical fertilizers (Gupta *et al.*, 2012). Greater sustainability in production on farms due to the integration of diverse enterprises of different economic importance. Recycling of wastes being built in the system helps to reduce dependence on external high-energy inputs thus conserving natural and scarce resources. The farming system provides a progress of money to the farmer in sustainable mean round the year by way of disposal of eggs, edible mushroom, milk, honey, silkworm cocoons, *etc.* This will help to reduce dependency on single enterprises and reduce the risk of money lenders also (Devendra, 2002). Integrated farming system is a resource management strategy to achieve economic and sustained production to meet the diverse requirement of farm household while preserving resource base. IFS can be practiced as micro business by farm youth for attaining regular income. IFS reduces the risk of failure as often one component or one crop based business leads to market instability. The other advantages of IFS include effective recycling of residues within the farm there by reducing the cost of production per unit area (Khan *et al.*, 2012).

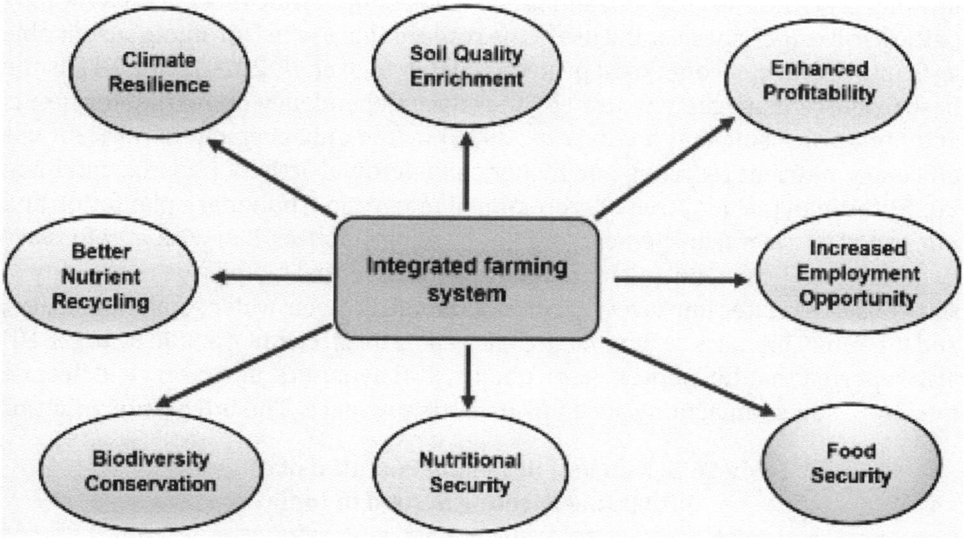

Figure 16.2: Ecosystem Services Provided by Integrated Farming System [*Source*: Paramesh *et al.* (2021)]

4. Employment Generation

The improvement in employment potential under IFS varied from 30 per cent to 485 per cent with a mean increase of 143 per cent over single enterprise (Shankar *et al.*, 2018; Behera and Mahapatra, 1999). The employment generation vary depending on the combination of enterprises chosen. The specialized agriculture practices and mono-cropping increased production cost, risk of crop failure, and lower market price (Manjunath *et al.*, 2017), due to which small and marginal farmers have migrated to neighbouring cities in search of jobs and livelihood (Arunachalam *et al.*, 2019). In this scenario, IFS will be a solution to reduce the economic risk with improved employment generation. The continuous labour requirement for multiple crops and livestock system provides an option for higher employment generation and keeps the farm families engaged in the farm activities. This holds good even during the COVID-19 pandemic for meeting the employment needs of reverse migrants (urban to rural). In IFS, the farm activities are continued round the year, thus, farm families are effectively engaged in farm *viz.,* daily care of animals, fodder block, agro-ecotourism activities, production of organic inputs and their marketing *etc.*

5. Residue Recycling and Soil Health

The intensive agriculture practices resulted in environmental degradation due to excessive use of high energy inputs such as fertilizers and pesticides. The use and recycling of locally available inputs and integrating them with the minimum needed quantities of external inputs would enhance the sustainability of the farming process. The crop and animal system exhibit a wide diversity in cropping patterns, livestock species, and use of the resource base which promote sustainable agriculture and environmental protection (Herridge *et al.*, 2019). Thus, IFS is the best resource management strategy to reduce dependency on market for inputs and to improve soil health (Hu *et al.*, 2016) by the enhancement of nutrient use efficiency, nutrient recycling and higher soil microbial activity (Sujatha and Bhat, 2015) through the adoption of crop-animal integration, boundary plantation and integrated nutrient management/organic farming practices. Likewise, agroforestry system/green leaf manuring in IFS also has the potential to upgrade the quality of soil, conserves water, improves carbon stock, total nitrogen, water stable aggregates, and microbial biomass carbon (Maughan, *et al.*, 2009). Further, Sainju *et al.* (2010) also reported that IFS improves soil quality, SOC dynamics, and crop yield despite moderate soil compaction caused from cattle presence. The integration of crops

Table 16.5: Nutrient Recycling Potential of different Integrated Farming System in India

Integrated Farming System	*Nutrient Recycling (kg ha^{-1})*			*References*
	N	*P_2O_5*	*K_2O*	
Field crops + Fish + Cattle	235.70	192.70	225.20	Kumar *et al.* (2011)
Field crops + Fish + Duck + Goat	110.40	58.70	68.10	
Crop + Fish + Poultry	192.50	119.70	77.80	
Cropping + Dairy + Fishery + Horticulture + Apiary	121.70	226.80	411.90	Singh *et al.* (2012)
Arecanut + Fodder + Dairy	218.00	51.80	33.00	Sujatha and Bhat (2010)
Crops + Dairy + Biogas + Vermicompost + Fishery + Horticulture + Agroforestry	112.16	53.50	114.70	Ravisankar *et al.* (2016)
Crop + Horticulture + Dairy + Vermicompost + Biogas + Fishery	64.00	36.00	41.00	
Crops + Horticulture + Cattle + Fishery + Poultry + Apiary	359.40	140.00	398.60	
Crop + Livestock (2 cows) + Fishery cum Duckery	38.40	33.10	43.00	
Crop + Horticulture + Dairy + Sheep + Poultry	91.00	42.00	75.00	Goverdhan *et al.* (2020)
Crop + Dairy + Fish + Poultry	55.00	17.00	76.00	Paramesh *et al.* (2021)

and livestock not only provides nutrient rich crop residues and animal manure but also reduces dependency on external purchase. Thus, IFS is an approach that produces residues and manure essential for crop growth and to maintain soil health and also helps in adoption of organic farming system.

6. Climate Resilience

The IFS systems of small and marginal farmers are relatively less depending on purchased inputs due to higher recycling potential, and thus, can be less vulnerable to climate change and crop failure due to higher diversity of farm by-products. The IFS system models from different parts of India also reported net negative emission of GHG emission due to higher carbon sequestration (Ravisankar *et al.*, 2019). All the IFS system having boundary plantation with perennial trees or horticulture (fruit crops) as a component emphasizes on increasing the residue recycling, where the tree components offset the negative effect of climate change by sequestrating more carbon into the soil and above-ground biomass of the trees and counterbalanced N_2O emissions. The IFS approach also has capacity to reduce CH_4 absorption through organic farming practices and may have further improved the abundance and activity of methanotrophs (Zhou *et al.*, 2008) and possibly decreased air diffusion that could have impaired CH_4 diffusion (CHEN *et al.*, 2011). So, the management practices such as nutrient management through composting and crop residues, and using legumes for nitrogen fixation, change in cultivation practices like direct seeded rice/SRI method of rice cultivation can increase the resilience of crops to changing climate and also reduces GHG emission. IFS is advocated as a promising strategy to increase agricultural production and rehabilitate degraded pastures while mitigating GHG emissions (Gil *et al.*, 2015). The improved agronomic management and conservation of biodiversity lead to resilient,

Table 16.6: Nutrient Recycling Potential of different Integrated Farming System in India

Location	*Components of Integrated Farming System*	*GHG Emission (kg CO_2 eq. ha^{-1})*
Palampur	Crops + Dairy + Horticulture + Fodder + Vermicompost + Boundary plantation + Kitchen garden	–1787
Jorhat	Crops + Dairy + Horticulture + Fishery + Poultry + Duckery + Goatry + Apiary + Vermicompost + Biogas + Liquid manure + FYM	–3175
Kalyani	Crops + Dairy + Horticulture + Vermicompost + Biogas + Fishery	–4517
Raipur	Crops + Dairy + Horticulture + Poultry + Fishery + Goat + Mushroom + Vermicompost + Boundary plantation + Kitchen garden	–7713
Hyderabad	Crops + Livestock + Horti-Pasture IFS model	–27036

Source: Ravisankar *et al.* (2019).

productive, and sustainable systems and can reduce environmental pollution. The combination of perennial forages with cropping, such as agroforestry, alley cropping, and intercropping, delivers different options for reducing the impact of climate change by improving carbon sequestration and nutrient availability (Bell *et al.*, 2014). Sunderland (2011) opined that the addition of multipurpose trees to the farming system provides both food and income to the small and marginal holders and acts a source of livelihood and sequesters atmospheric carbon.

7. Biodiversity Conservation

The widespread adoption of monoculture *viz.*, rice-wheat, rice-rice and rice-maize systems in India had severely affected the soil biology, increased genetic erosion, depleted groundwater availability and also caused several environmental problems. The small and marginal farmers choose crop diversity on their small farms through considering several factors including increased nutrition, market diversification as well as risk mitigation. The IFS have potential to enhance ecological function through biodiversity restoration as well as expanded whole-system economic and agronomic productivity. Agricultural diversification occurs when a farm or agricultural community adds more plants, plant varieties or animal breeds. The IFS promotes the cultivation of various crops together as intercrops, mixed crops, sequential crops *etc.* which may include annual, perennial and tree crops, and thereby provides ecosystem services from agriculture like nutrient recycling, improved soil quality, reduced economic losses due to crop failure, nitrogen fixation, water penetration and pollination. The farming system also stimulates soil microbial biodiversity with the production of planktons and macro-benthos in rice-fish-duck, rice-duck, and rice-fish system through the addition of compost or manure or by duck droppings over conventional rice production system (Nayak *et al.*, 2018). IFS encourages polyculture of annual and perennial crops, vegetables, flowers, spices, medicinal and aromatic plants *etc.*, integration of livestock or fish with crops, including cover cropping, fodder production, and rotational grazing at the field scale; and thus resulting in heterogeneous landscapes complementary which supports the whole agro-biodiversity (Perfecto *et al.*, 2005). The IFS also promotes non-crop biodiversity, first through ecological management strategies such as the use of minimal fertilizer and insecticides and then through the use of organic management practices. The addition of trees (fruit and timber trees) to the farming system able to provide both income and nutritional security to the farm family and also attracts honeybees, and other insects, in a broader set increases biodiversity, act as windshield, and adds aesthetic value to the farm (Sunderland, 2011). In climate changing scenario where the cultivation of irrigated rice, sugarcane, cotton and other intensive crops in certain parts of India will become risky due to the drought condition, the cultivation of drought-tolerant crops such as millets, short-duration pulses, vegetables, and root crops will become essential to achieve food security (Rufino *et al.*, 2013). So, the IFS with multiple enterprises encourages biodiversity conservation, provides feed, fodder, and fuel, and also reduces risks associated with crop failure.

Integrated Farming System Models Developed throughout India

An integrated farming system consists of a range of resource-saving practices that aim to achieve acceptable profits, high sustained levels of production and preserving the environment, while minimizing the negative effects of intensive farming. The integration has to be made in such a way that product of one component should be the input for other enterprises with high degree of complimentary effects on each other, which in turn reduces the cost of cultivation, provides a regular source of income and employment. IFS also reduces the dependence on chemical inputs and thus helps to cope with aberrant climatic conditions and ensures for the establishment of a climate smart agriculture and maintains sustainability on a long term basis.

Table 16.7: Various IFS Models Developed under AICRP on Integrated Farming System under different Agro-climatic Zones of India

Sl. No.	*NARP Zone (Location of the Centre)*	*Farming System*	*Area (ha)*	*Mean Production (Eq. Yield of Base Crop of the Region)*	*Mean Net Returns (₹)*
A.	**Western Himalayas**				
1.	Chhata (J&K)	Cropping system (0.49 ha) + Dairy (1 cow + 1 buffalo) + Horticulture (0.30 ha) + Apiary (0.001 ha) + Fishery (0.10 ha) + Poultry (0.001 ha) + Vermicompost (0.05 ha) + Biogas (2 m^3)	1.00	12.40	189502
2.	Palampur (HP)	Cropping system (0.75 ha) + Dairy (2 cows) + Horticulture (0.17 ha) + Vermicompost (0.01 ha)	1.00	16.50	78738
3.	Pantnagar (UK)	Cropping system (0.52 ha) + Dairy (2 cows) + Horticulture (0.19 ha) + Agroforestry (0.23 ha) + Vermicompost (0.005 ha)	1.00	24.20	147989
B.	**Eastern Himalayas**				
4.	Umiam (Meghalaya)	Cropping system (0.70 ha) + Horticulture (0.20 ha) + Poultry (600 Broiler + 50 Layer) + Piggery (3 nos.) + Fishery (0.05 ha)	1.00	20.60	155820
C.	**Lower Gangetic Plains**				
5.	Kalyani (WB)	Cropping system (0.42 ha) + Dairy (2 cows) + Horticulture (0.11 ha) + Fishery (0.09 ha) + Vermicompost (0.018 ha)	0.66	26.20	80128

Sl. No.	NARP Zone (Location of the Centre)	Farming System	Area (ha)	Mean Production (Eq. Yield of Base Crop of the Region)	Mean Net Returns (₹)
D.	**Middle Gangetic Plains**				
6.	Varanasi (UP)	Cropping system (0.81 ha) + Dairy (6 cows) + Horticulture (0.06 ha) + Fishery (0.10 ha) + Poultry (200 nos.) + Vermicompost (0.004 ha) + Mushroom (0.002 ha)	1.00	47.00	250465
7.	Patna (Bihar)	Cropping system (0.40 ha) + Dairy (2 cows) + Horticulture (0.10 ha) + Fishery (0.12 ha) + Duck (35 nos.) + Vermicompost (0.01 ha) + Biogas (2 m^3)	0.80	21.50	133367
E.	**Upper Gangetic Plains**				
8.	Modipuram (UP)	Cropping system (0.38 ha) + Dairy (2 buffaloes + 1 cow) + Horticulture (0.30 ha) + Vermicompost (0.01 ha) + Biogas (1.50 m^3) + Mushroom (0.002 ha)	0.70	31.90	192839
F.	**Trans Gangetic Plains**				
9.	Ludhiana (Punjab)	Cropping system (0.64 ha) + Dairy (2 cows) + Horticulture (0.20 ha) + Fishery (0.20 ha) + Agroforestry (0.03 ha) + Vermicompost (0.005 ha) + Apiary (5 boxes)	1.00	35.50	268106
10.	Hisar (Haryana)	Cropping system (0.90 ha) + Dairy (2 buffaloes) + Horticulture (0.06 ha) + Vermicompost (0.01 ha) + Mushroom (0.002 ha)	1.00	30.00	152555
G.	**Eastern Plateau and Hills**				
11.	Raipur (CG)	Cropping system (0.60 ha) + Dairy (2 cows) + Horticulture (0.22 ha) + Fishery (0.072 ha) + Poultry (200 nos.) + Vermicompost (0.003 ha) + Mushroom (0.003 ha) + Biogas (1.50 m^3)	1.00	16.50	144126
12.	Ranchi (JH)	Cropping system (0.80 ha) + Dairy (2 cows) + Fishery (0.10 ha) + Vermicompost (0.026 ha) + Mushroom (0.014 ha)	1.00	14.20	107909

Sl. No.	NARP Zone (Location of the Centre)	Farming System	Area (ha)	Mean Production (Eq. Yield of Base Crop of the Region)	Mean Net Returns (₹)
H.	**Central Plateau and Hills**				
13.	Jabalpur (MP)	Cropping system (0.90 ha) + Dairy (3 cows) + Horticulture (0.0024 ha) + Fishery (0.06 ha) + Poultry (300 nos.) + Vermicompost (0.0039 ha) + Mushroom (0.0024 ha)	1.00	16.05	80128
14.	Durgapura (RJ)	Cropping system (1.00 ha) + Dairy (2 cows) + Horticulture (0.25 ha) + Poultry (50 nos.) + Goat (6 nos.)	1.45	19.38	177839
I.	**Western Plateau and Hills**				
15.	Akola (MH)	Cropping system (0.70 ha) + Horticulture (0.25 ha) + Goat (12 nos.) + Mushroom (0.0024 ha) + Vermicompost (0.004 ha)	1.00	25.10	104858
16.	Rahuri (MH)	Cropping system (0.72 ha) + Dairy (2 cows) + Horticulture (0.20 ha) + Poultry (400 nos.) + Vermicompost (0.01 ha)	1.00	28.60	184417
17.	Parbhani (MH)	Cropping system (0.72 ha) + Dairy (2 cows + 1 buffalo) + Horticulture (0.20 ha) + Poultry (150 nos.) + Vermicompost (0.05 ha)	1.00	20.80	86236
J.	**Southern Plateau and Hills**				
18.	Rajender-nagar (TS)	Cropping system (0.70 ha) + Dairy (3 buffaloes) + Horticulture (0.20 ha) + Poultry (20 nos.) + Goat (35 nos.) + Vermicompost (0.015 ha)	1.00	26.50	97341
19.	Coimbatore (TN)	Cropping system (1.02 ha) + Dairy (2 buffaloes + 1 cow) + Horticulture (0.16 ha) + Vermicompost (0.005 ha)	1.20	27.20	241728
20.	Kathalgere (KA)	Cropping system (0.65 ha) + Dairy (1 buffalo + 2 cows) + Horticulture (0.15 ha) + Sheep (13+1 nos.) + Vermicompost (0.010 ha)	1.00	60.20	188608

Sl. No.	NARP Zone (Location of the Centre)	Farming System	Area (ha)	Mean Production (Eq. Yield of Base Crop of the Region)	Mean Net Returns (₹)
21.	Siruguppa (KA)	Cropping system (0.74 ha) + Dairy (2 buffaloes + 2 cows) + Horticulture (0.20 ha) + Fishery (0.045 ha.) + Vermicompost (0.0034 ha)	1.00	14.30	148091
K.	**East Coast Plains and Hills**				
22.	Bhubaneswar (OD)	Cropping system (0.32 ha) + Horticulture (0.19 ha) + Dairy (2 cows) + Fishery (0.33 ha) + Poultry (380 nos.) + Duck (20 nos.) + Agroforestry (0.094 ha) + Vermicompost (0.0033 ha) + Mushroom(0.010 ha) + Biogas (1.50 m^3) + Apiary (2 boxes)	1.25	22.80	75129
23.	Thanjavur (TN)	Cropping system (0.61 ha) + Horticulture (0.10 ha) + Dairy (1 cow + 1 buffalo) + Fishery (0.08 ha) + Poultry (150 nos.) + Vermicompost (0.002 ha)	0.80	15.00	138325
L.	**West Coast Plains and Hills**				
24.	Karjat (MH)	Cropping system (0.50 ha) + Dairy (3 cows) + Horticulture (0.40 ha) + Poultry (90 nos.) + Goat (6 nos.) + Vermicompost (0.0018 ha)	1.00	23.60	99000
25.	Karmana (Kerala)	Cropping system (0.20 ha) + Dairy (1 cow + 1 buffalo) + Duck (150 nos.) + Fishery (0.02 ha) + Vermicompost (0.0004 ha)	0.25	8.20	60555
27.	Goa (Goa)	Plantation crops (cashew, coconut, arecanut with intercrops) (0.70 ha) + Piggery + Poultry + Seedling production unit	0.79	24.00	86756
M.	**Gujarat Plains and Hills**				
28.	S.K. Nagar (Gujarat)	Cropping system (0.70 ha) + Dairy (2 buffaloes) + Horticulture (0.25 ha) + Vermicompost (0.01 ha)	1.00	36.90	237607

Sl. No.	NARP Zone (Location of the Centre)	Farming System	Area (ha)	Mean Production (Eq. Yield of Base Crop of the Region)	Mean Net Returns (₹)
N.	**Islands**				
29.	Port Blair (A&N)	Rice (0.50 ha) + Vegetables/ Flowers and Fish in BBF system (0.40 ha) + Livestock (2 cows) + Vermicompost	1.00	27.00	145000
30.	Port Blair (A&N)	Plantation crops (coconut) + Spices (clove, nutmeg) + Pineapple + Tapioca + Sweet potato (0.90 ha) + Piggery (4 nos.) + Fish (0.06 ha) + Poultry (25 nos.) + Lined pond for Azolla (0.01 ha) + Vermicompost	1.00	28.00	224000

Source: Panwar *et al.* (2018).

Conclusion

Integrated Farming System (IFS) is a holistic approach having interactions among the different enterprises. It combines different activities of agronomic crop farming with horticulture, animal husbandry, fisheries, forestry and other related farm activities on the same field to solve food and nutritional security problems resulting from decreasing food productivity and climate change. IFS itself is important for efficient management of available resources at the farm level, to generate adequate income and employment for the rural poor, protection of the environment, and livelihood security. As IFS relies more on farm and local resources for which IFS was found more sustainable and profitable. Organic residues when added in the form of plant and animal wastes helps to improve the soil health and hence productivity with lesser environmental hazards and increased margin for profit. IFS provides scope to accommodate more crops, livestock, trees, honeybee, *etc.* for which the carbon sink in the system is more and more resilient to climate vagaries and can be a potential approach to mitigate climate change. Providing awareness about benefits of IFS to farmers, government policy and subsidy support is essential to promote large-scale adoption of region specific IFS models.

REFERENCES

Altieri, M.A., Funes-Monzote, F.R. and Petersen, P. (2012). Agroecologically efficient agricultural systems for smallholder farmers: contributions to food sovereignty. *Agronomy for Sustainable Development.* **32**(1): 1-13.

Ansar, M., and Fathurrahman, F. (2018). Sustainable integrated farming system: A solution for national food security and sovereignty. IOP Conference Series: *Earth and Environmental Science*. **157**: 012061.

Balusamy, M., Shanmugham, P.M. and Baskaran, R. (2003). Mixed farming an ideal farming. *Intensive Agriculture*. **41**(11 and 12): 20-25.

Barbosa, F.A., Soares Filho, B.S., Merry, F.D., de Oliveira Azevedo, H., Costa, W.L.S., Coe, M.T., da Silveira Batista, E., Maciel, T.C., Sheepers, L.C. and de Oliveira, A.R. (2015). Cenários para a pecuária de corte amazônica. *IGC/UFMG*, 1-154.

Behera, U.K. and France, J. (2016). Integrated farming systems and the livelihood security of small and marginal farmers in India and other developing countries. *Advances in Agronomy*. **138:** 235-282.

Behera, U.K. and Mahapatra, I.C. (1999). Income and employment generation for small and marginal farmers through integrated farming systems. *Indian Journal of Agronomy*. **44**(3): 431-439.

Behera, U.K., Dass, A., Rautaray, S.K., Choudhary, A.K. and Rana, D.S. (2013). Integrated farming system research in India: an overview. *Integrated Farming Systems for Enhancing Livelihood of Small and Marginal Farmers*. Division of Agronomy, Indian Agricultural Research Institute, New Delhi, pp. 40-78.

Behera, U.K., Kaechele, H. and France, J. (2015). Integrated animal and cropping systems in single and multi-objective frameworks for enhancing the livelihood security of farmers and agricultural sustainability in Northern India. *Animal Production Science*. **55**(10): 1338-1349.

Bell, L.W. and Moore, A.D. (2012). Integrated crop-livestock systems in Australian agriculture: Trends, drivers and implications. *Agricultural Systems*. **111:** 1-12.

Bell, L.W., Moore, A.D. and Kirkegaard, J.A. (2014). Evolution in crop-livestock integration systems that improve farm productivity and environmental performance in Australia. *European Journal of Agronomy*. **57:** 10-20.

Bhargavi, B. and Behera, U.K. (2020). Securing the livelihood of small and marginal farmers by diversifying farming systems. *Current Science*. **119**(5): 854-860.

Bohra, J.S., Kushwaha, S., Singh, S.R., Singh, R.P., Singh, S.K., Singh, S., Mishra, P., Tripathi, A.K., Singh, N. and Singh, A. (2014). Watershed based integrated farming system modules for livelihood security of small and marginal farmers of *Vindhyan* region, In: *Proceedings of National Symposium on Agricultural Diversification for Sustainable Livelihood and Environmental Security*, held during 18-20 November 2014 at Ludhiana, Punjab, pp 547-548.

Channabasavanna, A.S. and Biradar, D.P. (2007). Relative performance of different rice-fish-poultry integrated farming system models with respect to system productivity and economics. *Karnataka Journal of Agricultural Sciences*. **20**(4): 706.

Chen, W., Wolf, B., Zheng, X., Yao, Z., Butterbach-Bahl, K., Brüggemann, N., Liu, C., Han, S. and Han, X. (2011). Annual methane uptake by temperate semi-arid steppes as regulated by stocking rates, aboveground plant biomass and top soil air permeability. *Global Change Biology*. **17**(9): 2803-2816.

Das, A., Choudhury, B.U., Ramkrushna, G.I., Tripathi, A.K., Singh, R.K., Ngachan, S.V., Patel, D.P., Layek, J. and Munda, G.C. (2013). Multiple use of pond water for enhancing water productivity and livelihood of small and marginal farmers. Indian Journal of Hill Farming. **26**(1): 29-36.

Das, A., Layek, J., Ramkrushna, G.I. and Babu, S. (2018). Integrated organic farming system in North East India. In: A. Das, K.P. Mohapatra, S.V. Ngachan, A. S. Panwar, D.J. Rajkhowa, G.I. Ramkrushna and L. Jayanta (Eds.), Conservation Agriculture for advancing food security in changing climate, pp. 1-30, Today and Tomorrow Printers and Publishers.

Devendra, C. (2002). Crop-animal systems in Asia: future perspectives. *Agricultural Systems*. **71**(1-2): 179-186.

Devendra, C. and Thomas, D. (2002). Small-holder farming systems in Asia. *Agricultural Systems*. **71**(1-2): 17-25.

Dey, A., Sarma, K., Kumar, U., Mohanty, S., Kumar, T. and Bhatt, B.P. (2019). Prospects of rice-fish farming system for low lying areas in Bihar, India. *Organic Agriculture*. **9**(1): 99-106.

Dong, Y., Zhang, S., Qi, Y., Chen, Z. and Geng, Y. (2000). Fluxes of CO_2, N_2O and CH_4 from a typical temperate grassland in Inner Mongolia and its daily variation. *Chinese Science Bulletin*. **45**(17): 1590-1594.

Ganesan, G., Chinnasamy, K.N., Bala Subramanian, A. and Manickasundram, P. (1990). Studies on rice-based farming system with duck cum fish culture in deltaic region of Thanjavur district, Tamil Nadu. *Farming Systems Newsletter*. **1**(2): 14.

Gil, J., Siebold, M. and Berger, T. (2015). Adoption and development of integrated crop-livestock-forestry systems in Mato Grosso, Brazil. *Agriculture, Ecosystems and Environment*. **199:** 394-406.

Gill, M.S., Samra, J.S. and Singh, G. (2005). Integrated farming system for realizing high productivity under shallow water table conditions. Research Bulletin of Punjab Agriculture University (Punjab Agricultural University, Ludhiana, Punjab, India).

Gill, M.S., Singh, J.P. and Gangwar, K.S. (2009). Integrated farming system and agriculture sustainability. *Indian Journal of Agronomy*. **54**(2): 128-139.

GOI. (2014). District Census Hand Book. Office of Registrar General and Census Commissioner of India, Ministry of Home Affair, Government of India.

Gopinath, K.A., Dixit, S., Ravindra Chary, G., Srinivasarao, C., Osman, M., Raju, B.M.K., Ramana, D.B.V., Venkatesh, G., Grover, M. and Maheswari, M. (2013). Improving the rainfed farming systems of small and marginal farmers in Anantapur and Adilabad Districts of Andhra Pradesh, Central Research Institute for Dryland Agriculture, p. 46.

Goverdhan, M., Kumari, C.P., Reddy, G.K., Sridevi, S., Alibaba, M.D., Chiranjeevi, K. and Kumar, M.S. (2020). Evaluation of integrated farming system model for resource recycling and livelihood security of small and marginal farmers of Telangana State, India. *Current Journal of Applied Science and Technology.* **39**(34): 17-26.

Hens, L. and Begossi, A. (2008). Diversity and management: from extractive to farming systems. *Environment, Development and Sustainability.* **10**(5): 559-563.

Herridge, D.F., Win, M.M., Nwe, K.M.M., Kyu, K.L., Win, S.S., Shwe, T., Min, Y.Y., Denton, M.D. and Cornish, P.S. (2019). The cropping systems of the Central Dry Zone of Myanmar: Productivity constraints and possible solutions. *Agricultural Systems.* **169:** 31-40.

Hu, L., Zhang, J., Ren, W., Guo, L., Cheng, Y., Li, J., Li, K., Zhu, Z., Zhang, J., Luo, S., Cheng, L., Tang, J. and Chen, X. (2016). Can the co-cultivation of rice and fish help sustain rice production? *Scientific Reports.* **6**(1): 28728.

Jayanthi, C., Balusamy, M., Chinnusamy, C. and Mythili, S. (2003). Integrated nutrient supply system of linked components in lowland integrated farming system. *Indian Journal of Agronomy.* **48**(4): 41-46.

Jayanthi, C., Mythili, S. and Chinnasamy, C. (2002). Integrated farming systems – A viable approach for sustainable productivity, profitability and resource recycling under low land farms. *Journal of Ecobiology.* **14**(2): 143-148.

Jhana, K.M. and Pems, D.E. (2011). The impact of integrated aquaculture-agriculture on small scale farm sustainability and farmers livelihoods: Experience from Bangladesh. *Agricultural Systems,* **104:** 392-402.

John, J. (2014). Homestead farming in Kerala: a multi-faceted land-use system. *Review of Agrarian Studies.* **4:** 80-105.

Korikanthimath, V.S. and Manjunath, B.L. (2009). Integrated farming systems for sustainability in agricultural production. *Indian Journal of Agronomy.* **54**(2): 140-148.

Kumar, A. and Ghosh, A. (2021). Integrated Farming System: An Overview. *Just Agriculture.* **2**(2): 7-8.

Kumar, R., Pandey, S. and Pandey, A. (2006). Plant roots and carbon sequestration. *Current Science.* **91**(7): 885-890.

Kumar, S., Bhatt, B.P., Dey, A., Shivani Kumar, U., Idris, M., Mishra, J.S. and Kumar, S. (2018). Integrated farming system in India: Current status, scope and future prospects in changing agricultural scenario. *Indian Journal of Agricultural Sciences.* **88**(11): 1661-1675.

Kumar, S., Bhatt, B.P., Dey, A., Shivani, A., Kumar, U., Idris, M., Mishra, J.S. and Kumar, S. (2018). Integrated farming system in India: Current status, scope and future prospects in changing agricultural scenario. *Indian Journal of Agricultural Sciences*. **88**(11): 13-27.

Kumar, S., Singh, S.S. and Dey, A. (2011). Integrated farming systems for Eastern India. *Indian Journal of Agronomy*. **56**(4): 297-304.

Kumar, S., Subash, N., Shivani, S., Singh, S. S., Dey, A., Kumar, S., Subash, N., Shivani, S., Singh, S.S. and Dey, A. (2012). Evaluation of different components under integrated farming system for small and marginal farmers under semi-humid climatic environment. *Experimental Agriculture*. **48**(3): 399.

Kumar, S.S. and Narayanagowda, K. (2017). Development of scale to measure attitude of schedule caste farmers towards integrated farming system in southern Karnataka and its application. *Mysore Journal of Agricultural Sciences*. **51**(2): 297-302.

Laxmi, L., Gupta, N., Singh, M. and Sharma, R.P. (2015). Evaluation of Integrated Horticulture-cum-Fish Farming in Malwa Region of Madhya Pradesh, India. *Current World Environment*. **10**(2): 667-671.

Liu, C., Holst, J., Brüggemann, N., Butterbach-Bahl, K., Yao, Z., Yue, J., Han, S., Han, X., Krümmelbein, J., Horn, R. and Zheng, X. (2007). Winter-grazing reduces methane uptake by soils of a typical semi-arid steppe in Inner Mongolia, China. *Atmospheric Environment*. **41**(28): 5948-5958.

Mahapatra, I.C. and Behera, U.K. (2004). Methodologies of farming systems research. *In*: Panda, D., Sasmal, S., Nayak, S.K., Singh, D.P., Saha, S. (Eds.), *Recent advances in rice-based farming systems*, 17-19th November 2004. Cuttack, Orissa, Central Rice Research Institute, pp. 79-113.

Mahapatra, I.C. and Behera, U.K. (2011). Rice (*Oryza sativa*)-based farming systems for livelihood improvement of Indian farmers. *Indian Journal of Agronomy*. **56**(1): 1- 19.

Manjunath, B. L. and Itnal, C.J. (2003). Farming system options for small and marginal holdings in different topographies of Goa. *Indian Journal of Agronomy*. **48**(1): 4-7.

Manjunath, B.L. and Singh, S.P. (2005). Systems approach in coconut for higher productivity and profitability. *Technical Bulletin No. 6*, ICAR Research Complex for Goa, pp. 1- 66.

Manjunath, B.L., Paramesh, V., Mahajan, G.R., Das, B., Reddy, K.V., Chakurkar, E.B. and Singh, N.P. (2017). Sustainability through resource recycling, soil fertility and carbon sequestration from integrated farming systems in west coast India. *The Bioscan*. **12:** 1-6.

Manjunath, B.L., Paramesh, V., Mahajan, G.R., Reddy, K.V., Das, B. and Singh, N.P. (2018). A five years study on the selection of rice-based cropping systems in

Goa for west coast region of India. *Journal of Environmental Biology*. **39**(3): 393-399.

Maughan, M.W., Flores, J.P.C., Anghinoni, I., Bollero, G., Fernández, F.G. and Tracy, B.F. (2009). Soil quality and corn yield under crop-livestock integration in Illinois. *Agronomy Journal*. **101**(6): 1503-1510.

Meena, M.S., Singh, K.M., Bhatt, B.P. and Kumar, U. (2013). Gender perspective in integrated farming system. ICAR Research Complex for Eastern Region.

Mohanty, R.K., Jena, S.K., Kumar, A., Sahoo, N. and Roy Chowdhury, S. (2008). Rice-fish culture: an ingenious agricultural heritage system, ICAR-Water Technology Centre for Eastern Region.

Nath, S.K. and Barik, K.C. (2013). Increasing incomes of resource poor farm families through an integrated farming system in the CDR ecosystem. *International Journal of Rural Studies*. **20**(1): 1-6.

Nayak, P.K., Nayak, A.K., Panda, B.B., Lal, B., Gautam, P., Poonam, A., Shahid, M., Tripathi, R., Kumar, U., Mohapatra, S.D. and Jambhulkar, N.N. (2018). Ecological mechanism and diversity in rice-based integrated farming system. Ecological Indicators. 91: 359-375.

Nedunchezhiyan, M., Jata, S. K., Gowda, K.H., Chauhan, V.B.S., Bansode, V.V. and Mukherjee, A. (2016). Livelihood improvement through tuber crops based integrated farming system. Root and Tuber Crops Based Integrated Farming System: A Way Forward to Address Climate Change and Livelihood Improvement, p. 22.

Panda, N. (2016). Backyard Poultry – A Viable option for Poverty Alleviation. Root and Tuber Crops Based Integrated Farming System: A Way Forward to Address Climate Change and Livelihood Improvement, p. 133.

Pandey, P.R., Gupta, J.K., Narvariya, R.K., Meena, S.C. and Narwariya, D. (2019). Constraints faced by farmers in adoption of integrated farming system in Vindhyan Plateau of Madhya Pradesh. Plant Archives. **19**(2): 512-514.

Pandey, R.N. and Bhogal, T.S. (1980). Prospects of increasing income and employment on mixed farms. *Indian Journal of Agricultural Economics*. **35:** 144-151.

Panwar, A., Ravisankar, N., Shamim, M. and Prusty, A. (2018). Integrated Farming Systems: A Viable Option for Doubling Farm Income of Small and Marginal Farmers. *Bulletin of the Indian Society of Soil Science*. **32:** 66-88.

Parajuli, R., Dalgaard, T. and Birkved, M. (2018). Can farmers mitigate environmental impacts through combined production of food, fuel and feed? A consequential life cycle assessment of integrated mixed crop-livestock system with a green biorefinery. *Science of the Total Environment*. **619-620:** 127-143.

Paramesh, V., Arunachalam, V. and Nath, A.J. (2019). Enhancing ecosystem services and energy use efficiency under organic and conventional nutrient management system to a sustainable arecanut based cropping system. *Energy*. **187:** 115902.

Paramesh, V., Chakurkar, E.B., Bhagat, T., Sreekanth, G.B., Kumar, H.B.C., Rajkumar, S., Gokuldas, P.P., Mahajan, G.R., Manohara, K.K. and Ravisankar, N. (2021). Impact of integrated farming system on residue recycling, nutrient budgeting and soil health. *Indian Journal of Agricultural Sciences*. **91**(1): 44-48.

Paramesh, V., Parajuli, R., Chakurkar, E.B., Sreekanth, G.B., Kumar, H.B.C., Gokuldas, P.P., Mahajan, G.R., Manohara, K.K., Viswanatha, R.K. and Ravisankar, N. (2019). Sustainability, energy budgeting, and life cycle assessment of crop-dairy-fish-poultry mixed farming system for coastal lowlands under humid tropic condition of India. *Energy*. **188:** 116101.

Paramesh, V., Ravisankar, N., Behera, U., Arunachalam, V., Kumar, P., Solomon Rajkumar, R., Dhar Misra, S., Mohan Kumar, R., Prusty, A.K., Jacob, D., Panwar, A.S., Mayenkar, T., Reddy, V.K. and Rajkumar, S. (2022). Integrated farming system approaches to achieve food and nutritional security for enhancing profitability, employment and climate resilience in India. *Food and Energy Security*. **11:** 1-16.

Paroda, R. (2012). Reorienting agricultural research for development to address emerging challenges in agriculture. *Journal of Research*. **49:** 134-138.

Patel, K.M., Patel, P.K., Desai, L.J., Patel, K.N., Patel, S.A. and Chaudhary, H.L. (2020). Integrated farming systems for livelihood security of small and marginal farmers. *Multilogic in Science*. **10**(33): 600-603.

Patra, S. and Samal, P. (2018). Integrated farming system in India: A holistic approach to magnify the economic status of innovative farmers. *Journal of Pharmacognosy and Phytochemistry*. **7**(3): 3632-3636.

Perfecto, I., Vandermeer, J., Mas, A. and Pinto, L.S. (2005). Biodiversity, yield and shade coffee certification. *Ecological Economics*. **54**(4): 435-446.

Purnomo, S.H., Sari, A.I., Emawati, S. and Rahayu, E.T. (2021). Factors influencing the adoption of integrated crop-livestock to support land conservation of organic agriculture in Mojosongo area, Karanganyar, Indonesia. *IOP Conference Series: Earth and Environmental Science*. **724**(1): 12049.

Pushpa, J. (2010). Constraints in various integrated farming systems. *Agriculture Update*. **5**(3/4): 370-374.

Radha, Y., Eshwaraprasad, Y. and Vijayabhinandana, B. (2000). Study on income and employment generation on agricultural based livestock farming system. VIII Annual Conference of AERA, pp. 28-29.

Ramrao, W. Y., Tiwari, S. P. and Singh, P. (2006). Crop-livestock integrated farming system for the Marginal farmers in rainfed regions of Chhattisgarh in Central India. *Livestock Research for Rural Development*. Volume 18, Article #102. Retrieved September 2, 2022, from http://www.lrrd.org/lrrd18/7/ramr18102.htm

Rao, C.A.R., Raju, B.M.K., Samuel, J., Dupdal, R., Reddy, P.S., Reddy, D.Y., Ravindranath, E., Rajeshwar, M. and Rao, C.S. (2017). Economic analysis of farming systems. *Agricultural Economics Research Review*. **30:** 37-45.

Rautaray, S.K., Das, P.C., Sinhababu, D.P., Sinhababu, D.P. and Singh, B.N. (2005). Increasing farm income through rice-fish based integrated farming system in rainfed lowlands of Assam. *Indian Journal of Agricultural Science*. **75**(2): 79-82.

Ravisankar, N., Singh, P., Mishra, R.P., Prusty, A.K., Shamim, M., Singh, R., Tripathi, D. and Mohan, B. (2016). Annual Report of AICRP on Integrated farming system.

Ravisankar, N., Singh, P., Mishra, R.P., Prusty, A.K., Shamim, M., Singh, R., Tripathi, D. and Mohan, B. (2019). Annual Report of AICRP on Integrated farming system.

Reddy, D.V.S. and Biddappa, C.C. (2000). Coconut-based cropping/farming systems in India. Journal of Plantation Crops. 28(1): 1-18.

Roy, S.K. (2014). Rural Development in India: What roles do NABARD and RRBs play? *Journal of Business Management Social Sciences Research*. **3**(9): 1-15.

Rufino, M.C., Thornton, P.K., Ng'ang'a, S.K., Mutie, I., Jones, P.G., van Wijk, M.T. and Herrero, M. (2013). Transitions in agro-pastoralist systems of East Africa: Impacts on food security and poverty. *Agriculture, Ecosystems and Environment*. **179:** 215-230.

Sachinkumar, T.N., Basavaraj, H., Kunnal, L.B., Kulkarni, G.N., Mahajanashetty, S.B., Hunsyal, C.S. and Hosamani, S.V. (2012). Economics of farming System in Northern Transitional Zone of Karnataka. *Karnataka Journal of Agricultural Sciences*. **25**(3): 350-358.

Sainju, U.M., Stevens, W.B., Caesar-TonThat, T. and Jabro, J.D. (2010). Land use and management practices impact on plant biomass carbon and soil carbon dioxide emission. *Soil Science Society of America Journal*. **74**(5): 1613-1622.

Salton, J.C., Mercante, F.M., Tomazi, M., Zanatta, J.A., Concenço, G., Silva, W.M. and Retore, M. (2014). Integrated crop-livestock system in tropical Brazil: Toward a sustainable production system. *Agriculture, Ecosystems and Environment*. **190:** 70-79.

Schönbach, P., Wolf, B., Dickhöfer, U., Wiesmeier, M., Chen, W., Wan, H., Gierus, M., Butterbach-Bahl, K., Kögel-Knabner, I., Susenbeth, A., Zheng, X. and Taube, F. (2012). Grazing effects on the greenhouse gas balance of a temperate steppe ecosystem. *Nutrient Cycling in Agroecosystems*. **93**(3): 357-371.

Shankar, D., Banjare, C. and Sahu, M.K. (2018). Tuber Crops Based Integrated Farming System Studies in Bastar and Kondagaon Districts of Chhattisgarh. *International Journal of Current Microbiology and Applied Sciences*. **7**(9): 1650-1658.

Shanmugasundaram, V.S., Baluswamy, M. and Rangaswamy, A. (1995). Integrated farming system research in Tamil Nadu. *Journal of Farming Systems Research and Development*. **1**(1): 1-9.

Sheikh, M.M., Riar, T.S., Kanak Pervez, A.K.M. (2021). Integrated farming systems: A review of farmers' friendly approaches. *Asian Journal of Agricultural Extension, Economics and Sociology*. **39**(4): 88-99.

Shekinah, D.E. and Sankaran, N. (2007). Productivity, profitability and employment generation in integrated farming systems for rainfed Vertisols of western zone of Tamil Nadu. Indian Journal of Agronomy. **52**(4): 275-278.

Shekinah, D.E., Jayanthi, C. and Sankaran, N. (2005). Physical indicators of sustainability – A farming systems approach for the small farmer in the rainfed Vertisols of the western zone of Tamil Nadu. *Journal of Sustainable Agriculture*. **25**(3): 43-65.

Sheokand R.S., Narinder, S. and Sheoran, O.P. (2000). Comparative economics of different farming systems under paddy-wheat farming sequence in rice-belt of Haryana. *Haryana Agricultural University Research Journal*, **30**(1/2): 49-53.

Siddeswaran, K., Sangetha, S.P. and Shanmugam, P.M. (2012). Integrated farming system for the small irrigated upland farmers of Tamil Nadu. Extended SummariesVolume 3: 3rd International Agronomy Congress held during 26-30 November 2012 at New Delhi, pp. 992-993.

Singh, A., Tanwar, S.P.S., Meghwal, P.R., Kumar, P. and Kumar, P. (2017). Exploring horticulture-based integrated farming for more return. *Indian Horticulture*. **65**(2): 3-7.

Singh, H., Burark, S.S., Sharma, S.K., Jajoria, D.K. and Sharma, R.P. (2017). Economic evaluation of farming systems for agricultural production in southern Rajasthan. *Economic Affairs*. 62(1): 47-53.

Singh, J.P., Gangwar, B., Kochewad, S.A. and Pandey, D.K. (2012). Integrated farming system for improving livelihood of small farmers of western plain zone of Uttar Pradesh, India. *SAARC Journal of Agriculture*. **10**(1): 45-53.

Singh, Rajender, Singh, N., Phogat, S.B., Sharma, U.K., Singh, R. and Singh, N. (1999). Income and employment potential of different farming system. *Haryana Agricultural University Journal of Research*, **29**(3-4): 143-145.

Singh, S.P., Gangwar, B. and Singh, M.P. (2008). Economics of sugarcane-based farming system in western Uttar Pradesh. *Agricultural Economics Research Review*. **21**(1): 109-117.

Solaiappan, U., Subramanian, V. and Sankar, G.R. (2007). Selection of suitable integrated farming system model for rainfed semi-arid vertic-inceptisols in Tamil Nadu. *Indian Journal of Agronomy*. **52**(3): 194-197.

Sujatha, S. and Bhat, R. (2010). Response of vanilla (Vanilla planifolia) intercropped in arecanut to irrigation and nutrition in humid tropics of India. *Agricultural Water Management*. **97**(7): 988-994.

Sujatha, S. and Bhat, R. (2015). Resource use and benefits of mixed farming approach in arecanut ecosystem in India. *Agricultural Systems*. **141:** 126-137.

Sunderland, T. C.H. (2011). Food security: Why is biodiversity important? *International Forestry Review*. 13(3): 265-274.

Surve, U.S., Patil, E.N., Shinde, J.B., Bodake, P.S. and Kadlag, A.D. (2014). Evaluation of different integrated farming systems under irrigated situations of Maharashtra. *Indian Journal of Agronomy*. **59**(4): 518-526.

Tiwari, S.P., Saxena, R.R., Nandeha, K.L., Vardia, H.K., Sharma, R.B. and Rajagopal, S. (1999). Augmentation of economic status of Bastar tribals through integrated (crop, livestock, poultry, duck, fish) farming system. *The Indian Journal of Animal Sciences*. **69**(6): 448-452.

Varughese, K. and Mathew, T. (2009). Integrated farming systems for sustainability in coastal ecosystem. *Indian Journal of Agronomy*. **54** (2): 120-127.

Vinodakumar, S.N., Desai, B.K., Channabasavanna, A.S., Rao, S., Patil, M.G. and Patil, S.S. (2017). Integrated farming system for livelihood security of small farmers of North-East Karnataka. *International Research Journal of Agricultural Economics and Statistics*. **8:** 216-221.

Walia, S.S., Aulakh, C.S., Gill, R.S., Dhawan, V. and Kaur, J. (2016). Intensive integrated farming system approach – A vaccination to cure agrarian crisis in the Punjab. *Indian Journal of Economics and Development*. **12**(1): 451-455.

Wezel, A., Casagrande, M., Celette, F., Vian, J.F., Ferrer, A. and Peigné, J. (2014). Agro-ecological practices for sustainable agriculture – A Review. *Agronomy for Sustainable Development*. **34**(1): 1-20.

Zhou, X.Q., Wang, Y.F., Huang, X.Z., Tian, J.Q. and Hao, Y.B. (2008). Effect of grazing intensities on the activity and community structure of methane-oxidizing bacteria of grassland soil in Inner Mongolia. *Nutrient Cycling in Agroecosystems*. **80**(2): 145-152.

Chapter 17

Integrated Farming Systems for Livelihood Security in Coastal and Island Regions

S. Dam Roy, T.P. Swarnam and A. Velmurugan*

ICAR-Central Island Agricultural Research Institute, Garacharama, Port Blair – 744 105, Andaman and Nicobar Islands
Corresponding author: sibnarayan@gmail.com

ABSTRACT

Integrated farming system provides a better solution to meet the demand for various agricultural commodities and ensure food security by efficient utilization of available resources. In India, especially in coastal plains and island region rice and coconut based systems are predominant without much diversification. In such areas farm diversification through IFS principles helps to minimize risk by the integration of livestock and aquaculture with rice or farming. Livestock with vegetables under coconut will be other viable options to increase the farm income. Thus, IFS helps to ensure entrepreneur opportunity and diversification of products by integration of different agricultural enterprises to maximize production, profitability and minimize the use of external inputs in the farm.

Keywords: *Efficient resource use, Enhanced production, Diversity, Entrepreneur opportunity.*

Introduction

The household livelihood security is defined as adequate and sustainable access to income and resources to meet basic needs including adequate access to food, potable water, health facilities, educational opportunities, housing, time for community participation and social integration (Frankenberger and McCaston, 1998). This includes household food, nutrition and income security. A variety of on-farm and off-farm activities provide a wide range of opportunities to make a livelihood. In this context, one or combinations of livelihood promotion, protection and provisioning strategies may enhance household food, nutrition and income security. The strategies for livelihood support includes, agricultural diversification, alternative income generating activities, promotion of suitable off farm activities, improving on-farm storage capacity and common property management through the people participation.

Diversification of production is one of the main household strategies to improve livelihoods. Diversification may be by taking up diverse species, varieties, livestock breeds or new enterprises for production in a given farm. It also includes landscape diversification *i.e.* inclusion of various crops and cropping systems interspersed in space and time. Besides, diversification may also occur when taking up non-farm activities or processing of farm products. Thus on-farm and non-farm diversification provides the buffering capacity against any household income fluctuations in short term and providing households with a wider options to address future challenges. The diversity can be created temporally as well as spatially, adding even greater functional diversity and resilience to the existing agricultural systems. In this regard, integrated farming system approach not only allows the diversification of existing farming system but also allows synergy between the components to increase production and productivity.

Changes in the farming techniques to maximize the production from the existing cropping pattern and optimal utilization of resources can be made possible by IFS. In this system the farm wastes are better recycled for productive purposes. IFS's activities are focused mainly round a few selected, interdependent, interrelated and often interlinking production systems based on a few crops, animals and related off-farm activities. This helps in harnessing the complementarities and synergies among different agricultural sub-systems/enterprises and augmenting the total productivity, sustainability and gainful employment. IFS strives to enhance food, nutritional, livelihood security and income of farmers in addition to reducing risk of total failure of crops in monocropped areas due to weather vagaries, pest and diseases and market fluctuations. El Titi (1992) defines integrated farming system as "a holistic pattern of land use which integrates natural regulation processes with farming activities to achieve maximum replacement of off-farm inputs and to sustain farm income". The advantages of IFS over conventional farming system are widely reported by several researchers that confirmed the reduction of cost of production and thus increasing productivity, income, nutrition and overall welfare besides increasing personal savings and health of farmers. Diversification of farming activities improved the utilization of labour; reduced unemployment in areas where there was a surplus of underutilized labour and provided a source of living and entrepreneurs activities particularly for small households (Van Brakel *et al.*, 2003).

Status of Agriculture in Coastal and Island Regions

India has 9000 km long coast and agriculture in these region is rich in agro-biodiversity and are fed by the river streams, back-water streams and other freshwater sources. In these areas the agricultural output is from crops, fisheries and aquaculture which provide scope for entrepreneur activities for several thousand people. Due to geographical conditions, rice occupies as the major staple crop. In addition, horticultural crops and plantations, especially coconut predominate coastal agriculture. The coastal and island regions of India is endowed with good amount of rainfall, surface and ground water resources, rich in biodiversity, mostly of deltaic soils and humid to sub-humid condition suitable for growing variety of crops.

The western coastal areas and islands are receiving copious amount of rainfall during south west monsoon (>200cm). Though the amount of rainfall received in east coastal region is less (50-150cm), the rivers bring water from areas of heavy rainfall in the west to areas of low rainfall in the east and deposit soil containing large quantities of clay and silt in the deltas. Traditionally, rice is the major crop in. Among the other crops, coconut, betelnut, rubber, cashew, ragi, cotton, pulses and ground nut are predominant in coastal areas. In many places along the coastal plain double cropping with rice as main crop is practiced. The major cropping systems found in different parts of coastal and island regions are given in Table 17.1. The western coastal plains are narrow, flat river valleys and areas near estuaries have good loamy soil, where rice is grown during kharif season. In some areas a second crop of rice is taken up by lifting water from wells, rivers and streams. In other areas, pulses, oilseeds and spices are planted. Plantation crops are very important component of western coast and islands. The alluvium rich east coastal plains are known as the rice bowl of Tamil Nadu, Andhra Pradesh and Odisha. It has several important rivers namely, Mahanadi, Godavari, Krishna, Cauvery and Vaigai flowing towards East to the Bay of Bengal, irrigating vast delta area and providing irrigation lifeline to the crops. The major crops grown in eastern coastal plains are rice, sugarcane, groundnut, mustard, sorghum, maize, other millets, pulses, vegetables, banana, coconut, mango, cashew nut and orchard crops. By virtue of its location, maritime fishing is another important occupation for the coastal people.

Table 17.1: Major Cropping Pattern in Coastal and Island Regions of India

States	*Major Cropping System*	*Constrains*
Tamil Nadu and Andhra coast	Rice – Rice, Rice – Vegetables, Rice – Rice – Pulse, Rice – Rice – Green manure, Sugarcane – Fallow	☆ As most of the rivers are either level or below sea level, drainage, salinity and floods are the major problem. ☆ Non availability of labour in time for agricultural operations.

States	*Major Cropping System*	*Constrains*
Odisha coast	Rice – Green gram, Rice/Maize – Cowpea, Jute – Paddy – Pulses, Paddy – Mustard, Rice – Potato – Rice, Early rice – Vegetables.	✰ Red and lateritic black, alluvial and saline soils. ✰ Waterlogging and salinity.
Gangetic delta	Rice – Rice, Rice – Vegetables.	✰ Arsenic pollution. ✰ Inundation of low and medium lands during early monsoon season.
Malabar coast	Rice – Rice – Fallow, Rice – Rice – Pulse/ Vegetables/Green Manure.	✰ Use of local or traditional varieties. ✰ Non-adoption of improved/new production practices by the farmers. ✰ Soil acidity, drainage congestion and high clay content.
Konkan coast	Rice – Rice, Rice – Pulse.	✰ Small and scattered land holdings, soil erosion. ✰ Low cropping intensity (115 per cent). ✰ Uncontrolled irrigation leading to flooding and less diversification.
Gujarat plains	Pearl millet – Mustard, Cluster bean – Mustard, Pulses – Mustard/Wheat – Summer pearl millet, Rice – Rice, Rice – Sugarcane, Groundnut – Wheat/Castor, Groundnut – Pearl millet, Cotton – Fallow (yield of seed cotton).	✰ Sandy soil, low rainfall. ✰ Higher water requirement of most of the crops resulted in low water use efficiency. ✰ Calcareous, alkaline soils, more clay.
Andaman and Nicobarand Lakshadweep	Rice – Pulse/Vegetable, Rice – Fallow, Plantations + Tuber Crops, Plantations Crops.	✰ Soil acidity, coastal salinity, acid sulphate soils and drainage congestion. ✰ Shallow, coarse soils, low in N and K and erosion.

Source: Modified from Gangwar and Singh, 2011.

Status of Coastal Aquaculture

The coastal aquaculture in these areas is highly competing with crop production. There is growing completion for land and water amongst different components of agricultural production systems that demands farmers to make choices based on rational division. Unlike East Asian countries, crop and fish integrated or simultaneous production system is limited in the coastal plains of India. At the same time, in coastal Andhra Pradesh, intensive aquaculture activities particularly shrimp farming resulted in salinization of coastal land. Besides, these issues the coastal aquaculture is relatively intensive and fetch good return to the farmers and practiced as specialized professional.

Constraints for Agricultural Production

The challenges originating from population growth, developmental needs and the burden of providing food to millions of people are the major anthropogenic factors facing coastal regions of our country. Besides, combinations of natural and hydrological factors are responsible for land degradation of different kinds. The key constraints affecting agriculture and agricultural development in coastal areas are their proximity to sea causing severe drainage and soil salinity problems due to sea water intrusion, stagnation of rain and runoff water and flooding from rivers or periodic storm surges. The wind storms or tidal waves also cause physical damage to the standing crops or agricultural trees resulting in huge losses. As most of the coastal plains are found in the river deltas, they have to depend on upstream reservoirs which influence the timing, quantity and quality of river flows thus affecting the agricultural activities in deltaic regions. Over the years it has been observed that the upstream dams and irrigation schemes deprived coastal areas of water for irrigation and, by removing silt and regulating floods, affects the fertility of coastal alluvial agricultural land. More than 20 per cent of Indian population and major metro cities are concentrated in coastal regions only. Urbanization resulted in space constraints and intense competition for land and water. In certain regions agriculture require extensive amounts of land to produce relatively low-value outputs. As competition for land intensifies particularly enterprise driver, grain crops and livestock grazing tend to be marginalized and replaced, either by non-agricultural activities or with higher-value outputs, which require higher capital inputs and less land (for example intensive livestock raising or horticulture). Due to which smallholders who are unable to invest and intensify, thus tend to lose their land and migrate to the coastal towns. Increasing water scarcity compounded by climate change may divert water away from agriculture to residential and industrial uses, and to other types of agriculture.

Impact of climate change is also a major issue in coastal and island regions. It is projected that water resources in the Indian subcontinent will come under increasing pressure due to the changing climate. It is observed that in Andaman and Nicobar islands over the years the total number of rainy days is increasing and the same is decreased during the crop season. Climate change is projected by many studies to cause sea-level rise, with significant consequences to the coastal agro-ecology and livelihoods of farmers and fishermen. Agriculture in coastal areas is fed by river streams, river deltas and backwater streams. These areas are not only suitable for crop production but also for fisheries and aquaculture. Further, climate change is projected to affect the yields of paddy, maize and coconut in the coastal areas. In general, climate change is projected to have a profound impact on the abundance and distribution of reef fishes particularly in six major regions along the Indian coastline *viz.* Gulf of Kachchh in Gujarat, the Malwan coast of Maharashtra, Lakshadweep islands, Gulf of Mannar and Palk Bay in Tamil Nadu and Andaman and Nicobar Islands.

Integrated Farming Systems for Livelihood Security and Entrepreneur Promotion

In India, especially in coastal plains and island region rice is grown under submerged conditions and water remained ponded (10-30 cm depth) for a long. In such areas farm diversification helps to minimize risk by the integration of aquaculture with rice farming. Rice-fish system is more popular in coastal lowlands where, fish in rice field utilize the untapped aquatic productivity of rice ecosystem (Table 17.2). It is observed that the rice bottom is highly fertilized due to the production of zoo and phytoplankton and these resources are fully utilized by the fish. The rice + *Azolla* + fish on an average gave Rs. 8817/ha more over the rice monoculture and ₹ 3219 ha^{-1} over the rice + fish. For successful rice-cum-fish farming in coastal lowlands, raised bunds are made around the fields to hold water for longer periods and to avoid over-flowing of flood water besides a suitable fish refuge/trench system. Raising of marketable fish and prawn in rainfed lowlands including coastal saline areas along with rice was achieved in West Bengal, Odisha, Andhra Pradesh and Andaman Islands. In shallow lowlands, 500 to 700 kg ha^{-1} of fish or prawn and 5 to 6 t ha^{-1} of rice can be obtained under such mixed cropping systems, while it is possible to increase the production of fish/prawn to 2000 kg ha^{-1} in deep water situations where water remains in the field for longer period (Figure 17.1). Vegetables and horticultural crops on raised bunds around the field with the soil excavated from the trenches further increase the land productivity to a great extent (Sinhababu and Venkateswarlu, 1998; Velmurugan *et al.*, 2015).

The coastal saline lands which grow *rice* during *kharif* or remain unproductive and these wetlands provide ample scope for development of rice-fish/prawn production system. Under such conditions, salt tolerant rice varieties and fresh water fish/prawn during *kharif* or sole cropping of rice during *kharif* followed by sole cropping of salt water fish/prawn in sequence during *rabi* can be taken up in rice-fish system. These wetlands can produce fish/prawn amounting to 400 to 600 kg ha^{-1} during three months of *rabi*, and another 500 to 600 kg ha^{-1} during the monsoon season in an integrated rice/fish/prawn culture (Ghosh, 1992). Rice-fish integrated farming has proved by many others as a viable technology for the low-lying rice fields in West Bengal and Kuttanad region of Kerala.

Table 17.2: Different Farming System Models for Lowlands of Coastal Region

Agro-ecological Region	*Prevailing System*	*Net Return (₹ ha^{-1} yr^{-1})*	*Integrated Farming Systems*	*Net Return (₹ ha^{-1} yr^{-1})*	*References*
Eastern coastal plains	Rice – Rice	15299	Rice – Rice – *Azolla*/ *Calotropis*+ Fish	17488	Shanmuga Sundaram and Balusamy (1993)
	Rice – Rice – Black gram	8,312	Rice – Rice – Cotton + Maize + Poultry/Fish	17209	
	Rice – Rice – Rice – Fallow – Pulses	13,790	Rice – Rice – Rice – Fallow – Cotton + Maize + Duck cum Fish	24117	Ganeshan *et al.* (1990)
	Rice	22,971	Rice + Fish	28569	Balusamy *et al.* (2003)
			Rice + *Azolla* + Fish	31788	
	Rice	-	Rice + Fruits + Vegetables + Flowers + Fish + Poultry + Duckery + Mushroom + Apiary + Biogas	58360	Behera and Mahapatra (1999)
	Rice – Rice	-	Rice – Fish + Vegetables + Fruits + Dairy + Poultry	187175	Reddy *et al.* (2018)
Western coastal plains	Rice – Rice – Fallow	12,167	Rice + Fish	15375	Varughese and Thomas (2009)
	Rice	19210	Rice – Brinjal + Mushroom + Poultry	77310	Korikanthimath and Manjunath (2009)
	Coconut	6230	Coconut + Fodder + Dairy	32340	
	Coconut	-	Coconut + Fodder + Dairy + Poultry cum Fish	66, 678	Reddy and Bidappa (2000)

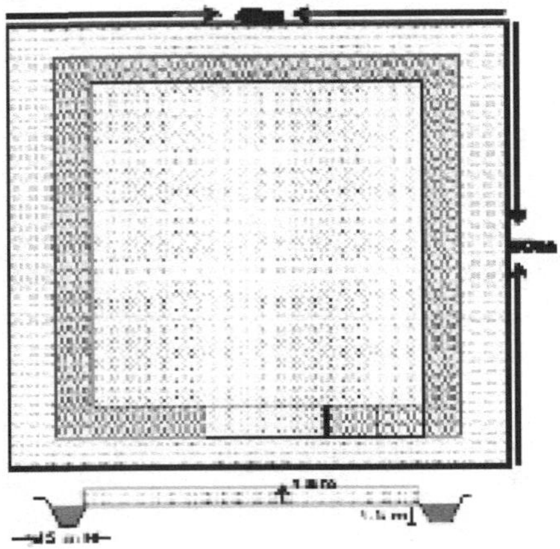

Schematic diagram of paddy-cum-fish with border bed technique

Excavation of paddy-cum-fish technique with border bed at farmers' field

Crop cultivation under paddy-cum-fish with border bed technique at farmers' field

Figure 17.1: Rice-Fish System in Andaman Islands.

The coastal regions receive high rainfall during monsoon season and farm ponds on 20 per cent of land area helps in harvesting rainwater which would help in raising a second crop after rice and also for providing protective irrigation to rice during *kharif* and part of the land can be elevated with the soil excavated from the ponds which provided better land situation for raising a good crop of rice having lower salt concentration and shallower depth of standing water than the original fields. The coastal saline soils generally remain flooded throughout the crop growth period. The bunding of fields to prevent entry of brackish water can facilitate growing of fish with rice. In highly saline conditions, where saline water regularly enters the ponds from estuarine/creeks, the main consideration is not for raising rice crop but for prawn and brackish-water fish because these areas do not support a successful crop of rice (Sinhababu and Venkateswarlu, 1995).

Farming system is a resource management strategy to avail maximum efficiency of a particular system. The study of energy use efficiency of IFS by Manjunath (2002) revealed the variation of the mean total energy input. Integration of poultry and mushroom enterprise with rice-brinjal system required highest energy input (52,030 MJ ha^{-1}) and followed by rice – groundnut system integrated with mushroom and poultry (46,077 MJ ha^{-1}). However, rice cropping alone or enterprises recorded the least requirement of energy. The energy output was maximum (1,65,334 MJ ha^{-1}) under rice - brinjal + mush room + poultry with 3.18 system energy efficiency mainly due to the lesser energy input involved in contrast to energy rich output enterprises indicating efficient utilization of scarce and costly resource in integrated faming system approach. The integrated farming system has the potential for employment generation in agriculture. The study by Korikanthimath and Manjunath (2009) indicated that rice-brinjal system integrated with mushroom and poultry generated 392 man days/ha compared to rice alone (Table 17.3).

Table 17.3: Employment Generation of Rice Based IFS

Farming System	*Employment Generation (man-days ha^{-1} $year^{-1}$)*		
	Men	*Women*	*Total*
Rice alone	24	90	110
Rice – Groundnut + Mushroom + Poultry	58	292	350
Rice – Cowpea + Mushroom + Poultry	55	290	345
Rice – Brinjal + Mushroom + Poultry	78	314	392
Rice – Sunnhemp + Mushroom + Poultry	52	257	309

Source: Korikanthimath and Manjunath, 2009.

The coastal regions plantation based farming systems is an important farming practice among the small and marginal farmers to maximize land use, farm productivity, and profitability. The suitability of diversified farming system in coconut plantations was confirmed by many researchers. Korikanthimath and Manjunath (2009) developed a coconut based farming system involving coconut, fodder, and dairy for coastal zones of Goa (Table 17.4). The productivity levels of the farming system revealed that the base crop coconut yielded 4,700 nuts ha^{-1} $year^{-1}$ in the system. The intercropped forage grass-legume mixture with hybrid Napier (PBN-16) and *Centrosema* yielded 82.6 t ha^{-1} $year^{-1}$ of coconut garden. Further, the milk yield from the integrated dairy unit was 1,648 litres $year^{-1}$. Besides, the incidental byproducts of coconut (5.74 t ha^{-1}), litter (3.2 t ha^{-1}) and cow dung (8.69 t ha^{-1}) from the dairy were additional inputs available for recycling in the system.

Table 17.4: Economics of Coconut-based Farming Systems

Component	*Cost of Maintenance (₹)*	*Net Returns (₹)*	*B : C Ratio*	*Generation of Employment (man-days ha^{-1} yr^{-1})*		
				Men	*Women*	*Total*
Coconut alone	12150	6230	0.51	72	5	77
Coconut + Fodder	28995	25580	0.88	91	46	137
Coconut + Fodder + Dairy	62535	32340	0.52	248	108	356

Composting of these recyclable manurial resources produced 4.70 t of coconut compost, which was available for recycling of nutrients within the system. The economic analysis of the integrated system indicated net returns of ₹ 32335 year^{-1} during the period as compared to ₹ 6225 ha^{-1} from cultivation of coconut alone. High yielding nature of hybrid Napier used for intercropping coupled with on-farm generation of inputs *i.e.* green fodder for cattle and manure for the crop resulted in higher net returns by reducing input cost. Monocropping of coconut resulted in employment potential of only 77 man-days, the majority of the works being men oriented (93.5 per cent). Further, intercropping of forage grasses in coconut depicted a potential of 142 man days of employment, 60 per cent being men oriented works. In total, forage intercropping in coconut provided an employment potential of 219 man-days. Compared to a monocropped situation, intercropping added an additional 142 man-days of employment. Besides, maintenance of two milch cow's right through the year required an additional 137 man-days. Thus, in an integrated system, nearly 356 man-days of employment was generated. In contrast to monocropping of coconut (77 man-days), the integrated system provides an additional (279 man-days) employment, which, could lead to an additional income of ₹ 23780 ha^{-1} for the family labour. Further, integrated dairy provided alternating returns and regular employment throughout the year. Thus, it is evident that the integrated system provides for a higher degree of resource use efficiency, productivity, and sustainability.

The coconut-based IFS studies conducted at CPCRI, Kasaragod and Kayamkulam revealed that the integration of coconut with dairy, poultry and fishery is an ecologically sustainable system which fetches higher income to cultivators when compared to coconut grown as monocropping. To assess the economic viability of CBIFS, the average cost and returns for a period of two years (2011 -2013) were analyzed. The total cost involved in maintaining the system under CBIFS with integrated nutrient management practices was ₹ 389700.00. The total returns realized from the system was ₹ 675212.00 and net returns was ₹ 285512.00 (Subramanian *et al.,* 2014). Besides improved net returns and employment generation, the IFS system also prevents soil erosion and helps in nutrient recycling. The tree components by virtue of their deep roots intercept absorb and recycle nutrients that would have been otherwise lost by leaching, thereby making nutrient cycle more closed. The dense canopy cover provided by tree crops protects soil against the direct action of rain thus reducing surface runoff. A dynamic equilibrium can be expected with respect of organic matter and plant nutrients on the garden floor due to the continuous addition of leaf litter and its constant removal through decomposition.

The orientation and structure of arecanut canopy permits 40 per cent of incident radiation to penetrate down and thus it doesn't properly utilize the available soil, space and light. Studies on the arecanut rooting pattern revealed that palms planted at 2.7 × 2.7 m could use effectively only 30 per cent of land area. Therefore, the arecanut plantation is more ideal for inter cropping and multistoried cropping than coconut plantations. In other words the compact nature of arecanut crown allows more sunlight to pass down to ground and maintain high humidity which in turn favours growth of shade loving shrubs/trees, vines (betel leaf and pepper) and variety of short statured biennials and annuals such as pineapple, ginger, turmeric and tuber crops. Cardamom has been one of the time tested most remunerative spice crop which provide scopes for entrepreneur opportunities. Net profit of ₹ 161850.00 ha^{-1} could be obtained by mixed cropping of arecanut and cardamom as against ₹ 103624.00 ha^{-1} (mean of 8 years) in case of sole arecanut. In recent years, Vanilla as mixed crop with arecanut has become the most profitable spice (Korikanthimath and Manjunath, 2009).

Farming System Models for Island Region

The major land forms in Andaman Islands are longitudinal hills, intermountain valley and coastal plains (Figure 17.2). Agriculture is confined to valley and coastal plains while plantations of coconut and areca nut are found in hill slopes adjoining the valley or coastal plains. Forests are found in the higher slopes and hilltops, while mangroves are scattered in the coastal region. Rainfed rice is grown in 8400 ha mainly in valley and coastal plains due to water logging caused by heavy intensive rainfall of around 2500 mm during June to November accentuated by poor drainage

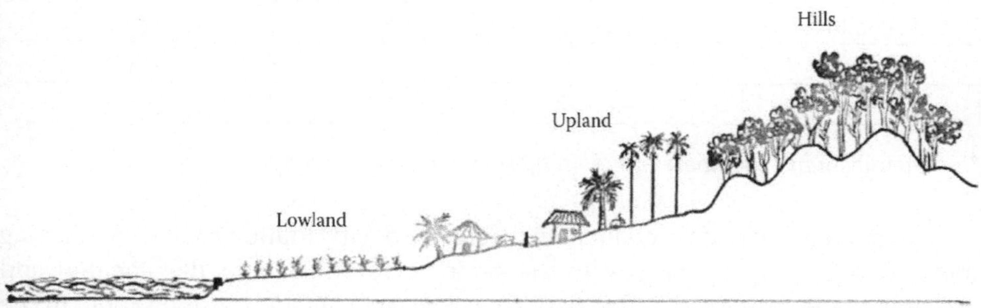

Figure 17.2: Typical Landscape of the Island Region.

and lower slope (2 per cent). The productivity is very low due to poor management. Only long duration rice is grown in many parts of the Island and some places short duration rice followed by vegetables were also being practiced.

In coastal and valley plains of the islands major constraints are waterlogging and water scarcity during different seasons. To tackle the twin problems of waterlogging and water scarcity raised bed and Furrow (BBF) system was made mainly to produce vegetables, rice and fish together in the water logged or degraded areas through land manipulation. A farming system model was evaluated in an area of 0.75 ha to increase the farm production and to ensure food and nutritional security to small farmers of the island. This involves land manipulation in an area of 0.30 ha where raised beds and furrows were alternately made to facilitate growing of vegetables in beds and rice – fish in the furrows. In remaining 0.35 ha area rice based cropping systems were followed. The crops were selected in such a way to meet the family nutritional requirement in terms of cereals, pulses, oilseeds, vegetables *etc.*, and dairy component was also integrated to achieve maximum returns. The results indicated that crop + dairy based integrated farming system (IFS) can yield ₹ 2.03 lakhs of total net income from a total area of 0.75 ha with employment generation of 302 man-days year^{-1}. The dairy component contributed 54 per cent of net income while generating 182 man-days year^{-1} followed by crops (40 per cent) with an employment generation of 114 man-days year^{-1}. The system could able to meet the basic food requirement of a family of five members (Table 17.5). The surplus produce of vegetables and milk can be sold in the market for income generation and can be used for purchasing the deficit items. The other advantage of integration is the income generation/money flow throughout the year. The sale of milk and vegetables ensure income to the household throughout the year.

Table 17.5: Production of Various Essential Food Items from the IFS Model of 0.75 ha

Item	*Demand (kg)**	*Supply (kg)*	*Surplus/Deficit (kg)*
Cereals	1900	2100	200
Pulses	170	50	–120
Oilseeds	376	316	-60
Fruits	170	200	30
Vegetables	750	2904	2154
Milk (L)	950	2863	1903
Fish	—	20	—

*for a family of five members based on ICMR recommendations.

In Andaman Islands coconut plantations are mostly found on hill slopes having more than 25 per cent slopes. In the slopes plantation crops like coconut and arecanut are dominant with few fruit trees like jackfruit, guava, and sapota in the homestead gardens. The soils are poorly fertile and subject to soil erosion. A farming system model was evaluated at Farmers filed involving plantations + poultry and livestock. The normal cropping pattern followed by farmer was arecanut, coconut + sapota and banana as sole crop. In the integrated approach, a modification was made in the existing cropping pattern by integrating black pepper with arecanut. Cropping was undertaken in 0.90 ha and 0.10 ha was allotted for backyard poultry and livestock. Vanraja birds (20 nos.) and ducklings (5 nos.) were supplied to the farmer for rearing as backyard poultry. Birds were fed with grains, broken rice, wheat, coconut waste, vegetable waste *etc.* that are available within the system. Dung served as source of FYM for fruit and other plantation crops thereby reducing the cost of production. By integrated approach the net farm income increased to ₹ 3.94 lakhs ha^{-1} $year^{-1}$ as compared to plantations alone. The contribution from crop component is 75 per cent towards net returns followed by cattle (21 per cent) and poultry (5 per cent). The employment generation increased to 438 man-days ha^{-1} $year^{-1}$ by integrating backyard poultry and cattle rearing with cropping compared to cropping alone (240 man-days ha^{-1} $year^{-1}$). An employment of 61 man-days from poultry and 137 man-days from cattle could be generated. Cattle and buffalo rearing involved intensive use of family labour and provided significant employment opportunity for small and marginal farmers (Rao, 1983). Recycling of resources in terms of energy is one of the main beneficial effects of integrated farming system. The total output energy was 133018.00 MJ as against total input energy of 6030.00 MJ indicating an efficient conversion of input energy into by products (Ravisankar *et al.,* 2007).

Farming System Models for Nicobar Islands

The Nicobar Islands are situated in the South-east of the Bay of Bengal and separated from the Andaman group of Islands by 10° channel. In Nicobar Islands, the plantation crops of coconut and areca nut occupy 84 per cent of the area under agriculture (Handbook of Statistics 2011). The crops like sweet potato, tapioca, greater yam, pine apple, banana, papaya and pandanus are grown in home gardens to a limited extent. Rearing pig and back yard poultry are the traditional activity found in the tribal areas. This is reflected in their food consumption pattern where these crops along with coconut and pig play an important role in their daily diet. It is observed that their daily diet doesn't constitute fruits and vegetables and the average egg consumption is far below the recommended level due to non-availability and higher cost. With increasing population and the need to ensure household food and nutritional security, a suitable strategy was evaluated for augmenting the production of fruit and vegetable crops, livestock and poultry within the limited resources of farm families in different islands in a participatory approach to enhance the food and nutritional security of the tribal farm households through homestead based integrated farming system comprising diverse interventions like home garden, backyard poultry, goat rearing and composting (Figure 17.3).

Figure 17.3: Homestead IFS for Tribal Areas of Nicobar Region.

Components in Farming System

1. **Home gardens (400 m^2):** In the home gardens 150 m^{2}area is allotted for growing seasonal vegetables *viz.* okra, brinjal, tomato, green amaranth, cucumber, bitter gourd and bottle gourd, sweet corn. In another 125 m^2 area fruit crops like banana, pine apple, papaya can be grown. Grow tuber crops *viz.* tapioca, sweet potato, greater yam and colocasia in an area of 100 m^2. Besides, plant fruit trees of *Pandanus sp.* and *Morinda citrifolia* and in corners of the field and *Sesbania grandiflora* (agathi) and *Gliricidia sepium* (Jindapalli) as border trees to act a biofence besides serving as green fodder and green leaf manure.
2. **Livestock:** The livestock component in the present system includes goat and poultry.
   a) **Goat –** A total of 2 adults goats are reared which are fed on the green fodder grown as bio-fence and cassava tubers planted in home gardens.
   b) **Poultry –** A farm family can maintain a dual purpose flock of Vanaraja or local Nicobari fowl under backyard condition. A total of 10-13 birds comprising 5 to 7 adult layers, 1 adult rooster, 4 to 5 young pullets and a young rooster and 1 hen with chicks will allow farm family to

produce an average of 25 to 33 eggs per week while maintaining the flock at no cost. The birds are normally fed with coconut from their garden, morinda and other leguminous leaves. The local hen is used mainly as brood to set the eggs. In general, 14 to 21 eggs would be set in each quarter to maintain the stock size.

c) **Compost** – To provide organic manure by utilizing farm wastes and or crop residues compost tanks were made at a corner of the home garden. Above the composting unit, pandal was made with local materials and perennial crops like coccinia, greater yam were grown. The manure obtained from goat and poultry wastes along with crop residue were used for making the compost.

Food and Nutritional Security

From the system about 300 kg of seasonal vegetables, 117 kg of greens, 214 kg of tubers, and 200 kg of fruits were produced from an area of 400 m^2. The egg production has also increased resulting in improved consumption. After the intervention the consumption of vegetables including greens, fruits and egg increased significantly from 50 to 250 per cent due to on farm production and availability. The consumption of tubers decreased by 25 per cent with increased consumption of other food items indicating a change in dietary pattern more towards balanced nutrition (Table 17.6). The diet diversity also indicates the food and nutritional security where the diet which includes diverse types of foods is internationally considered as a healthy one. The number of food items consumed by the tribal family has increased by inclusion of variety of vegetables, greens, fruits and egg indicating more diversified food consumption ensuring more food and nutritional security.

Table 17.6: Change in Average Daily Consumption of Specific Food Items due to Intervention

Food Item	*RDI Person^{-1} (g)*	*RDI Family^{-1} (g)#*	*Consumption Quantity (g family^{-1} day^{-1})*			*Deficit/ Surplus*	
			Before	*After*	*Change*	*Before*	*After*
Cereals	400	2000	1360	1625	265	-640	-375
Pulses	80	400	190	265	75	-210	-135
Vegetable	150	750	255	625	370	-495	-125
Greens	50	250	50	175	125	-200	-75
Fruits	100	500	265	480	215	-235	-20
Tuber	100	500	750	560	-190	250	60
Fish, Meat	50	250	200	300	100	-50	50
Egg (week)	3.5	17.5	0.5	12	11.5	-17	-5.5

RDI – Recommended Dietary Allowance as per ICMR standards for a family of 5 members.

The surplus produce of vegetables and egg was sold in the neighbourhood earning additional income of about ₹ 2750.00 and ₹ 5000.00 from the sale of kids of resulting in a total supplementary income of ₹ 7750.00 from the farming system model within the limited resources available. The additional income was reflected in the increased consumption of cereals and pulses enabling the household to improve the share of cereals and pulses in their daily food intake resulting in a more balanced consumption of food. A total of 95 man-days were generated by the system *viz.* 52 man-days in home garden, 40 man-days in livestock rearing and 3 man-days in composting spread throughout the year. Composting of organic residues on farm resulted in effective recycling of farm waste for crop production.

Conclusion

The integrated farming system approach will play a major role to sustain livelihood security especially of small and marginal farmers and conserve the resource base through efficient resource recycling within the system. Declining size of landholdings without any alternative opportunity for income augmentation is resulting in fall in farm income, and causing agrarian distress. The smallholders prefer to integrate the various enterprises to reduce the risk of specialist agriculture without assured market and income reliability is high. Thus the integration of components enhances efficiencies in family labour and farm nutrient recycling. The IFS models developed on ecosystems and sub-systems can be fine-tuned through farmer participatory approach based on site specific requirements and dissemination of such models will help in enhancing farm income and ensure livelihood security of marginal and small farmers and anchoring sustainability in agriculture.

REFERENCES

Balusamy, M., Shanmugham, P.M. and Baskaran, R. (2003). Mixed farming an ideal farming. *Intensive Agriculture* **41** (11–12): 20–25.

Behera, U.K. and Mahapatra, I.C. (1999). Income and employment generation of small and marginal farmers through integrated farming systems. *Indian Journal of Agronomy* **44**(3): 431- 439.

El, T. (1992). Integrated farming: An ecological farming approach in European Agriculture. *Outlook Agriculture.* **21**: 33-39.

Frankenberger, T.R. and McCaston, M.K. (1998). The household livelihood security concept. *Food, Nutrition and Agriculture.* **22**:30-36.

Ganesan, G., Chinnasamy, K.N., Bala Subramanian, A. and Manickasundram, P. (1990). Studies on rice based farming system with duck cum fish culture in deltaic region of Thanjavur district, Tamil Nadu. *Farming Systems Newsletter* **1**(2): 14.

Gangwar, B. and Singh, A.K. (2011). Efficient Alternative Cropping Systems. Project Directorate for Farming Systems Research, Modipuram, Meerut, India. pp. 339.

Ghosh, A. (1992). Rice – fish farming development in India: past, present and future. *In*: *ICLARM Conference Proceedings on Rice-fish Research and Development in Asia*, dela Cruz, C.R., Lightfoot, C. Costa-Pierce, B.A., Carangal, V.R. andBimbao, M.P. (Eds.) pp. 27–43.

Korikanthimath, V.S. and Manjunath, B. (2009). Integrated farming systems for sustainability in agricultural production. *Indian Journal of Agronomy* **54**(2): 140-148.

Manjunath, B.L. (2002). Integrated farming systems for coastal region of Goa. *Ph.D. Thesis* Department of Agronomy, University of Agricultural Sciences, Dharwad.

Rao, M.V. (1983). Helping them to buy and sell nutrition. *Intensive Agriculture.* **21**(8):17-20.

Ravisankar, N., Pramanik, S.C., Jeyakumar, S., Singh, D.R., Bibi, N., Nawaz S. and Biswas, T.K. (2007). Study on Integrated Farming System (IFS) under different resource conditions of Island Ecosystem, *Journal of Farming System Research and Development.* **13** (1): 1-9.

Reddy, D.V.S. and Biddappa, C.C. (2000). Coconut based cropping/farming practices in India. *Journal of Plant and Crops,* **28**(1):1–18.

Reddy, R.V.S.K., Sree, E.K., Reddy, A.D., Deepthi, V., Nirmala, T.V., Raju, S.G.S, Subbaiah, K.V. and Prasad, J.V. (2018). An evaluation study on viable integrated farming system (IFS) model in Godavari delta of Andhra Pradesh. *Journal of Pharmacognosy and Phytochemistry.* **7**(1S): 2361-2368.

Shanmugasundaram, V.S. and Baluswamy, M. (1993). Rice-fish-Azolla - An Integrated Farming System in Low Land Wetlands Wetlands. *Farming Systems.* **9**: 105-107.

Sinhababu, D.P. and Venkateswarlu (1995). Rice – Fish in rainfed lowlands. *CRRI Bulletin,* Central Rice Research Institute, Cuttack, India p. 14.

Sinhababu, D.P. and Venkateswarlu, B. (1998). Modern Frontiers of rice-fish systems. *In*: Advances in fisheries and fish production. Ahmed, S.H. (Ed). Hindustan Publishing Corporation, New Delhi. pp. 206–228 and p. 309.

Subramanian, P., Dhanapal, R., George, V.T., Maheswarrapa H.P. and Thamban C. (2014). Coconut based integrated farming system. *Technical Bulletin No.78,* CPCRI, Kasaragod, Kerala, India.p.32.

Van Brakel, M.L., Morales, E.J., Turingruang, D. and Little, D.C. (2003). Livelihood improving functions of pond based integrated agriculture and aquaculture systems. *MRC Fisheries Programme* (*FP*). Institute of Aquaculture, University of Stirling, Scotland, UK.

Varughese, K. and Mathew, T. (2009). Integrated farming systems for sustainability in coastal ecosystem. *Indian Journal of Agronomy*, **54**(2):120-127.

Velmurugan, A., Swarnam, T.P. and Lal R. (2015). Effect of land shaping on soil properties and crop yield in tsunami inundated coastal soils of Southern Andaman Island. *Agriculture, Ecosystems and Environment*, **206**: 1-9.

Chapter 18

Waste-Wealth Relationship through Integrated Fish-based Farming Approaches

Jham Lal, Nitesh Kumar Yadav, Soibam Khogen Singh and Pradyut Biswas

Department of Aquaculture, College of Fisheries, Central Agricultural University (Imphal), Lembucherra, Tripura – 799210

ABSTRACT

In addition to meeting the demand, the modern technologies that are being adopted to increase agricultural production for the nutrition and food security of the rapidly growing world population will lead to environmental degradation. Integrated aquaculture (IA) can ensure as well as sustain sufficient food safety while also being environmentally friendly. Integrated fish farming is a diverse and collaborative method of cultivating that focuses on fish as well as other farm products. This farming system is the recycling of animal manure (faeces, urine, and also uneaten feeds) to be used as fertilizers, and occasionally as food for increasing fish in ponds, enclosures as well as cages. Supplementary feed, manuring, as well as fertilization, comprise the main input expense in aquaculture which can be diminished through sustained natural live food (plankton) production obtained by recycling organic wastes from several livestock husbandry systems including such as cattle, pigs, poultry, ducks, etc. The Integrated Fish-based Farming also increased the income and reduce cost of system. So, the Adoption of Integrated fish farming (IFF) is one of the great future option for sustainable production, regular income source of the farming community and also supported management of agricultural and livestock waste in India as well as other developing nations of the world.

Keywords: *Integrated fish Farming, Waste management, Income generation, Ecological reclamation.*

Introduction

Aquaculture is a diverse production sector with various output strategies as well as practices. Aquaculture continues to grow faster than all other animal food-producing sectors combined. Sustainable aquaculture is dependent on eco-friendly, economically viable, and socially feasible cultivation systems. Supplementary feed, manuring, as well as fertilization, comprise the main input expense (around 50-60 per cent) in aquaculture which can be diminished through sustained natural live food (plankton) production obtained by recycling organic wastes from several livestock husbandry systems including such as cattle, pigs, poultry, ducks, *etc.* Integrated fish farming is a mixture of two or additional cultivation systems including fish-livestock farming fish-agriculture/horticulture farming *etc.* The fundamental principle encompassed in integrated farming is the complete utilization of farm wastes, which are recycled into the fish pond (as manure/fertilizer) to improve its primary productivity (Pandey *et al.,* 2016).

Fish farming can be associated with agricultural production, livestock farming, and irrigation practices to improve resource utilization and, ultimately, increase production as well as net incomes. An integrated farming system is comprised of several resource-saving strategies that aim to accomplish appropriate incomes and long-term production levels while decreasing the adverse consequences of intensive farming as well as protecting the habitat (Ahmed and Garnett, 2010). Integrated fish farming is a sustainable-agriculture technology cultivated widely in Asia such as India as well as further nations of the planet (Ahmed and Garnett, 2011). Integrated fish farming is a substantiated environment sustainable and economically feasible technology that incorporates reasonable resource utilization. This farming system has enormous potential for improved resource utilization and productivity. It is a low-input technique for attaining tolerable biomass crop production per unit area while also assisting in the reduction of environmental pollution (Pandey *et al.,* 2016). IFF is a long-standing practice in China, Hungary, the Philippines, Taiwan, Japan, North Korea, Malaysia, Indonesia, Vietnam, Pakistan, as well as several other developing nations, such as India (Bisht *et al.,* 2009).

Integrated fish farming is the simultaneous cultivation of crops, and livestock as well as the cultivation of fish in the same place. The extensively well-known integrated farming strategies are expressed in words of "Fish-Cum," which means "Fish and" others as the case may be (Prein, 2002; Singh *et al.,* 1991). In this technique, nothing is expended as well as ecological equilibrium is strengthened. Recycling organic waste material for fish culture helps to clean the environment while also offering economic advantages (Shyam *et al.,* 2012). The recycling of animal dung/wastes in aquaculture ponds is critical for natural fish farming, which benefits sustainable aqua farming while also lowering costs for supplementary feeds as well as fertilizers (Kalita *et al.,* 2015). The major potential links between both farm animals as well as fish farming consider the use of nutrient content, specifically the recycling and composting of animal waste for fish farming. The word nutrients to all aspects including nitrogen (N) as well as phosphorous (P) that act as fertilizers to enhance natural food webs instead than traditional farmed animals nutrition utilization including feed ingredients, though solid slaughtering waste material consumed to carnivorous fish fall into the latter category. There are several consequences for the use of many resources such as capital, labor, as well as water. Several factors influence prospective livestock-fish production links (Little and Edwards, 2003).

A Definition of Integrated Farming

The term "integrated" comes from the Latin word "integrare," which denotes "to earn entire, "to complete by the addition of parts," or "to incorporate portions into an entirety. The crop, animal husbandry, and fish subsystems may function unassisted in particular farming strategies, as well as their commodities are just additive. However, an outcome from one subsystem in an integrated farming system that would oppositely be wasted evolves an intake to another farming, leading to increased effectiveness of the outcome of desired commodities from the land/ water region under an agriculturalist's supervision. There is a synergistic effect in integrated farming because the combined effects of the subsystems are tremendous than the sum of their particular consequences (Edwards *et al.,* 1988) and some examples were given in Table 18.1 (Kumar *et al.,* 2012a).

Table 18.1: One and Two Acre Integrated Farming System Module

Sl.No.	*Land Allocation*	*One Acre Integrated Farming System*	*Two Acre Integrated Farming System*
1.	Components	Crop + Goat + Poultry	Crop + Livestock + Fishery
2.	Allied	Mushroom/Goat manure/ Vermi-composting	Duckery/Vermi-composting/Bee keeping/ FYM
3.	Cereal crops	50 per cent area	50 per cent area
4.	Horticultural crops (Fruits + vegetables)	22.5 per cent area	12.5 per cent area
5.	Cucurbits or seasonal vegetables	4 per cent of total area	Making wire fences
6.	Livestock	2.5 per cent area	1.80 per cent area
7.	Poultry (100-200 birds)	225 sq. ft. area	Fish + Duck integration17.8 per cent area
8.	Mushroom (200 mushroom bags)	25 X 20 area	Duck (1000 m^2)
9.	Fodder production	12.5 per cent area	12.5 per cent area
10.	Spices	Turmeric, ginger or guinea grass	Turmeric, ginger or guinea grass
11.	FYM/vermin-composting pits	1.4 per cent area	1.4 per cent area

The basic principles of integrated farming include utilizing the synergistic effects of interconnected farm operations as well as sustainability, such as the proper utilization of livestock manure (Dalsgaard *et al.,* 1995). The following are the goals of integrated fish farming systems were following and systemic diagram shown in Figure 18.1:

- Beneficial assimilation of crops and animals through purposeful farm utilization
- Improving agriculture as well as farm animals' stability
- Production of higher-quality and more diverse food products
- Environmental security through effective as well as careful utilization of natural resources.

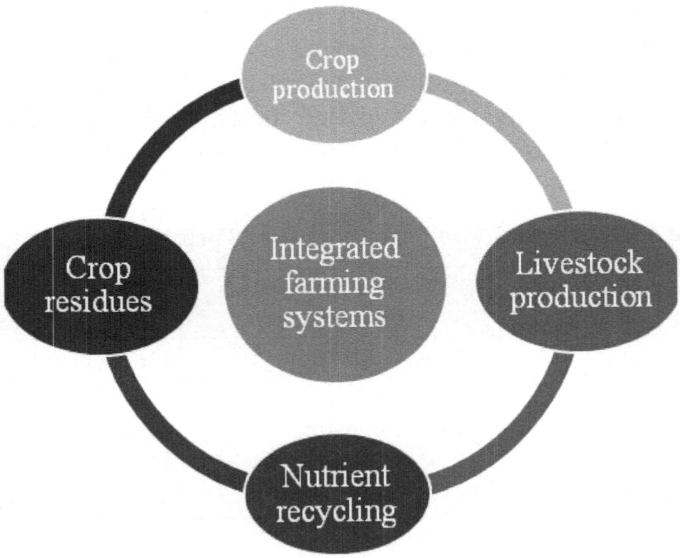

Figure 18.1: The Important Factors of Integrated Farming Strategies.

Origin of Integrated Fish Farming System

Integrated farming is thought to have originated over 2,400 years ago in China (Hu and Zhou, 1989), and it includes a dynamic model of fish poly-culture (a standardized method of rearing fish), poultry, farmed animals, as well as agricultural production, and the "integrated" (cumulative) utilization manure, grass, and other crops as feeds and fertilizers (NACA, 1989). IFFS, also known as Agro-pisciculture or combined agriculture-aquaculture, has a lengthy record in Asia, dating back over 1500 years in India (Coche, 1967). In this context, an ancient technique known as combined agribusiness aquaculture strategies may carry considerable commitment for ASEAN members. Created in China more than 2,000 years ago,

IAA has traditionally performed as a direction for farm homes to enhance their foods, improve earnings, as well as enrich the soil even with restricted reserves by converting nutrients (Yang *et al.*, 2001).

Principles of Integrated Fish Farming

The integrated fish farming principle focuses on raising fish alongside animals as well as crops. Integrated aquaculture farming is broadly defined as the continuous or consecutive integration of two or several human-induced ecosystems, at least one of which is aquaculture. However, the links involving fish farming as well as anthropogenic impacts encompass responsibilities in cleanliness (night soil or even other types of human faecal matter re-use, waste treatment), nutrients retrieval (aquaponics, breweries), as well as energy recovery (farming in warmed industrial effluent of power stations, dairy, and so on) (Prein, 2002). This cultivation maximizes resource utilization through efficient recycling and reuse of by-products from one system. It also allows for the efficient use of accessible cultivating space in order to maximize production (AIFP, 2005). Integrated fish cultivation, as well as living creature domestication, are predicated on the assumption that 'there is no waste,' and effluent is simply a misplaced resource that can be recycled into a useful resource for another commodity.

Types of Integrated Fish Farming System

The integrated fish farming system categorized three kinds of crop-animal integrated strategies based on the usage of agribusiness strategies the local agro-ecological circumstances, and the level of integration among yield as well production (Rota, 2009):

- ☆ Pasture cultivation technique (cultivating annual grain crops into annual pasture)
- ☆ Cropping system without chemical inputs (similar to pasture cropping technique without any usage of fertilizers, herbicides, insecticides as well as pesticides)
- ☆ Holistic management strategy (comparable to pasture cultivating with animal feeding incorporated into the farming methods) (Reddy, 2016).

Integrated fish farming systems are categorized mainly 3 types according to their settings.

(i) Complete or Vertical Integration

Comprehensive or Vertical integration (cultivation of animals instantly above the fish pond so that livestock faeces can fall over immediately into the cultivation pond, rising vegetables on the dyke or midst of the pond where it can easily utilize the pond water).

(ii) Semi-integration

This is only a portion of the integration. For example, a fishpond integrated with a dairy farmhouse is situated a significant distance from the water body, with disposal transmitted towards the fish pond.

(iii) Free Integration

The procedures are independent in this case, for example, the fish pond is integrated with the supermarket. These are not associated and thus operate independently. As a result, no energy flows between one and the other. However, funds from either local or supermarket sources are used to establish fish ponds as well as create jobs (Eyo *et al.,* 2004).

The integration of different factors affected integrated fish farming system in tropical and non tropical region were mentioned in Table 18.2 and Figure 18.2.

Table 18.2: Factor Offsetting Integrate Fish Farming System in Tropical and Non Tropical Region

Factor	*Farming Systems*	
Climate	Tropical	Non-tropical
Types of water	Freshwater	Brackish water
Water supply	Irrigated	Rainfed
Farm size	Small scale	Large scale
Farming system	Crops	Livestock
Species of crops	Rice	Maize
Species of live stock	Ruminants	Monogastries
Ruminants	Large scale(Cattle, Buffalo)	Small scale(Goat, Sheep)
Monogastries	Pig	Poultry

Source: Edwards *et al.* (1988).

Utilizing Wastes in Several Ways of Recycling

A combined fish farming technique is one of the most effective strategies for improving livestock and plant protein generation while minimizing the usage of soil, water, as well as waste resources. The major biological function of an integrated agribusiness method is metabolic end recycling; however, enhanced area utilization, in which two subsystems inhabit a component or all of the area needed for one subsystem, may be an essential consideration for improved productivity (Edwards *et al.,* 1988). The animal droppings alone can be effectively manipulated by utilizing numerous strategies and techniques and nutrient content of different animal waste were mentioned in table 3 (Gaddie and Douglas, 1975).

☆ Fresh animal waste can pertain instantly to the fish rearing ponds. Pig ties, chicken crops, and enclosures for ducks as well as geese can be

Figure 18.2: Different of Types of Integrated Fish Farming System.

established on the dikes or above the ponds. Fresh feces thus arrive at the ponds immediately, preventing power losses due to the processing of dung as well as transport. The feedstuff of animals is not completely absorbed as well as then can be instantly employed by the fish. As a result, the number of organisms should be equivalent to the component of the water surface.

- ☆ The leftovers, as well as the liquid portion of anaerobic fermentation to produce biogas, are sometimes used to fertilize water systems. Composting from aerobic fermentation can also be used for fish farming.
- ☆ Livestock waste can be utilized informally throughout one or two trophic phases of the nutrition chain, including cultivating fodder grain for fodder herbivores fish, raising earthworms or additional animals directly to feed carnivorous fish, or using it as an ingredient of pelleted feeds.
- ☆ Chicken waste can be utilized to nourish, as well as pig dung, which can, in turn, be utilized as fertilizers for the fish pond (Long, 2012).

Table 18.2: Nutrients of different Farming Waste Material

Types of Farming System	*Moisture (per cent)*	*Organic Material (per cent)*	*Nitrogen (per cent)*	*Phosphorus (per cent)*	*Potassium (per cent)*
Cow manure	80-85	14.6	0.30-0.45	0.15-0.25	0.05-0.15
Pig dung	85	15.0	0.50-0.60	0.45-0.60	0.35-0.50
Poultry dropping	78	50	1	1.25	0.90
Goat		60	17.31	1.78	2.88
Rabbit		50	12.50	1.33	1.20
Duck	56.6	25.2	1	1.4	0.62

Ecological Sustainability of Integrated Fish Farming

Every recent, large-scale diet system has visible environmental as well as social consequences. Even the long-term viability of recent, large-scale organic agribusiness has been called into question. The sustainable development of ecological parameters in an agro-ecosystem is critical to the cultivating venture's success. The Brundtland Commission (WCED, 1987) defined long-term sustainability as the capacity to join the requirements of the existing without compromising future generations' capabilities to convene their requirements. Several aspects of combined fish farming improvement have had significant social as well as environmental consequences (Black, 2000). The deteriorated state of the greatest aquatic ecosystems, especially in combination with social problems about subtracting "new" sources of water contamination to already overloaded ecological systems, will necessitate fish farming developing a new ecological perspective as well as sustainable operational methods (Akpan and Okafor, 1997). Inadequately governed combined strategies generally have higher nutrient loading, which causes cyanobacterial bloom (Pearl and Tucker, 1995), whereas well-managed systems have inadequate nutrient loading. In a well-managed system, waste emission is minimal to non-existent. However, ecologically long-term fish farming is the creation of aquatic farming systems that protect and improve the structures and functions of their natural and social environments. It entails reasonable ecological perspectives that will create fish farming strategies for numerous integration methods by utilizing available resources, beneficial recycling of manure and substances that can deteriorate natural ecosystems, as well as adequate career creation as well as trade techniques that will be generally recognized as gathering both financial and ecological appreciations.

Nutrient Dynamics in Integrated Fish Farming Systems

Nutrient dynamics in any aquatic ecosystem is the investigation of important plant nutrient levels as they move through one compartment to another and from one transition stage to another. There are two types of nutrient availability in any aquatic environment: (i) internal-autochthonous and (ii) external-autochthonous. Internal factors come from either the sediments phase or natural fixing procedures (this applies only to the major plant nutrient nitrogen as it commonly emerges from nitrogen fixers utilizing the atmospheric nitrogen as well as which is relatively common in the freshwater ecosystem). Carbon, nitrogen, potassium, as well as phosphorus, are important plant nutrients that promote phytoplankton growth in any aquatic environment. Because potassium is abundant in several aquatic environments, the dynamics of potassium do not require as much caution as they do in terrestrial ecosystems. Fertilization with organic or inorganic nutrients is required to maintain sufficient phytoplankton as natural food to promote optimum fish production. In theory, pond fertilization should be based on Liebig's law of the minimum, which indicates that biological production is restricted by the nutrient content in the lowest supply. Under scarce nutrient supply, the percentage of nutrient availability by phytoplankton is dependent on the concentration, as well as total phytoplankton production is straight proportional to the initial composition of limiting nutrients. However, in practice, optimal pond fertilization is a very complicated consequence due to the dynamics of intrinsic as well as extrinsic factors in pond ecosystems. For example, the difference in nutrient prerequisites of individual algal species governs the necessity to procure a proportional supply of important nutrients (Ahilan *et al.,* 2011).

Ecological Reclamation of Organic Waste

Waste reclamation refers to the regeneration of a waste constituent for usage in a manner other than its original purpose; re-use refers to numerous uses of a specified material or commodity; as well as reclaiming refers to regaining from manure the dispute from which a commodity was created as well as re-introducing it into the manufacturing operation to the reproduction of the similar commodity (Ademoroti, 1996). In integrated fish farming, various organic manure is being recovered, and converted, as well as re-used. This farming strategy has been recommended as an environmentally peaceful method of converting manure, particularly those generated by the farming of higher trophic status fish varieties that necessitate an external energy supply (Buschmann, 2000). Reclamation, conversation, as well as re-use aid in the resolution of waste disposal and resource material scarcity issues. Combined fish husbandry is a conventional technique that has contributed to the development of fish farming in China (Hu and Zhou, 1989). In several of the integration processes in Asia as well as other fractions of improving nations, fish crops continued as a greatly significant workout. However, marketable pig/chicken/duck cultivation has been followed by fish as well as crop generation, primarily through utilizing animal manure as fish diet as well as crop dung, respectively (Mukherjee, 1995).

Advantage of Integrated Fish Farming

The advantages of integrated fish farming are not short term as a food security and income generation but also provided sustainability (Figure 18.3) and described below.

(i) Food Security

Integrated fish farming provides opportunities in this regard because it continues to serve as a food-production base that manages to combine crop cultivation, livestock raising, as well as fish cultivation. The potential of integrated aquaculture can be quite broad. The fish farm does not exclusively provide sufficient fertilizer to generate a huge number of fish, but it further generates flesh, milk, eggs, vegetables, and other products by fully utilizing the aquatic ecosystems, water surface, land, as well as pond silt to enhance the food required for human ingestion (Eyo *et al.*, 2004).

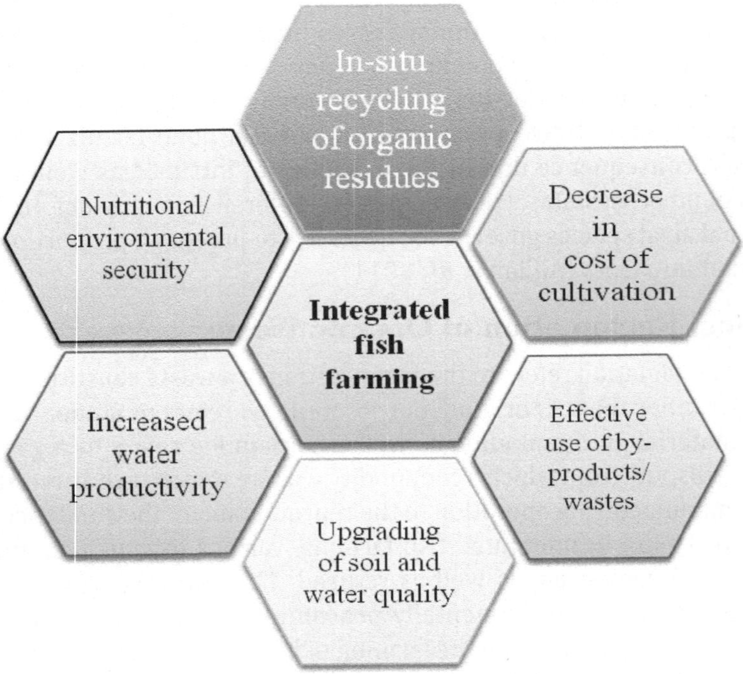

Figure 18.3: Advantage of Integrated Fish Farming.

(ii) Increase Food Supply

A wide range of products or products derived from livestock farming practices. Crop productions, as well as fish farming, have been accomplished. This increases the availability of protein-rich foods. Using only pelleted feeds as well as livestock protein origins for fish feed is not cost-effective and minimizes the amount of nutrition accessible to human utilization (Eyo *et al.,* 2004). The productivity of various elements of the cultivation strategy, such as paddy, milk, animal, chicken, pig, goat keeping, duckery, apiculture, sericulture, horticulture, mushroom farming, and so on, is required in their integration when using the agribusiness strategy. The integration is designed in an extremely a direction that the output of one constituent becomes the intake for another, with a higher level of complementary impacts on one another. The cattle are nourished fodder, which produces milk. The feces, urine, as well as litter, generate farmhouse manure as well as energy for fish ponds. The siltation of fish ponds is used as crop waste. Farmyard organic matter can replace approximately 25 per cent of the approved N, P, as well as K for crop productions while al enhancing the physicochemical and natural properties of the soil (Gill *et al.,* 2009).

(iii) Regular Source of Income

A generates a year-round framework for a consistent source of money from numerous farm commodities. For example, in chicken, vegetables, as well as aquaculture, before harvesting the fish, which may withstand several months, the fisherman can sell the eggs and veggies, generating income for some time (Onada, and Ogunola, 2016). Some examples of economics of integrated fish farming system were depicted in Table 18.3.

Table 18.3: Economic Viability of Integrated Farming System Models

States	*Farming System*	*Net Profit/ha/year*	*Reference*
Assam	Rice + fish	27896	Goswami *et al.* (2004)
Bihar	Rice + wheat + dairy	86105	Kumar *et al.* (2014)
	Crop + fish + poultry	2253	Kumar *et al.* (2012b)
Chhattisgarh	Crop + dairy + goat + poultry + duck	58456	Ramrao *et al.* (2006)
	Crop + dairy + goat + poultry + duck	452096	Sharma *et al.* (2017)
Tamil Nadu	Goat + fish + cropping applied	Rs. 131118	Jayanthi *et al.* (2003)
	Cropping + Fish +Pigeon	98778/ha/annum	Balasamy *et al.* (2003)
	Rice + fish + poultry	97731	Manjunatha *et al.* (2014)

States	Farming System	Net Profit/ha/year	Reference
	Rice + poultry + fish + mushroom	11755	Rangaswamy *et al.* (1995)
	Cropping + Fish + Goat	131118 ha^{-1} $year^{-1}$	Balasamy *et al.* (2003)
	Cropping + fish + poultry	114665	Jayanthi *et al.* (2003)
	Rice + Fish + Azolla	31788 ha^{-1} $year^{-1}$	Balasamy *et al.* (2003)
Goa	Rice + Fish	28569	Meena *et al.* (2022)
	Rice + Azolla + Fish	32140	Meena *et al.* (2022)
Kerala	Rice + groundnut + mushroom + poultry	102857	Meena *et al.* (2022)
	Rice + cowpea + mushroom + poultry	105627	Meena *et al.* (2022)
	Rice + brinjal + mushroom + poultry	165334	Meena *et al.* (2022)
Karnataka	Rice + fish + poultry	62977	Channabasavanna and Biradar, (2007)
Odisha	Crops + poultry	37320	Mohanty *et al.* (2010)
Punjab	Fish + piggery	113200 ha^{-1} $year^{-1}$	Meena *et al.* (2022)
Rajasthan	Crop + goat + sheep + dairy	15013	Kumar and Upadhyay, (2009)
Madhya Pradesh	Dairy + 15 goats+10 poultry + 10duck + fish	44913 ha^{-1} $year^{-1}$	Meena *et al.* (2022)
Uttar Pradesh	Crop + dairy + goat + fish + apiary	135820	Singh *et al.* (2011)

(iv) Self-sufficiency

Integrated is appropriate for underprivileged farmers with unusually poor purchasing behavior and consistent low spending on diet as well as additional dietary requirements (Ayinla, 2003). The diverse environment of an integrated fish farm creates better job opportunities than the single unified fish cultivation (Huazhu and Baotony, 1989). The strategy is all-encompassing because the period is nicely employed in other farm operations depending on the nature of the integrated implicated. Crops such as vegetables are obtained continuously even during the dry season in fish cum agricultural production because water from the pond is utilized to adequately water the farms. Farmers are involved in one or more cultivation activities year-round, making them self-sufficient as well as productive all year (Gabriel *et al.*, 2007).

(v) Biological Productivity

The natural productivity of any aquatic body is commonly assessed by estimating the qualitative as well as the quantitative abundance of planktons, which are the biological nutrition of fish. Livestock waste increases pond biological productivity via a variety of pathways, resulting in increased fish production. The biomass obtained in kg/ha is referred to as productive capacity, and it is used to calculate the system's "net community production." This capability indicates the quality of the underlying soil as well as the quality of the pond water; it is also dependent on the characteristics of waste used. The community structure of plankton is affected by the nature of manure. The proportion of biomass endorsed per unit of power flow in a specific ecosystem helps determine how mature the strategy is. This reveals the technique's capacity to transform the power available into something beneficial (Gebru, 2021).

(vi) Energy Production

Animal excrement can be used to generate biogas and fuel for residential home uses including cooking and lighting, as well as rural enterprises such as powering mills and water pumps. Biogas can start replacing the usage of droppings cakes, charcoal, as well as timber as fuel. 25 kg of completely new cow dung can generate one cubic meter of biogas, which could be utilized to continue providing power for light, warmth, or motive power (Reddy, 2016).

Constraints of Integrated Fish Farming

A consequence of climate change includes such water availability as well as weather patterns such as excessive rain, and exacerbates eutrophication and stratification in stationary waters, resulting in crop failures.

(i) Despite technical assistance, scientific fish cultivation has yet to gain widespread acceptance. State Fisheries departments' understaffed extension divisions have been unable to perform the Aesthetics role. The prevailing information gap between the scientists as well as the farmer requires to be bridged by reorienting the occurring extension setup.

(ii) The accessibility of creation technology is not exclusively characteristic. Affect manufacturing judgments. The commodity should be effectively traded. Makers' cooperatives in regions with multipurpose objectives can move a lengthy path toward achieving sustainable output programs. In a declaration to join the short-term prerequisites of budgets for buying core intakes, a burden can be generated by the hypothecation of production (Livestock as well as fish). The current practice of insisting on security as well as marketing materials may be improved better to benefit manufacturers.

(iii) In the case of personal waters body, structural stiffness such as numerous proprietorships, public leasing rights, as well as various water uses have frequently hampered the improvement of fish farming. There is a necessity for cultivation and a reasonable residence regulation comparable to agricultural land.

(iv) Poaching (criminal withdrawal of fish) is another deterrent to output. This, however, is legislation as well as an order issue. The emerging prerequisite in the legislation of the country could be strengthened.

Conclusion

This chapter concluded that integrated approach is improved technology as other mono-cropping system or sole fish farming. Integrated fish farming practices reduced problem of safe disposal of organic waste which were produced through agriculture or livestock. The organic waste increased the biological productivity, consequently enhancing fish production in a limited resources. The integrated fish farming also reduced the malnutrition problem and help for nutritional security. The integrated fish farming reduced the cost of cultivation and improved the income as well as environmental safety.

REFERENCES

Ademoroti, C.M.A. (1996). Environmental Chemistry and Toxicology. Floudex press Ltd., Ibadan, Nigeria. pp.-217

Ahilan, B., Ravaneshwaran, K. and Kumaravel, P. (2011). Integrated Aquaculture. Daya publishing house, ISBN: 9789383048137. pp-105-114.

Ahmed, N. and Garnett, S. T. (2010). Sustainability of freshwater prawn farming in rice fields in southwest Bangladesh. *Sustainable Agriculture*, **34**: 659–679.

Ahmed, N. and Garnett, S.T. (2011). Integrated Rice-Fish Farming in Bangladesh: Meeting the Challenges of Food Security. *Food Security*, **3**(1): 81-92.

AIFP. (2005). Farming in Nigeria waters. Newsletter of the Aquaculture and Inland Fisheries Project (AIFP) of the National special Programme for food security in Nigeria. FAO Office Abuja, Nigeria. **3**(4): 2-4.55.

Akpan, E.R. and Okafor, N. (1997).On Organic Fertilization and Plankton Development in Two Experimental Fresh Water Ponds of Nigeria. *Journal of Aquaculture in the Tropics,* **12**: 147-154.

Ayinla, O.A. (2003). Integrated fish farming: A veritable tool for poverty alleviation/ Hunger eradication in the Niger Delta Region. *In*: Eyo, A.A and Atanda, J.O (eds). Conference Proceedings of Fisheries Society of Nigeria, Owerri, Nigeria. pp. 40-41

Balusamy, M., Shanmugham, P.M. and Baskaran, R. (2003). Mixed farming an ideal farming. Intensive Agriculture, **41**: 20 – 25.

Bisht, P., Oli, B.P. and Shah, K. (2009). Integrated poultry cum fish farming in Central Himalaya. *Environment Conservation Journal,* **10**(3): 25-30.

Black, K.D. (2000). Environmental Impacts of Aquaculture. CRC Press, Boca Raton, 36.

Buschmann, A.H. (2000). Integrated cultivation of salmon and seaweed: Productive, Environmental and Economic issues. Aquaculture Canada (2000) Abstracts.

Channabasavanna, A.S. and Biradar, D.P. (2007). Relative performance of different rice-fish-poultry integrated farming system models with respect to system productivity and economics. *Karnataka Journal of Agricultural Sciences*, **20**: 706-709.

Coche, A.G. (1967). Fish culture in rice fields a world-wide synthesis. *Hydrobiologia*, **30**: 1-44.

Dalsgaard, J.P.T., Light foot, C. and Christensen, V. (1995). Towards quantification of ecological sustainability in farming systems analysis. *Ecological Engineering*, **4**: 181-189.

Edwards, P., Pullin, R.S.V. and Gartner, J.A. (1988). Research and education for the development of integrated cmp-livestock-fish farming systems in the tropics. ICLARM (International Center for Living Aquatic Resources Management, Manila, Philippines) Studies and Reviews, pp 16-53.

Eyo, A.A., Ayanda, J.O., Falayi, B.A. and Adelowo, E.O. (2004). Economic prospects of investment in fish cum livestock farming. *In*: 18th Annual Conference of the Fisheries Society of Nigeria (FISON) at Owerri, Nigeria on dated 8-12 December, 2003.pp 82-104.

Gabriel, U. U., Akinrotimi, O. A., Bekibele, D. O., Anyanwu, P. E. and Onunkwo, D. N. (2007). Economic benefit and ecological efficiency of integrated fish farming in Nigeria. *Scientific Research and Essay*, **2**(8): 302-308.

Gaddie, E.R. and D.E. Douglas (1975). Earthworm for ecology and profit. Vol. 1, Scientific Earthworm farming, Bookworm publishing company, Ontario, California, USA.

Gebru, T. (2021). Integrated Aquaculture with Special Reference to Fish Integration with Animal Husbandry to Enhance Production and Productivity. *Journal of Fisheries and Livestock Production*, **9**: 306.

Gill, M. S., Singh, J. P. and Gangwar, K. S. (2009). Integrated farming system and agriculture sustainability. *Indian Journal of Agronomy*, **54**(2): 128-139.

Goswami, M., Biradar, R.S. and Sathiadhas, R. (2004). Techno-economic viability of rice-fish culture in Assam. *Aquaculture Economics and Management*, **8**: 309-317.

Hu, B. and Zhou, E. (1989). Review on development of integrated fish farming in China. In: Selected Papers of Freshwater Fisheries Research. *Scientific Press, Beijing*, pp. 77-110.

Huazhu, Y. and Baotony, (1989). Integrated fish farming in China, NACA Technical Manual Bangkock, Thailand. pp. 153-209.

Jayanthi, C., Baluswamy, M., Chinnusamy, C. and Mythily, S. (2003). Integrated nutrient supply system of linked components in lowland integrated farming system. *Indian Journal of Agronomy*, **48**: 241-246.

Kalita, H. K., Talukdar, B., Baishya, R. A., Basumatary, S., Das, J., Dutta, A., Ch. Barman, R. and Sarma, D. (2015). Role of Duck Droppings on Pond Productivity Through Fish-Cum-Duck Integrated Farming System in Agro Climatic Condition, Assam. *Nature Environment and Pollution Technology*, **14**(3): 541.

Kumar, S. and Upadhyay, A.D. (2009). Goat farmers' coping strategy for sustainable livelihood security in arid rajasthan: An empirical analysis. *Agricultural Economics Research Review,* **22**: 281-290.

Kumar, S., Bhatt, B.P., Dey, A., Kumar, U., Chandra, N. and Shivani, D. (2014). Sustainable integrated farming system models for small farm holders of Bihar. Research in Farming Systems. Today and Tomorrow's Printers and Publishers, New Delhi, India.

Kumar, S., Dey, A., Kumar, U., Chandra, N. and Bhatt, B. P. (2012a). Integrated Farming System for Improving Agricultural Productivity. In: Status of Agricultural Development in Eastern India. Eds: Bhatt, B.P., Sikka, A.K., Mukherjee, J., Islam, A. and Dey, A. Published by ICAR Research Complex for Eastern Region. pp. 205-230.

Kumar, S., Subash, N., Singh, S.S. and Dey, A. (2012b). Evaluation of different integrated farming system (IFS) for small and marginal farmers under semihumid climatic environment. *Experimental Agriculture,* **48**: 399 –413.

Little, D. and Edwards, P. (2003). Integrated livestock-fish farming systems. Published by Food and Agriculture Organization of the United Nations, Rome 2003. ISBN 92-5-105055-4.

Long, A.C. (2012). Fish feeding in integrated fish farming. Cyber tech publications. pp. 188-189.

Manjunatha, S.B., Shivmurthy, D., Satyareddi, S.A., Nagaraj, M.V. and Basavesha, K.N. (2014). Integrated farming system - an holistic approach: A review. *Journal of Agriculture and Allied Sciences,* **3**: 30 – 38.

Meena, L.R., Subash, N., Anjali, Kumar, D. and Nirmal, (2022). Integrated Farming System Models Development for Small and Marginal Households for Sustainable Production and Livelihood Improvement in India: An Overview. *Medicon Agriculture and Environmental Sciences,* **3**(1): 05-18.

Mohanty, D., Patniak, S.C., Das, P.J., Parida, N.K. and Nedunchezhiyan, M. (2010). Integrated farming system for sustainable livelihood: a success story of a tribal Farmer. *Odisha Review,* **5**: 41 – 43.

Mukherjee, T.K. (1995). Integrated crop-livestock-Fish production systems for maximizing productivity and economic efficiency of small holders farms. In: Symoens, J.J. and Micha, J,C. Seminar proceedings RAOS/GA/FAO on the Management of integrated freshwater Agro piscicultural Ecosystems in Tropical Areas (Brussels, 16-19 May, 1994) pp. 121-143.

NACA, (1989). Integrated fish farming in China. Network of Aquaculture Centers in Asia-Pacific (NACA) Technical manual 7. A world food day, publication of the network of aquaculture Centres in Asia and the pacific. Bangkok, Thailand.

Onada, O.A. and Ogunola, O.S. (2016). Improving Food Security in an Eco-Friendly Manner through Integrated Aquaculture. *Open Access Library Journal*, **3**: e2476.

Pandey, A., Dhawan, A., Srivastava, A. and Singh, A. (2016). Integrated fish-cum-duck farming: productivity enhancement through technological interventions. *Indian Journal of Ecology*, **43**: 894-896.

Pearl, H.W. and Tucker, O.S. (1995). Ecology of Blue-Green Algae in Aquaculture Ponds. *Journal of the World Aquaculture Society*, **26**: 109-131.

Prein, M. (2002). Integration of aquaculture into crop-animal systems in Asia. *Agricultural Systems*, 71: 127–145.

Ramrao, W.Y., Tiwari, S.P. and Singh, P. (2006). Crop livestock integrated farming system for the marginal farmers in rain fed regions of Chhattisgarh in Central India. *Livestock Research for Rural Development*, **18**: Article #102.

Rangaswamy, Venkittasamy, R., Premsekar, N., Jayanthi, C., Purusothaman, S. and Palaniappan, S.P. (1995). Integrated farming system for rice based ecosystem. *Madras Agricultural Journal*, **82**: 287-290.

Reddy, P.P. (2016). Integrated crop-livestock farming systems. *In*: Sustainable intensification of crop production, Springer: Singapore, ISBN : 978-981-10-2701-7. pp. 357-370.

Rota, A. (2009). Integration of crop and livestock production in conservation agriculture: guidelines for project design. International Fund for Agricultural Development, Rome, Italy

Sharma, R.L., Abraham, S., Bhagat, R. and Prakash, O. (2017). Comparative performance of integrated farming system models in Gariyaband region under rainfed and irrigated conditions. *Indian Journal of Agricultural Research*, **51**: 64-68.

Shyam, R., Shaha, G. S. and Dey, M.K. (2012). Aquaculture success stories. Central Institute of Freshwater Aquaculture (CIFA). Bhubaneswar.

Singh, J.P., Gangwar, B., Pandey, D.K. and Kochewad, S.A. (2011). Integrated Farming System Model for Small Farm Holders of Western Plain Zone of Uttar Pradesh. PDFSR Bull No. 05. Project Directorate for Farming Systems Research (PDFSR), Modipuram, Meerut, India. p-58.

Singh, S., Singh, P., Verma, N. and Kumar, D. (1991). Integrated fish farming-rationale and scope. College of Fisheries, G.B. Pant University of Agriculture and Technology, Pantnagar, India.

WCED (1987). Our Common Future. Report of the World Commission on Environment and Development (WCED), published by the Oxford University Press, Oxford, the United Nations.

Yang, H., Fang, Y. and Chen, Z. (2001). Integrated Grass-Fish Farming Systems in China. In: Integrated Agriculture-Aquaculture: A Primer, FAO Fisheries Technical Paper 407. FAO, Rome, Italy.

Chapter 19

Non-Conventional Energy Sources for Integrated Farming System in Irrigation and Water Management

Deepak Kumar[1], Rashmi[2] and Utkarsh Kumar[3]*

[1]Ph.D. Scholar, Department of Soil and Water Conservation Engineering, College of Agricultural Engineering and Technology, Junagadh Agricultural University, Junagadh – 362001, Gujarat
[2]Research Scholar, Department of Agricultural Economics, Institute of Agricultural Sciences, Banaras Hindu University, Varanasi – 221005, Uttar Pradesh
[3]Crop Production Division, ICAR-Vivekananda Parvatiya Krishi Anusandhan Sansthan, Almora – 263601, Uttarakhand
**Corresponding author E-mail: utkarsh.kumar@icar.gov.in*

ABSTRACT

India and the rest of the globe are dealing with the depressing prospect of an altered climate as a result of the ongoing increase in the use of fossil fuels. The diesel and electric water pumps used for irrigation in India are a significant consumer of fossil fuel-based energy. Using diesel and electric water pumps results in an annual increase in CO2 emissions of up to 45 million tonnes, or between 8 and 12 percent of all greenhouse gas emissions. The Indian government initiated a solar pumping initiative for agriculture and drinking water in 2014–2015 with an ambitious objective of installing 0.1 million Solar Photovoltaic Water Pumps (SPVWP) through 2020–2021 due to its established benefits globally. The government's initiatives and systems of support for the expansion of SPVWP in India were thoroughly examined in this study. Updates on solar water pumping technology, performance analysis, optimal sizing, the deterioration of PV generators used to power the pumps, economic and *environmental considerations, and developments in PV material science and efficiency are the main topics of the study. This article provides an update on the development and application of solar water pumping technology. Areas for potential follow-up research are also noted. A model to describe the motor-pumps subsystems utilised in PV pumping installations is also included in this paper. For various total heads, the model explicitly describes the water flow output (Q) as a function of the electrical power input (P) to the motor-pump. A method to calculate the amount of carbon dioxide (CO_2) emissions saved by using water pumping facilities powered by a solar array instead of diesel fuelled generators is provided based on the motor-pump subsystem model.*

Introduction

India's GDP (Gross Domestic Product) growth is significantly influenced by the agriculture sector. According to the World Bank's Employment in Agriculture report, the agriculture industry directly employs around 70 per cent of the entire population, which has led to an ongoing increase in energy consumption in recent years. According to Central Statistics Office data for energy statistics 2016, the agriculture sector had the highest growth rate in power consumption among all sectors, using 18 per cent of the total power generated in 2014–2015 (Ministry of statistics and programme implementation, Central Statistics Office, Energy Statistics, 2016). This is mostly due to the ongoing rise in the use of fossil-based water pumps for irrigation, including both electric and diesel ones (International Energy Agency and India Energy Outlook, 2015, Reddy, 1999). There are an estimated 21 million irrigation pumps in India, 9 million of which run on diesel and 12 million on electricity, and they are primarily responsible for the agricultural sector's high energy consumption (Narale and Rathore, 2015). Furthermore, groundwater is the primary source of irrigation in India. As a result, according to the World Bank's 2012 report (World Bank, Water Paper. Managing the Invisible: Understanding and Improving Groundwater Governance, 2012) India is the world's largest irrigation groundwater user.

Because millions of water pumps require energy in the form of diesel or electricity, India's agriculture sector consumes 85 million tonnes of coal each year. Furthermore, diesel-powered pump sets consume 4 billion litres of diesel (Shakti Foundation, Feasibility Analysis for Solar Agriculture Water Pumps in India, 2014). Fuel-powered groundwater pumping accounts for 8 per cent –12 per cent of total greenhouse gas emissions (Shah, 2009). The combustion of fossil fuels increases CO2 emissions, which is a major contributor to the current climate change crisis. The reliance on fossil-based energy must be reduced in order to progress toward sustainable development. Replacing an estimated 21 million pump sets used in agriculture with energy-efficient pump sets would result in 131.96 billion kWh of annual energy savings. This would also result in a 45 million tonnes CO2 reduction in annual greenhouse gas emissions (Press information bureau, Ministry of Power).

Aside from that, particularly in India, fossil fuel-based water pumping has several limitations, including a lack of power grid availability, a continuous increase in the price of diesel, and increased maintenance costs. Access to electricity as an energy source for agricultural use is a major issue in India, forcing farmers to rely on fossil fuel-based pumping systems. Furthermore, more than 60 per cent of India's electricity is generated using fossil fuels. Furthermore, according to International Energy Agency data (IEA, 2016), approximately 244 million people in India still do not have access to electricity (International Energy Agency and World Energy Outlook, 2016). Considering all of these factors, as well as the critical need for energy in the agricultural sector, a solar photovoltaic water pumping (SPVWP) system may be a suitable and sustainable replacement for fossil fuel-powered water pumps.

An Overview of Solar Photovoltaic Water Pumping Technology in India

In India, the Ministry of Renewable Energy Resources (MNRE) is in charge of the expansion and development of SPVWP for irrigation and drinking water across the country. According to the MNRE guidelines for SPVWP systems released in 2015 (Ministry of New and Renewable Energy and Solar Photovoltaic Water Pumping System, 2015), the PV module capacity of the SPVWP system must be in the range of 900 Watt to 9 KWP and must operate under daily average solar irradiation of 7.15 KWh/m^2 on the surface of the PV array (*i.e.* coplanar with the PV modules). Table 19.1 contains the detailed technical specifications of the SPVWP system released by MNRE and followed throughout the country, while Figure 19.1 depicts a typical SPVWP system installation (Climate Technology Centre and Network. UNFCC Technology Mechanism).

Table 19.1: Technical Specification of DC and AC Solar Photovoltaic Water Pumping system in India (Ministry of New and Renewable Energy and Solar Photovoltaic Water Pumping System, 2015)

Sl. No.	*Description*	*PV Array (W_p)*	*Motor Capacity*	*Shut Off Dynamic Head (Meters)*	*Water Output (L/day)*
1	Model-I* (DC)	900	1 HP	12	90,000
2	Model-II* (DC)	1800	2 HP	12	180,000
3	Model-III* (DC)	2700	3 HP	25	135,000
4	Model-I (DC)	1200	1 HP submersible with controller	45	42,000
5	Model–II (DC)	1800	2 HP submersible with controller	45	63,000
6	Model-V (DC)	3000	3 HP submersible with controller	100	42,000
7	Model-VIII (DC)	4800	5 HP submersible with controller	150	45,600

Sl. No.	Description	PV Array (W_p)	Motor Capacity	Shut Off Dynamic Head (Meters)	Water Output (L/day)
8	Model-IX (DC)	6750	7.5 HP submersible with controller	70	141,750
9	Model-XIII (DC)	9000	10 HP submersible with controller	100	126,000
10	Model-I* (AC)	900	1 HP submersible with controller	12	81,000
11	Model-II* (AC)	1800	2 HP submersible with controller	15	162,000
12	Model-III* (AC)	2700	3 HP submersible with controller	15	243,000
13	Model-V* (AC)	4800	5 HP submersible with controller	15	432,000
14	Model –I (AC)	1200	1 HP submersible with controller	45	38,400
15	Model – II (AC)	1800	2 HP submersible with controller	45	57,600
16	Model – V (AC)	3000	3 HP submersible with controller	100	39,000
17	Model – VIII (AC)	4800	5 HP submersible with controller	150	40,800
18	Model – IX (AC)	6750	7.5 HP submersible with controller	70	128,250
19	Model-XIII (AC)	9000	10 HP submersible with controller	100	117,000

* SPVWP models for shallow well (surface).

A brief detail of SPVWP system components are given below:

I. **Solar Panels**: The SPVWP should be run with a PV array capacity ranging from 200 WP to 10000 WP (Ministry of New and Renewable energy and Solar Photovoltaic Water Pumping System, 2015). To achieve the desired power, the solar panels will be connected in series and parallel.

II. **Pump:** The SPVWP system in India uses any of the following pumps:

a) Surface mounted pumps

b) Submersible pumps

c) Floating motor pumps

d) Any other pump set after approval from the ministry

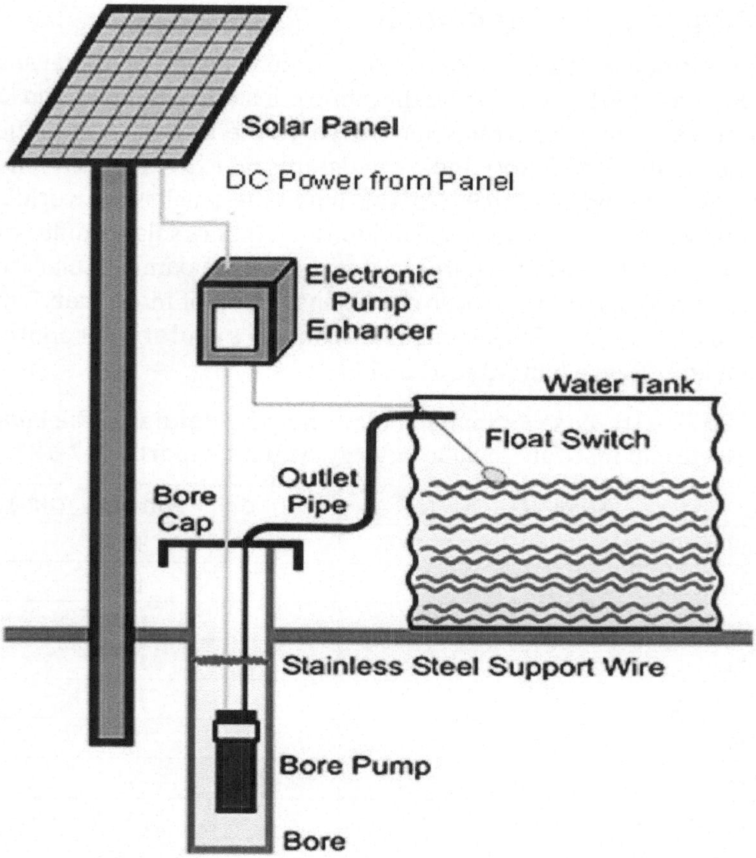

Figure 19.1: A Typical Solar Water Pump Installation (Climate Technology Centre and Network. UNFCC Technology Mechanism).

III. **Electric motor:** Following are the two types of motors available in India:

a) Inverters are required to convert DC to AC for AC motors. Solar pumping systems make use of variable-frequency inverters, which optimise current by matching it between the panel and the pump.

b) Permanent magnet DC motors are generally more efficient. They can be equipped with or without carbon brushes. The carbon brushes in DC motors should be replaced every two years. As a result, brushless DC motors are becoming increasingly popular in solar water pump systems.

IV. **Electronics**: Major electronic components used in SPVWP are Maximum Power Point Tracker (MPPT), Inverter (if A.C pumps are used) and a Controller.

Availability of Solar Irradiation

The performance of solar water pumps is determined by the availability of solar irradiation at the location where the pump is installed (Sontake and Kalamkar, 2016). An increase in solar irradiation increases the flow rate and efficiency of SPVWP (Hamrouni *et al.* 2008). India's geographical extent is between 8°4' and 37°6' north latitude and 68°7' to 97°25' east longitude, and it is the world's seventh largest country, with a landmass of 2.9 million km^2. India's solar profile is very rich. India's geographical location ensures that it receives maximum solar irradiation all year. Many regions in India receive solar insolation of more than 5 KWh/m^2/day and account for nearly 58 percent of the country's land area, or approximately 1.89 million km^2 (Ramachandra *et al.* 2011).

Table 19.2: State-Wise Estimated Solar Energy Potential in the Country (National Institute of Solar Energy, Annual Report 2017-18)

Sl.No.	*State/UT*	*Solar Potential (GW$_p$) #*
1	Andhra Pradesh	38.44
2	Arunachal Pradesh	8.65
3	Assam	13.76
4	Bihar	11.20
5	Chhattisgarh	18.27
6	Delhi	2.05
7	Goa	0.88
8	Gujarat	35.77
9	Haryana	4.56
10	Himachal Pradesh	33.84
11	Jammu and Kashmir	111.05
12	Jharkhand	18.18
13	Karnataka	24.70
14	Kerala	6.11
15	Madhya Pradesh	61.66
16	Maharashtra	64.32
17	Manipur	10.63
18	Meghalaya	5.86
19	Mizoram	9.09
20	Nagaland	7.29
21	Odisha	25.78
22	Punjab	2.81
23	Rajasthan	142.31
24	Sikkim	4.94

Sl.No.	State/UT	Solar Potential (GW_p) #
25	Tamil Nadu	17.67
26	Telangana	20.41
27	Tripura	2.08
28	Uttar Pradesh	22.83
29	Uttarakhand	16.80
30	West Bengal	6.26
31	UTs	0.79
	Total	**748.98**

The time period and water requirement of major crops in India is shown in Table 19.3 (Bhave, 1994).

Table 19.3. Comparative Chart of Water Requirement and Availability of Solar Radiation (Bhave, 1994)

Field Crop Season	*Time Period (Month)*	*Water Requirement (mm)*	*Average Solar Radiation Availability (KWh/m²/day)*	*Water Source*
Kharif (Major crop is Rice)	July-October	900-2500	Maximum (>5)	Rain, Tube well
Rabi (Major crop is Wheat)	November-March	450-650	Normal (>4)	Canals, Tube wells
Zaid (Major crop is sugarcane)	March-June	1500-2500	Maximum (>5)	Tube wells

The period when the major crops that require a lot of water are grown coincides with the period when India receives the most solar irradiation, which is from March to November. As a result, Indian farmers can use solar water pumps to replace existing fossil-fueled water pumps for irrigation.

Economics of SPVWP Systems

The Indian market is extremely cost-conscious. As a result, in order to successfully introduce any new technology into this market, the long-term cost (capital cost and operation and maintenance) must be as low as possible. Because SPVWP systems must compete with well-established fossil-based water pumping technology, they must be economically and financially appealing to the end user. In this regard, various dominating factors such as annual increases in the cost of diesel and fossil-based electricity, frequent power outages, and the inability of electric water pumps to connect to the power grid are making solar water pumps more economically appealing.

Table 19.4: Shuba *et al.* (2010) Shows a Comparison of the Cost of Water Pumping Incurred by using different Water Pumping Sources of 5 HP Pump Capacity for Irrigation in India

Factors	*SPVWP*	*Diesel Powered Pumps*	*Grid Connected Pumps*
Capital cost (INR)	5,00,000 (system cost)	50,000 (Pump cost)	1,50,000 (initial connection cost) and 50, 000 (Pump cost)
Fuel consumption	Nill	1.23 L/hour	4.5 KWh/hour
Fuel price	Nill	58 Rupees/Liter	6 Rupees/KWh
Fluctuation in fuel price	Nill	Increasing	Increasing
Discount rate	10 per cent	10 per cent	10 per cent
Hours of operation per year (6h/day for 200 days)	1200	1200	1200
Net present value (NPV)Over 25 years (INR)	5,16,300	9,27,400	5,60,000
Cost of pumping per hours (INR)	17.2	18.7	30.9

In comparison to diesel and electric pumps, the SPVWP is clearly more cost effective. As a result, it can serve as a better replacement option for fossil-fueled water pumps in India.

Policies and Status of SPVWP in India

In India, the Ministry of New and Renewable Energy (MNRE) is in charge of the growth and development of solar water pumping, and it was in 1993 that the MNRE launched a programme to install 50,000 solar PV water pumping systems across the country. This programme is being carried out with the assistance of several organisations, including the Indian Renewable Energy Development Agency (IREDA) and state nodal agencies (SNAs). Between 1990 and 2000, the high cost of PV silicon solar cells resulted in low penetration of SPVWP across the country. As a result, only 13,964 SPVWP were installed until 2014, compared to the targeted 50,000. (Ministry of New and Renewable Energy, 2014). Despite the fact that the government provided a variety of financial incentives such as subsidies, low-interest rates, and soft loans through IREDA and SNAs, SPVWP growth lagged far behind the stated target.

The Indian government launched the Solar Pumping Program for Irrigation and Drinking Water in 2014. The program's goal was to install 0.1 million pumps in 2014–2015, and it is expected that by 2020–21, at least 1 million SPVWP would be installed across the country (Ministry of New and Renewable Energy, 2014). To promote this programme and ensure its proper implementation on the ground, the MNRE will provide financial assistance through SNAs and the National Bank for Agriculture and Rural Development (NABARD) in accordance with the schemes listed in Table 19.5 (Press information bureau. Ministry of New and Renewable Energy, Government of India Pattern of Central Financial Assistance (CFA) for Off-grid Solar PV Pumps (2015).

Table 19.5: Subsidy Pattern for SPVWP Program in India

Sl.No.	*SPVWP System*	*Capacity*	*Maximum Subsidy in INR (per HP)*
1	DC Pumps	Upto 2 HP	57,600
		>2 HP to 5 HP	54,000
2	AC Pumps	Upto 2 HP	50,400
		>2 HP to 5 HP	43,200
3	DC/AC pumps	>5 HP to 10 HP	194,400

Source: Chandel *et al.* (2017).

The release of the subsidy is conditional on all of SPVWP's systems and sub-systems confirming the minimum technical requirements as laid out by MNRE guidelines from time to time. Beneficiaries must spend 20 per cent of the total system cost of SPVWP, plus bank loans at standard interest rates, over a 10-year repayment period (Ministry of New and Renewable Energy, 2014).

Key Barriers in the Growth of SPVWP in India

It has been more than two decades since the Government of India launched a programme to promote and develop SPVWP throughout the country. Unfortunately, the SPVWP market has not grown in line with expectations, with only 25,000 SPVWP installations by 2016. To improve the SPVWP scenario, the MNRE announced in 2014 a target of installing 1 million SPVWP by 2021. However, the solar pump market in India continues to struggle. Policies for the growth and development of renewable energy face a difficult implementation bottleneck (Beck and Martinot, 2004). Economic and financial barriers, policy and regulatory practises, a lack of education, and technological barriers are all examples of bottlenecks. These barriers become even more critical in countries such as India, where the entire solar PV market must contend with them (Rathore *et al.* 2017).

Discussion and Policy Recommendation

Because of the use of diesel and electric water pumps for irrigation, Indian agriculture has one of the highest carbon footprints in the world. In India, an estimated 21 million fossil fuel-powered water pumps are currently in use for irrigation. This situation is becoming more critical, as nearly 244 million Indians lack access to electricity, and air quality is deteriorating day by day due to increased use of fossil-based energy. SPVWP can provide an environmentally sustainable, dependable, and cost-effective irrigation water pumping solution in India.

Literature Survey of PV Water Pumping Systems

Table 19.6: Summary of PV Water Pumping System Performance Evaluation Studies

Sl.No.	*Reference*	*Country*	*Application*	*Research Finding*
1	Kou *et al.* (1988)	USA	Domestic	Predicted monthly water pumped by a system within 6 per cent of TRNSYS predicted based on hourly data.
2	Hadj Arab *et al.* (1991)	Spain	Domestic	Optimized a proposed PV water pumping system by studying individual requirements with a simulation program.
3	Loxsom and Veroj (1994)	Thailand	Irrigation	Algorithm was developed to estimate the water pumped as per insolation.
4	Alghuwainem *et al.* (1996)	Saudi Arabia	Irrigation	Self excited induction generator utilization avoids need for matching devices and tracking system.
5	Katan *et al.* (1996)	Australia	Domestic	System efficiency increases with MPPT and sun tracker.
6	Pande *et al.* (2003)	India	Irrigation	Reported 6 year pay-back period including subsidies on PV modules.
7	Mohanlal *et al.* (2004)	Egypt	Irrigation	System efficiency is increased upto 20 per cent by manually tracking thrice in a day.
8	Gad (2009)	Egypt	Domestic	Computer simulation program is used to simulate the performance of a proposed PV water pumping system.
9	Mokeddem *et al.* (2011)	Algeria	Irrigation	System efficiency increased by orientation and sizing of PV array and motor pump system.
10	Khan *et al.* (2012)	Bangladesh	Rural water supply	System efficiency is increased by adding DC-DC buck converter for a direct coupled PV water pumping system.

Sl.No.	Reference	Country	Application	Research Finding
11	Atlam and Kolhe (2013)	Turkey	Domestic	System performance and efficiency can be improved by matching the output characteristics.
12	Benghanem *et al.* (2013)	Saudi Arabia	Irrigation	Electronic array configuration should be included in order to match maximum power points of PV array with pumps.
13	Reddy and Reddy (2014)	India	Domestic	Configuration of the photovoltaic system can be improved with dynamic models for inverter, single phase induction motor and Neural network based maximum power point tracking.
14	Setiawan *et al.* (2014)	India	Irrigation	Two important design aspects for PV water pumping system are identified; analyzing piping system to determine the type of pump to be used and power system planning

Source: Chandel *et al.* (2015).

Table 19.7: Summary Highlights of Optimal Sizing of PV Water Pumping Systems

Sl. No.	Reference	System Type	Optimal Sizing Technique	Research Finding
1	Argaw (1994)	Direct coupled interfaced with PWM DC/AC inverter	Non-linear optimization	Optimum matching factor of 0.74 and 0.55 are found using DC/AC inverter facing
2	Wagdy *et al.* (1994)	Direct coupled and battery buffered	Switched mode	Optimization provides maximum utilization of available solar radiation to minimize cost.
3	Cuadros *et al.* (2004)	Inverter coupled with AC pump	Multi-Step optimization	Output yield of crop can be improved by photo irrigation.
4	Firatoglu and Yesilata (2004)	Direct coupled	Multi-Step optimization	System performance is found better for lower photovoltaic array area.

Sl. No.	Reference	System Type	Optimal Sizing Technique	Research Finding
5	Zvonimir *et al.* (2007)	Inverter coupled with AC pump	Hybrid simulation	Electrical power of the PV generator, obtained by the new optimization method is relatively smaller as compared to usual method.
6	Hamidat and Benyoucef (2008)	Inverter coupled with AC pump	Conventional Method	DC engine and a positive displacement pump shows best performance when compared with AC engine centrifugal pump.
7	Kaldellis *et al.* (2009)	Battery based	Two-level analysis	Energy storage capability of a properly designed PV system determined analytically and experimentally to meet the electricity and water needs.
8	Yahia *et al.* (2011)	Direct coupled	Loss of power supply probability (LPSP)	Program developed for relationship between system power reliability and system configurations.

Source: Chandel *et al.* (2015).

Table 19.8: Techniques Used for Improving Efficiency of PV Water Pumping Systems

Sl. No.	Reference	Motor-Pump Set Type	Type of Technique Used	Research Finding
1	Ziyad and Dagher (1990)	DC	Electrical Array Reconfiguration Controller (EARC)	System improves pump performance during cloudy conditions and provides extra pumping hours.
2	Eduard (1997)	DC	Six-step-square-wave-inverter	System can be implemented by a simple micro controller
3	Odeh *et al.* (2006)	AC	TRANSYS modeling	Increasing PV array size improves water output volume and subsystem efficiency and decreases PV efficiency.
4	Abdolzadeh and Ameri (2009)	DC	Spraying water over PV modules	Spraying water over the PV modules strongly improves the system and sub-system efficiencies.

Sl. No.	*Reference*	*Motor-Pump Set Type*	*Type of Technique Used*	*Research Finding*
5	Azadeh (2010)	DC	Covering the surface of PV array by a thin film of water	Reduces system costs and provides needed power with lower array nominal power.
6	Abdolzadeh *et al.* (2011)	DC	Spraying water on PV module	Spraying water on the PV modules decreases modules temperature and increase modules' performance and pump flow rate.
7	Kim *et al.* (2011)	DC	Optimum power point tracking (OPPT) algorithm	System designed is suitable for indoor solar energy harvesting under dim lighting conditions.
8	Joao and Luis (2014)	AC	Two-inductor boost converter (TIBC) and Voltage source inverter (VSI)	Proposed solution is a viable option for more reliability.

Source: Chandel *et al.* (2015).

Table 19.9: Economic and Environmental Aspects of PV Water Pumping Systems

Sl. No.	*Reference*	*Country*	*Study Type*	*Research Finding*
1	Foster and Hanley (1998)	Mexico	Economic analysis	Economic viable PV water pumping systems gained foot hold and changing the face of water pumping in Mexico.
2	Hamidat (1999)	Algeria	Economic analysis	PV surface pumps to supply water can contribute to socio-economic development in remote Sahara regions.
3	Odeh *et al.* (2006)	Ireland	Economic viability	Mismatch between water demand and supply patterns have a major effect on economic viability of the PV pumping.
4	Kumar and Kandpal (2007)	India	Environmental and economic analysis	Capital cost of PV pump, its useful life, price of fuel substituted, and discount rate on the unit cost of CO_2. Emission mitigation are of importance for solar pumping promotion.

Sl. No.	Reference	Country	Study Type	Research Finding
5	Meah *et al.* (2008)	USA	Rural water supply	PV water pumping system reduce CO_2 emission considerably over its 25 – year life span.
6	Fedrizzi *et al.* (2009)	Brazil	Irrigation	Negligence of local specificities and technology transfer methods cause PV water pumping system failure
7	Kaldellis *et al.* (2011)	Greece	Economic and Environmental analysis	PV pumping systems are economical viable options for water consumption need of remote communities.
8	Purohit and Kandpal (2011)	India	Financial Evaluation	PV pumping systems are viable option when sufficient incentives are provided by Government.
9	Jamil *et al.* (2012)	India	Techno-economic analysis	Payback period of less than 4 years with huge saving over 16 years.
10	Foster *et al.* (2013)	USA	Rural water supply	Investment payback for PV water pumping system is averaged about 5-6 years.
11	Rezae and Gholamian (2013)	Iran	Economic analysis	Considerable savings are observed in PV water pumping system as compared to conventional system.

Source: Chandel *et al.* (2015).

Conclusions

PV water pumping technology is a dependable and cost-effective alternative to electric and diesel water pumps for crop irrigation. PV water pumping for urban, rural, and community water supplies and institutions is another potentially viable but underutilised sector. Remote, inaccessible areas with no grid electricity require special attention as well. These industries continue to rely on traditional electricity or diesel-powered pumping systems, resulting in higher recurring costs for users. Given the high installation costs of solar water pumps, particularly for large irrigation and water supply systems, governments must provide more incentives to make the technology a more appealing alternative to diesel and electrical water pumping. Factors influencing performance and efficiency improving techniques, use of highly efficient PV modules including bifacial modules, and degradation of PV generator are areas for further research in order to reduce costs, improve performance, and extend the life of pumping systems.

REFERENCES

Abdolzadeh, M. and Ameri, M. (2009) Improving the effectiveness of a photovoltaic water pumping system by spraying water over the front of photovoltaic cells. *Renewable Energy*. 34: 91–6.

Abdolzadeh, M., Ameri, M. and Mehrabian, M. A. (2011) Effects of water spray over the photovoltaic modules on the performance of a photovoltaic water pumping system under different operating conditions. *Energy Sources*.33: 1546–55.

Alghuwainem, S. M. (1996) Performance analysis of a PV powered dc motor driving a 3 phase self-excited induction generator. *IEEE Transactions on Energy Conversion*. 11: 1.

Argaw, N. (1994) Optimization of photovoltaic water pumps coupled with an interfacing pulse width modulated dc/ac inverter power conditioning devices. *(IEEE First WCPEC; December 5–9, 1994; Hawaii)*. p. 1165–8.

Atlam, O. and Kolhe, M. (2013) Performance evaluation of directly photovoltaic powered DC PM (direct current permanent magnet) motor – propeller thrust system. *Energy*.57: 692–8.

Azadeh, K. (2010) The effects of nominal power of array and system head on the operation of photovoltaic water pumping set with array surface covered by a film of water. *Renewable Energy*. 35: 1098–102.

Beck, F. and Martinot, E. (2004) Renewable energy policies and barriers. *Encyclopedia of energy*. 5 (7): 365–383.

Benghanem, M., Daffallah, K. O., Joraid, A., Alamri and Jaber, A. (2013) Performances of solar water pumping system using helical pump for a deep well: a case study for Madinah, Saudi. Arabia. *Energy Conversion and Management*. 6: 50–6.

Bhave, A. G. (1994) Potential for solar water-pumping systems in India. *Applied Energy*. 48 (3): 197–200.

Chandel, S. S., Naik, M. N. and Chandel, R. (2015) Review of solar photovoltaic water pumping system technology for irrigation and community drinking water supplies. *Renewable and Sustainable Energy Reviews*. 49: 1084-1099.

Chandel, S. S., Naik, M. N. and Chandel, R. (2017) Review of performance studies of direct coupled photovoltaic water pumping systems and case study. *Renewable and Sustainable Energy Reviews*. 76: 163–175.

Climate Technology Centre and Network. UNFCC Technology Mechanism. *https://www.ctc-n.org/technology/library/renewableenergy/solar-water-pumps*.

Cuadros, F., Rodrýguez, F. L., Marcos A, Coello, J. (2004) A procedure to size solar powered irrigation (photo irrigation) schemes. *Solar Energy*.76: 465–73.

Eduard, M. (1997) Water PV Pumping with a peak-power tracker using a simple six step square-wave inverter. *IEEE Transactions on Industry Applications*.3.

Fedrizzi, M. C., Ribeiro, F. S. and Zilles, R. (2009) Lessons from field experiences with photovoltaic pumping systems in traditional communities. *Energy for Sustainable Development*. 13: 64–70.

Firatoglu, Z. and Yesilata, B. (2004) New approaches on the optimization of directly coupled PV pumping systems. *Solar Energy*. 77: 81–93.

Foster, R. and Hanley, C. (1998) Life cycle cost analysis for photovoltaic water pumping systems in Mexico. *In: Proceedings of the 2nd world conference on photovoltaic energy conversion. Vienna, Austria*.

Foster, R. E., Alma, D., Ross, M. P. and Avilez, O. M. (2013) Ten-year reliability assessment of photovoltaic water pumping systems in Mexico. Sandia National Laboratories.

Gad, H. E. (2009) Performance prediction of a proposed photovoltaic water pumping system at South Sinai, Egypt climate conditions. *In: Proceedings of the thirteenth international water technology conference. Hurghada, Egypt*. p. 739–52.

Hadj, A. A., Chenlo, F., Mukadam, K. and Balenzategui, J. L. (1991) Performance of PV water pumping systems. *Renewable Energy*. 18: 191–204.

Hamidat, A. (1999) Simulation of the performance and cost calculations of the surface pump. *Renewable Energy*. 18: 383–92.

Hamidat, A. and Benyoucef, B. (2008) Mathematic models of photovoltaic motor-pump systems. *Renewable Energy*. 33 (5): 933–42.

Hamrouni, N., Jraidi, M. and Chérif, A. (2008) Solar radiation and ambient temperature effects on the performances of a PV pumping system. *Revue des Energies Renouvelables*. 11 (1): 95–106.

International Energy Agency, India Energy Outlook (2015) *Available from: https://www.iea.org/publications/freepublications/publication/IndiaEnergyOutlook_WEO2015.pdf.*

International Energy Agency, World Energy Outlook, (2016) *Accessed from: http://www.iea.org/newsroom/news/2016/november/world-energy-outlook-2016.html.*

Jamil, M., Ahmed, S. A. and Rizwan, M. (2012) SPV based water pumping system for an academic institution. *American Journal of Electrical Power and Energy Systems*. 1(1): 1–7.

Joao, V. M. C. and Luis, F. M. T. (2014) Implementation of a high-efficiency, high-lifetime, and low-cost converter for an autonomous photovoltaic water pumping system. IEEE *Transactions on Industry Applications*.

Kaldellis, J. K., Meidanis, E. and Zafirakis, D. (2011) Experimental energy analysis of a standalone photovoltaic-based water pumping installation. *Applied Energy*. 88: 4556–62.

Kaldellis, J. K., Spyropoulos, G. C., Kavadias, K. A. and Koronaki, I. P. (2009) Experimental validation of autonomous PV-based water pumping system optimum sizing. Renewable Energy 34: 1106–13.

Katan, R.E., Agelidis, V.G., Nayar, C.V. (1996) Performance analysis of a solar water pumping system. *In: Proceedings of the 1996 IEEE international conference on power electronics, drives, and energy systems for industrial growth (PEDES)*. p. 81–7.

Kim, J., Kim, C. and Kim, J. (2011) A regulated charge pump with a low-power integrated optimum power point tracking algorithm for indoor solar energy harvesting. *IEEE Transactions on Circuits and System*.58: 12.

Kou, Q., Klein, S. A. and Beckman, W.A. (1988) A method for estimating the long-term performance of direct-coupled PV pumping systems. *Solar Energy*. 64: 33–40.

Kumar, A. and Kandpal, T. C. (2007) Potential and cost of CO_2 emissions mitigation by using solar photovoltaic pumps in India. *International Journal of Sustainable Energy*. 26: 159–66.

Loxsom, F. and Veroj, P. D. (1994) Estimating the performance of a photovoltaic water pumping system. *Solar Energy*. 52: 215–9.

Meah, K., Fletcher, S. and Ula, S. (2008) Solar photovoltaic water pumping for remote locations. *Renewable and Sustainable Energy Reviews*.12 (2): 472–487.

Ministry of New and Renewable Energy (2014) *Available from: http://mnre.gov.in/file-manager/offgrid-solar-schemes/NABARD-operational-Guidelines-of-Solar-Water-Pumping Programme.pdf.*

Ministry of New and Renewable energy, Solar Photovoltaic Water Pumping System (2015) Available from: *https://mnre.gov.in/file-manager/UserFiles/technicalspecification_ spwps_2013_14.pdf.*

Ministry of statistics and program implementation, Central Statistics Office, Energy Statistics (2016) Accessed from: mospiold.nic.in/Mospi_New/upload/Energy_statistics_2016.pdf.

Mohanlal, K., Joshi, J. C. and Kothari, D. P. (2004) Performance analysis of a directly coupled photovoltaic water-pumping system. *IEEE Transactions on Energy Conversion*. 19: 3.

Mokeddem, A., Midoun, A., Kadri, D., Said, H., Iftikhar, R. A. (2011) Performance of a directly-coupled PV water pumping system. *Energy Conversion and Management*. 52: 3089–95.

Narale, P. and Rathore, N. S. (2015) Solar Water Pumping System for Agriculture. *Energetica.*

Odeh, I. (2014) Economic feasibility of large scale PV water pumping applications utilizing real field data for a case study in Jordan. *International Journal of Renewable Energy Development*. 3 (2): 107.

Odeh, I., Yohanis, Y. G. and Norton, B. (2006) Influence of pumping head, insolation and PV array size on PV water pumping system performance. *Solar Energy*.80: 51–64.

Pande, P. C., Singh, A. K., Ansari, S., Vyas, S. K. and Dave, B. K. (2003) Design development and testing of a solar PV pump based drip system for orchards. *Renewable Energy*. 28: 385–96.

Purohit, P. and Kandpal, T. C. (2005) Solar photovoltaic water pumping in India: a financial evaluation. *International Journal of Ambient Energy*. 26 (3): 135–146.

Ramachandra, T. V., Jain, R. and Krishnadas, G. (2011) Hotspots of solar potential in India. *Renewable and Sustainable Energy Reviews*.15 (6): 3178–3186.

Rathore, P. K., Rathore, S., Singh, R. P. and Agnihotri, S. (2017) Solar power utility sector in India: challenges and opportunities. *Renewable and Sustainable Energy Reviews*. 81: 2703–2713.

Reddy, P. K. R. and Reddy, J. N. (2014) Photovoltaic energy conversion system for water pumping application. *International Journal of Emerging Trends in Electrical and Electronics*. 10 (2): 2320–69.

Reddy, S. K. R. (1999) Energy Demand in Indian Agriculture: a Post Green Revolution Experience (Doctoral Dissertation), University of Hyderabad.

Rezae, A. and Gholamian, A. (2013) Technical and financial analysis of photovoltaic water pumping system for Gorgan. *International Journal on Cybernetics and Informatics*. 2: 2.

Setiawan, A., Purwanto, D. H., Pamuji, D. S. and Nurul, H. N. (2014) Development of a solar water pumping system in Karsts Rural Area Tepus, Gunungkidul through student community services. *Energy Procedia*. 47: 7–14.

Shah, T. (2009) Climate change and groundwater: India's opportunities for mitigation and adaptation. *Environmental Research Letters*. 4 (3): 035005.

Shakti Foundation, Feasibility Analysis for Solar Agriculture Water Pumps in India (2014) *Accessed from: http://shaktifoundation.in/initiative/feasibility-analysissolar-agricultural-water-pumps-india/.*

Shuba, V. R., Bharadwaj, A., Thatte, A. A., Harish, S., Iychettira, K. K., Perumal, R. and Nayak, G. (2010) Harnessing Solar Energy: Options for India, Center for Study of Science, Technology and Policy, Bangalore. *Accessed from: http://shaktifoundation.in/wp-content/uploads/2014/02/cstep harnessing per cent 20solar per cent 20energy per cent 20- per cent 20final per cent 20report.pdf.*

Sontake, V. C. and Kalamkar, V. R. (2016) Solar photovoltaic water pumping system-A comprehensive review. *Renewable and Sustainable Energy Reviews*. 59: 1038–1067.

Wagdy, R. and Nourb, M. A. (1994) Optimum design of a photovoltaic powered pumping system. *Journal of Power Sources*.50: l–9.

World bank, Water Paper. Managing the Invisible: understanding and Improving the Groundwater Governance (2012) *Energetica*. India.

Yahia, B., Arab, A. H. and Azoui, B. (2011) Optimal sizing of photovoltaic pumping system with water tank storage using LPSP concept. *Solar Energy*. 85: 288–94.

Ziyad, M. and Dagher, F. (1990) The effect of electrical array reconfiguration water pump on the performance of a PV-powered volumetric water pump. *IEEE Transactions on Energy Conversion*. 5: 4.

Zvonimir, G. and Margeta, J. (2007) A model for optimal sizing of photovoltaic irrigation water pumping systems. Solar Energy.81: 904–16.

Chapter 20

Biotechnology Interventions in Integrated Farming Systems: Entrepreneurship Opportunities in Climate Change Scenario

Aditya Pratap Singh[1], Y. Linthoingambi Devi[2], Pramod Kumar Pandey[3] and Telem M. Chanu[4]*

[1]Research Scholar, Department of Genetics and Plant Breeding, Bidhan Chandra Krishi Viswavidyalaya, Mohanpur – 741252, Nadia, West Bengal
[2]Research Scholar, Department of Genetics and Plant Breeding, Indira Gandhi Krishi Viswavidyalaya, Raipur – 492012, Chhattisgarh
[3]Assistant Maize Breeder, AICRP on Maize, School of Crop Improvement, College of Post Graduate Studies in Agricultural Sciences, Central agricultural University (Imphal), Umiam – 793103, Meghalaya
[4]Assistant Professor, Department of Social Science, College of Horticulture and Forestry, Central Agricultural University (Imphal), Pasighat – 791102, Arunachal Pradesh
Corresponding author: adityapratapbckv@gmail.com

ABSTRACT

In the present scenario of farming system where farmers face economic uncertainty owing to monoculture; integrated farming system with the intervention of biotechnology tend to empower them to sustain their livelihood. The integrated farming system integrates farming practices with livestock or fish farming using the natural resources on sustainable basis. Biotechnology entrepreneurship contributes in the development, creation and marketing of biotechnology products. In general, biotechnology intervention in agriculture has provided us an enhance in productivity, improved nutritional and processing quality, improved flavor of products, biofertilizers and so on. Thus with the intervention of biotechnology, several opportunities of entrepreneurship has opened up in the field of agriculture which can be embedded in integrated farming systems.

Keywords: *Integrated farming system, Biotechnology, Entrepreneurship.*

Introduction

Farmers have been experiencing frequent crop failures as a result of intensive monoculture systems that lead to pests and diseases and are prone to soil exhaustion. Extreme weather variability has exacerbated the problem. There is also a great deal of market risk. Farmers were forced for many years to sell their produce at distress prices. Because of the steep rise in input costs and unfavorable market prices, profit margins also started to decrease in regular years. Market prices are erratic, often recording very low prices shortly after harvest. As a result, farmers are dissatisfied, and younger farmers are moving from agriculture to urban areas. Youth in rural areas do not consider agriculture an assured source of income that can provide them with a livelihood. The only ray of hope in these circumstances comes from Integrated Farming Systems, which combine a diverse set of complementary and supplementary enterprises to achieve optimal sustainable use of natural resources, effective use of available byproducts, and profitability throughout the year for all farm members. This results in increased income generation and high level of income stability and prosperity of the farmers. Natural and market risks have a very high level of influence on the crop production activity, which is major activity in most of the farms. It is possible to minimize this risk by combining dairy production, sheep raising, and other businesses that have a steady market and provide income. Dairy enterprises provide income on a daily/weekly basis, while crop enterprises provide income on a seasonal basis (Wieczorek, 2003).

Both the agricultural and therapeutic biotechnology sectors share similar scientific fundamentals, including the long time to market of emerging products. Agricultural applications of biotechnology, however, present very different challenges, goals, and opportunities that provide entrepreneurs and innovators with a very different context. The concept of innovation refers to anything new, whether it's a way of doing something new or a new process. All integrated processes that contribute to the creation, development, and marketing of biotechnology products are included in biotechnology entrepreneurship. Entrepreneurs in these sectors have the ability to think about disruptive businesses, not just disruptive technologies, like the creation of agricultural biotechnology traits. We discuss the prospects of agricultural biotechnology in different components of integrated farming system in this chapter.

Biotechnological Interventions

Through the use of biotechnology (genetic engineering), crops have been designed to be become disease resistant and drought resistant, thus increasing yields. Biotechnological tools have empowered scientists to select gene sequences for disease resistance from other species and transfer them to high-value crops. Researchers at Cornell and the University of Hawaii developed two varieties of papaya resistant to papaya ringspot virus by introducing a different gene from the virus into papaya to create resistance (Gonsalves, 1998). Under licensing agreements, papaya growers have been receiving seeds for these two varieties, named 'SunUp' and 'Rainbow', since 1998. Another example comes from dry climates, where crops need to conserve water. Many crop varieties can be made drought-tolerant by using genes from naturally drought-resistant plants.

What are the Benefits of Genetic Engineering in Agriculture?

The risks and benefits of genetic engineering are the same as any other aspect of life. Although genetic engineering technology has been discussed much, so far there are no strong evidences to suggest medical risks associated with consumption of genetically engineered products. It is possible that transgenic organisms will provide benefits over and above those gained from traditional agricultural biotechnology. The following examples illustrate how agricultural biotechnology can benefit from applying currently available genetic engineering techniques.

1. Enhanced Crop Protection

A crop-protection technology provides farmers with cost-effective solutions to pest problems, which, if left uncontrolled, would cause severe yield losses. As we discussed above, genetic engineering has been successfully used to create corn, cotton, and potato plants that produce a protein that kills some insects when they eat the plants. An active ingredient in some "natural" insecticides is a protein derived from the soil bacterium *Bacillus thuringiensis*. Transgenic crop protection can be an effective way to control pests in some cases. *Bt* engineered into a corn crop, for example, produces a resistant plant, not just the part of the plant that has been sprayed with insecticide. This is further discussed in 'protection against pathogens' section. Whenever a new technology provides more effective controls, yields increase. Some new technologies are adopted because they are more affordable than current ones that offer an equivalent level of control. Some new technologies are not adopted for one reason or another because they do not compete with the existing technology. *Bt* as part of 'natural' insecticide is used by organic farmers to control insect pests in their crops, but these farmers may be against transgenic Bt crops.

2. Improvements in Food Processing

A genetically modified enzyme, chymosin, was the first food product to receive regulatory approval in 1990. 60 percent of all cheese is now made with this rennet instead of calf rennet. Purity, reliable supply, upto 50 percent cost reduction, and high cheese yield efficiency are a few of its benefits.

3. Improved Nutritional Value

It is possible to improve the taste, texture or nutritional quality of food using genetic engineering. Transgenic crops producing beta-carotene, a precursor to vitamin A, helps to prevent blindness in people with nutritionally inadequate diets. A few of transgenic crops possess higher levels of their principle nutrients, such as soybeans with higher protein content, potatoes with more starch and amino acids, beans with more essential amino acids, and rice with beta-carotene.

4. Better flavor

Aroma precursors can be transformed into flavoring compounds by increasing enzyme activity in plants. There are currently trials taking place with transgenic peppers and melons.

5. Fresher Produce

Genetic engineering has the potential to improve the keeping properties of fresh produce to make it easier to transport, thus allowing consumers to access nutrient-dense whole foods and prevent decay, damage, and nutrient loss. It is possible to ship vine-ripened tomatoes that will not bruise while being shipped. It is being investigated whether broccoli, celery, carrots, melons, and raspberries can be modified similarly. The shelf life of some processed foods has also been extended by modifying the fatty acid composition of ingredients such as peanuts.

Tissue Culture and Micropropagation

Plants can be regenerated via tissue culture using the principle of totipotency; first proposed by Haberlandt in 1902 (see Vasil, 2008). Micropropagation was first attempted by culturing excised root parts and shoot tips or axillary buds. Following this, it was possible to regenerate whole plants from cultured carrot callus tissue and tobacco stem cells through somatic embryogenesis. Miller *et al.* (1955) reported that plant regeneration can be induced in culture by supplying the right ratio of auxin and cytokinin to the medium. Murashige and Skoog (1962) analysed the composition of tobacco leaves to formulate a mineral nutrient formulation that sustained tobacco cell growth and division. As it is well known, MS media (Murashige and Skoog's media) has become a significant contribution to plant tissue culture. A method of transferring genes using Agrobacterium into crops and regenerating transgenic plants is widely popularized in current time. This method has proven useful for introducing desired agronomic traits into transgenic crops. Plant tissue culture was firmly established as a useful research tool, as well as a technique for improving crops through biotechnology. Plant tissue culture is the process of excising plant tissues from plants and growing them in nutrient media. Several variations are referred to under the general term "tissue culture". A few examples include meristem culture for propagating virus-free plants (*e.g.*, orchids, strawberries, grapes, and potatoes), protoplast culture, cell suspension culture, tissue culture, and pollen culture to produce haploid plants (Guha and

Maheshwari, 1966). Tissue culture could generally be used to improve crops in the following ways:

1. Establishment and initiation of cultures.
2. Performing specific modifications such as propagation of clones, elimination of viruses, selection of variants, genetic transformations, and so on.
3. Restoration of modified plants

Agricultural biotechnology has benefited enormously from plant tissue culture in the past. Plants grown from seed vary considerably in growth, flower characteristics, yield, disease resistance, resistance to environmental stress, and so on (Thorpe, 2007). As such, selecting those that possess desirable properties for vegetative multiplication would be valuable. A large number of comparable plants (clones) can be propagated through tissue culture at a large scale. Also, slow-growing orchids are desirable for horticulture, as are fruit crops such as strawberries and grapes. Upon co-cultivation with Agrobacterium or bombardment with transgenes, the transgene is introduced into only a limited number of cells. It is necessary to regenerate plants only from these specific populations of cells to recover transgenic plants. Consequently, the key to recovering transgenic plants from crop biotechnology is plant tissue culture. In over 40 countries, transgenic crops are planted on thousands of hectares that would not have been possible without tissue culture technology. The other applications of tissue culture include establishing variants or mutants (somaclonal variation, mutation), cryopreservation, producing dihaploids and producing useful chemicals (*e.g.,* Taxol). Plants can be tested for their toxicity to trace elements using liquid cultures (Kopittke *et al.,* 2010).

Artificial Seed

Somatic embryos can be embedded in a suitable matrix such as agarose or calcium alginate to produce artificial seeds. In 1992, alfalfa artificial seeds were successfully planted in the field after being encapsulated within calcium alginate. Somatic embryos are well suited for the purpose of creating "clonal seeds", which was the main aim of artificial seeds development. Khor and Loh (2005) suggested that the objective for the generation of artificial seeds may need to expand, since the concept is now extended to include the encapsulation of other tissue cultured or in vitro prepared materials (*e.g.,* protocorm-like bodies, shoot buds). The same is already under practice by many scientists. This concept has also been extended to include encapsulating non-endospermic seeds or protocorms of commercially valuable species, including orchids.

Biofertilizers

Insecticides, fungicides, and pesticides have been extensively used to increase production. Unfortunately, these chemicals deplete the soil of essential minerals, seriously impacting crops. Consequently, biofertilizers have been developed, which are basically cultures of microorganisms contained in a carrier material. Live or latent cells of microorganisms that easily solublize phosphate, fix nitrogen, and break down organic matter are included in biofertilizers to supplement seeds, soil, and composting. By using biofertilizers, the goal is to increase the number of such micro-organisms and accelerate their growth so that they can provide plants with nutrients that are more readily assimilated by them. Moreover, they contribute to the improvement of soil fertility by fixing atmospheric nitrogen and producing plant growth substances. In addition, they stimulate root growth by releasing hormones and antimetabolites. This results in soil mineralization and nutrient degradation (García-Fraile, 2015). Using them as a complement to chemical fertilizers is cost-effective. In addition to microorganisms like bacteria, fungi, and blue-green algae being used as fertilizers as such, carriers such as peat and lignite powder are also used to prolong their shelf life. In the context of agricultural productivity and the environment, biofertilizers are an essential component. There are several types of biofertilizers including:

- Biofertilizers that fix nitrogen
- Biofertilizers that solubilize phosphate
- Biofertilizers that mobilize phosphorus
- Micronutrient-rich biofertilizers
- Biofertilizers that promote plant growth

Protection against Insect Pests

In nature, there are a number of insecticidal proteins and molecules. These molecules can be harnessed and deployed in target crops by using molecular biology and genetic engineering. Marc Van Montagu, Mary-Dell Chilton, and Robert Fraley's pioneering work, using recombinant DNA and Agrobacterium mediated gene transfer approaches to transform plant cells, set the stage for genetically modified crops. Most commonly used genes for making transgenic plants contain genes for endotoxins, lectins, alpha-amylase inhibitors, and insect chitinases, as well as plant metabolic enzymes and insecticidal viruses. As a result of the development of Bt cotton, many organizations, both private and public, have come up with Bt transgenic versions of rice, tomato, brinjal, *etc.*, exploiting diverse Bt genes sourced from a variety of sources (Liu *et al.*, 2010). Maize, cotton and potato are among the crops that exhibit the highest number of transgenic events for insecticidal resistance. Researchers looked for genes/QTLs for resistance to different pests so that they could easily be introduced into elite cultivars by Marker Assisted Selection (MAS) strategy (Gatehouse, 2013). MAS was used, however, to introduce three Bt genes into 9311, a highly selective parental rice line in China used for hybrid seed production, from the donor parents Minghui63 (*cry*1Ac), Minghui63 (*cry*2A), and Minghui63 (*cry*1C). Agronomic traits were not significantly compromised by the improved lines in terms of pest resistance.

Bioprotection or Biological Control

For decades, synthetic pesticides dominated pest control, but today people are returning to biological pest control methods due to their specificity and environmental safety. A biological control is an intervention that inhibits a specific organism's growth or impact, reducing its abundance or damaging potential (Bhardwaj *et al.,* 2014). The use of insect predators to control diverse types of insects is being investigated. Various biological control agents have been identified worldwide. A microbial pesticide is one of the most widely used biological agents to control pests. Among them are crystallized (cry) proteins produced by *Bacillus thuringiensis* (Eilenberg *et al.,* 2001). Hormones, pheromones, growth regulators, and pheromones are examples of biological agents. For resource-poor farming systems, these agents are very beneficial because they eliminate the need for chemical pesticides, which are not only expensive, but also unsustainable. Fungi are also regarded as potential pesticide replacements. In North Africa, desert locust is a sporadic pest that can be controlled using *Metarhizium* fungi in combination with Phenylacetonitrile, a hormonal compound that affects how locusts swarm. Furthermore, *Metarhizium* strains have shown to be effective in controlling rhinoceros beetles. In addition to whole organisms, biocontrol agents also include their metabolites and macromolecules that are used to control insects. It is likely that entomopathogenic fungi are one of the most effective methods of controlling insects. When the spores and conidia of such fungi come in contact with the insect cuticle, they germinate, penetrate and proliferate in the insect tissue by absorbing its nutrients. Insect control agents, such as Bio-power, Bio-Catch, and Bio-Magic, have been formulated using spores or conidia. The most important entomopathogenic fungi include *Metarhizium anisopliae, Beauvaria bassiana, Nomuraea rileyi, Trichoderma viride*, and *Lecanicillium muscarium*, among others. Additionally, insect pathogenic viruses can also be used to control insect pests. *Helicoverpa armigera, Spodoptera litura*, and *Hyblaea puera* are infected by Nucleopolyhedrosis (NPV), while *Chilo infuscatellus* and *Plutella xylostella* are infected by Granulo viruses. Parasitoids such as *Trichogramma chilonis* and entomopathogenic nematodes like *Steinernema carpocapsae, S. bicornutum* and *Heterorhabditis indica* are also powerful insecticidal biocontrol agents. *Photorhabdus luminescens*, a gammaproteobacterium which produces insecticidal toxin complex A (Tc A) harbored by the entomopathogenic nematode *H. indica*, has been evaluated and registered at the International Centre for Genetic Engineering and Biotechnology (ICGEB), New Delhi. Bioprahar is the trade name under which this formulation has been commercialized. Furthermore, RNAi (RNA interference) mediated approaches have also been explored for controlling the insect pests. Cotton bollworm larvae are effectively controlled by transgenic plants expressing double-stranded RNA of a cytochrome p450 gene. A variety of plant extracts containing azadirachtin, limonoids and their metabolites and marketed as 'Nimbecidine EC' and 'Margosom EC' have been developed to control whiteflies, thrips and aphids.

Protection against Pathogens

Pathogens employ different molecular strategies to combat diseases. Especially when it comes to fungal and bacterial diseases, race-specific resistance is available, such as with rusts in wheat (*Puccinia tritici* causes black rust, *P. recondita* causes brown rust and *P. striiformis* causes yellow rust) and blast (*Magnaporthe grisea*) and BLB (*Xanthomonas oryzae*) in rice (Wally *et al.*, 2010). Gene pyramiding using race-specific markers and different harmonious combinations of resistance genes provide a durable, high level of resistance and a wide spectrum. When Xa21 is combined with critical resistance genes like Xa4, Xa5, Xa13, Xa14, *etc.*, it provides high protection against Bacterial Leaf Blight (BLB) in rice (Pal and Gardener, 2006). A similar phenomenon has also been observed in the case of rice blast in which staking together of region-specific harmonious resistance genes provided very high and durable resistance to the pathogen (Suh *et al.*, 2013). The MAS technique has been used by Indian scientists to develop varieties resistant to many diseases. In Pearl millet and rice, three resistant varieties have been released, namely HHB 67-2 for powdery mildew, RPBIO226 and Pusa 1460 for BLB. A wide range of genetic engineering strategies have been adopted to combat disease-causing agents that lack resistance sources in the gene pool. These efforts are based on a growing understanding of the molecular events that occur during plant-pathogen interactions (Wally and Punja, 2010). To combat fungal pathogens, strategies have included manipulating antifungal compounds, which broadly include proteins related to pathogenesis (PR), proteins that are involved in ribosome inactivation (RI), and metabolites produced by plants, such as phytoalexins and resistance genes (R genes).

Vermicomposting Biotechnology

Since ancient times, earthworms have been used to improve soil structure, aeration, and nutrient content. Even so, the idea of utilizing earthworms for composting a wide range of organic materials has been around since the mid-20th century (Senapati, 1993), as all the information about the efficiency of certain groups of earthworms in decomposing different organic materials was only discovered later. On the basis of their ecological niches, earthworms fall into three categories: endogeics, anecics, and epigeics. Deep in the soil live geophagous earthworms, also called endogeic earthworms. The anecics live just below the ground surface, feeding on organic materials mixed with the soil. In addition, epigeic organisms live on top of soils, feeding primarily on plant material and other organic matter. Due to their ability to consume organic wastes, these earthworms make excellent organic manures since they are able to consume a wide variety of organic materials (Lavelle, P., and Martin, A., 1992). Epigeic earthworms are voracious feeders capable of consuming large amounts of organic matter. Even though they consume a lot of food, they use only very small amounts for body synthesis and excrete 90-95 per cent of what they ingest in the form of vermicast. Because earthworm intestines have high concentrations of microorganisms, these vermicasts also have rich population of these organisms (Atlavinyte, 1982). The gut contents of earthworms have been found to contain a wide variety of symbionts including microfungi, bacteria, protozoa, actinomycetes, *etc.* (Dash and Senapati, 1986). These microorganisms thrive in earthworm intestines because they provide conducive environmental conditions for their proliferation (Ponomarera, 1962). Because earthworm food contains a greater concentration of different microorganisms, which are subject to microbial activity in the alimentary canal, as well as finely meshed food particles in the gizzard, earthworm food particles are subjected to more intense microbial activity. Besides microbes, earthworm guts contain various enzymes that aid in such processes. In the intestinal fluid of the worms, these enzymes operate in a pH range from near neutral to slightly alkaline (Senapati 1993). Earthworm guts have also been reported to contain hormones, antibiotics, *etc.* In the intestines of the earthworms, all of these components mix with the food materials to produce vermicast, which is released with the excreta. In the presence of different microbes, these vermicasts undergo rapid aerobic decomposition in which enzymes accelerate the process and growth-promoting substances help the process along. In shorter time periods, aerobic microorganisms decompose vermicasts more rapidly than anaerobic microorganisms, yielding nutrient-rich, well-humified organic manure known as 'vermicompost'. Some of the successful example of improvement in farming system through biotechnological intervention is given in Table 20.1.

Table 20.1: Examples of Improvement in Farming System through Biotechnological Intervention

Improvement in Integrated Farming System through Biotechnology	*Example*	*Reference*
Increased crop productivity	Bt engineered into a corn crop	Liu *et al.*, 2010
Artificial seed	Encapsulation of protocorm-like bodies, shoot buds other tissue culture or in vitro prepared materials	Loh, 2005
Biological control/ bioprotection	*Bacillus thuringiensis*	Eilenberg *et al.*, 2001
Protection against pathogen	Gene stacking for blast resistance	Suh *et al.*, 2013
Vermicomposting Biotechnology	Vermicast	Atlavinyte, 1982
GM Crops	Papaya ringspot virus resistant papaya	Wu *et al.*, 2018

Commercially Available Crops Improved through Genetic Engineering

Since 1996, the global use of transgenic or GM crops has increased. On the ISAAA website - http://www.isaaa.org - you will find the ISAAA's Annual Global Status Report. The report discusses how many countries are planting genetically modified crops, how many acres are planted, the benefits derived from the crop, farmer perspectives, and future prospects for the technology. Many transgenic genotypes are in use as commercial crops: alfalfa, canola, beans, carrots, chicory, cotton, creeping bent grass, eggplant, flax, corn, melon, papaya, petunia, plum, Polish canola, rice, rose, soybean, squash, sugar beet, sugarcane, sweet pepper, tobacco, tomato, and wheat *etc.* With genetic engineering, a plant can acquire multiple traits. These are known as stacked traits. Herbicides and insects can currently be tolerated by corn, cotton, and soybean crops. There have been many transgenic crops commercialized, including herbicide-resistant maize and insect-resistant cotton. An alternative strategic approach to sustainable agriculture is the refuge. The crop will be more able to resist pests and diseases and to withstand more herbicides when many genes are stacked for one trait. Several specific transformation events have been found to be effective refuges. To determine the effectiveness of these methods, weeds and insects resistant to them are closely monitored.

New and Future Initiatives in Crop Genetic Engineering

Commercial GM crops have benefitted crop production to date, but new products are being developed that are expected to improve food safety, the environment, and livestock feed. The following products may be deemed as examples of these: rice in iron and beta carotene (a micronutrient that is converted into vitamin A by the body; long-term bananas that ripen on the tree faster and can thus be harvested earlier; maize with improved feed value; delayed-ripening papaya; papayas resistant to papaya ringspot virus, *etc.* Among others, a tomato offering high amounts of flavonols, a powerful antioxidant, is available as well as drought-tolerant maize and wheat, maize with improved phosphorus availability, insect-resistant eggplant and rice, edible vaccines made from fruit and vegetables, and trees with low levels of lignin for making paper.

1. Applications in Dairy Industry

The applications of biotechnological interventions in dairy industry are given in Figure 20.1.

Some of the main points are discussed as follows-

- ☆ **Selecting dairy cattle with higher resistance to mastitis and other diseases using marker-assisted selection:** Genetic diseases can be detected using restriction fragment length polymorphism (RFLP), a technique used in prenatal diagnosis. PCR (Polymerase Chain Reaction) can be used to detect pathogens easily and quickly for checking of milk adulteration (mixing of milk from different livestock).
- ☆ **Development of cattle that are genetically superior:** Breeders use artificial insemination (using sperm from superior males to inseminate large numbers of females) to improve livestock. Embryo transfer involves the transfer of an embryo from a superior female animal to an inferior female animal's reproductive tract for the purpose of developing genetically superior high yielding cattle.

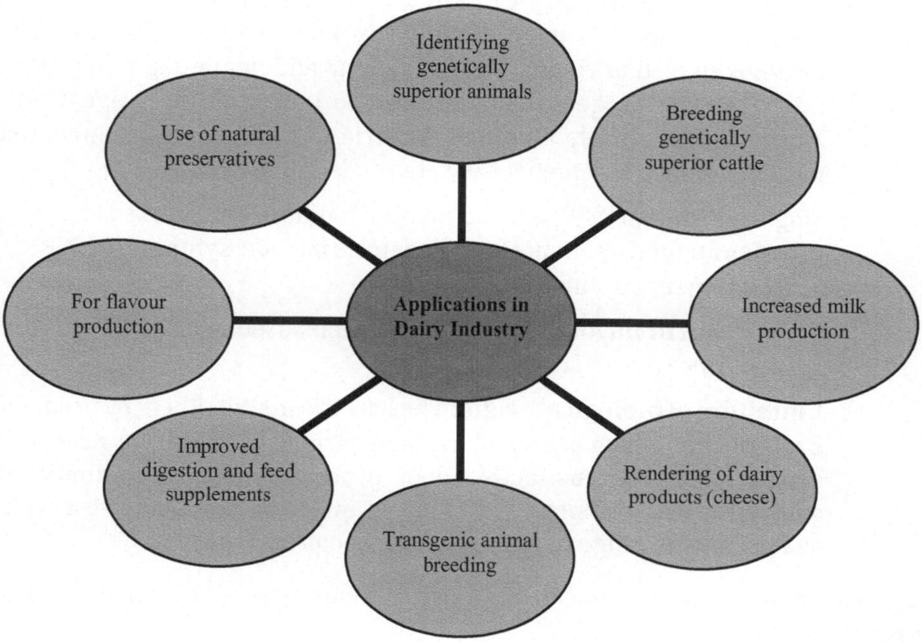

Figure 20.1: Improvement in Dairy Industry through Biotechnological Intervention.

- **Increase in milk production rate:** Number of secretory cells in the udder determines milk production rate, and BST, a recombinant DNA growth hormone, increases cell productivity. This results in increased milk production as a result of the application of BST. Designer milk with higher levels of desirable components, like casein, and lower levels of less desirable ones.
- **Cheese production and ripening:** Cheese production requires the enzyme rennet. For commercial cheese production, recombinant chymosin is produced using genetic engineering and cloned in several microorganisms in order to replace calf rennet. Cheese traditionally ripens over a long period of time; therefore, GM bacteria are used to synthesize enzymes that speed up the process. This includes proteinase, lipase, and D-galactosidase. It is possible to directly inject synthetic genes into fertilised eggs in order to produce transgenic animals.
- **Development of transgenic disease-resistant cows.** A transgenic cow that produces biopharmaceuticals (such as hormones and blood-clotting factors) for medical treatment.
- **Development of new vaccines for the treatment and prevention of dairy-cattle diseases.**

- **Improved digestibility of feed and feed supplements:** It may be possible to modify rumen microbes to increase fibre digestion and/or detoxify feed in ruminants. Cellulolytic and lignin-degrading rumen microbes can be genetically altered to increase fibre digestibility. Moreover, genetically modified bacteria may be able to remove toxic particles from the rumen if they are seeded in the rumen.
- Single cell protein (SCP) is a microbial cell culture product used to produce proteins for food or animal feed. Examples include: synthesis of SCP from *Saccharomyces* sp. and *Kluyeromyces* sp.
- **Production of flavor:** Important organisms used in dairy flavoring are *Leuconostoc* sp. and *Lactococcus lactis*
- **Inhibitors of biological origin:** The bacteria inhibit others by producing Bacteriocins, which are small proteins produced by starter strains. For example. Lactococcus lactis subsp. produces nisin, an antimicrobial substance. The compound is found in milk and acts as a preservative against a wide range of gram-positive bacteria.

In the dairy industry, some of the more common types of enzymes are listed as follows:

Rennet

Milk's liquid form is maintained by its proteins, including casein. Enzymes, such as proteases, are added to milk throughout the cheese-making process in order to hydrolyze caseins, which stabilize micelle arrangements, prevent coagulation. The most familiar enzyme is chymosin, which comes from rennet. Bio-engineered chymosin is becoming increasingly important in the manufacture of cheese products, including hard cheese, chedder, and other types of hard cheese. Other sources of chymosin include microbial, and animal sources. Besides caseins, milk contains a number of different types of proteins. Proteases are used to denaturize whey proteins, like lactalbumin and lactoglobulin, to produce creamier yogurt.

Lactase

Lactase breaks lactose down into galactose and glucose, which are the two essential sugars in milk. In addition to organizing lactose-free products, lactase is used in making ice cream, to make manufactured goods creamier and sweeter for lactose-intolerant individuals. Lactase is as frequently organized from *Kluyveromyces* sp. of yeast and *Aspergillus* sp. fungi.

Catalase

A cheese making process that requires the enzyme catalase has been established partially. Bovine livers contain catalase enzymes that convert hydrogen peroxide to water and molecular oxygen.

Lipases

Cheese is flavoured with lipase, which is used to breakdown milk fats. Hydrolyzed milk fats create free fatty acids that are responsible for the flavor. Shorter chain fats are preferred since they result in the flavor of many cheeses, while the longer chain fats can produce soapiness or no flavor at all.

2. Gene Probes

Detection of pathogens in low concentrations in foods is made possible with gene probes and immunological assays based on polyclonal or monoclonal antibodies targeting specific pathogens. There are more companies developing kits based on gene probes and immunological values for the detection of specific pathogens in foods. They are already available commercially.

Aside from these efforts, researchers are also working to develop designer animals and humanized milk from animals using CRISPER- and TALEN-based genome editing tools. The use of stem cells as an alternative to antibiotic treatment for diseased udders, skin, and hoof wounds has emerged in the dairy industry despite a high use of antibiotics. Animal diseases can now be treated with mesenchymal stem cells.

Conclusion

The biotechnology sector has attracted many investors as well as entrepreneurs; however, the old biotech models have not been very successful in creating self-sustaining profitable businesses. A biotech company has lost capital and assets due to long development cycles and underestimating risk. Businesses focus on achieving product development, financing, and strategic alliance milestones, losing sight of what could be the end goal of a business venture: profit. Even in the midst of a new start-up wave that appears to improve on the previous one, biotech companies consistently underestimate their projected revenues and breakeven time. Even so, given increased investments in R and D and recent policy reforms, including those aimed at biopharmacies, bioservices, and agriculture, there is an immense amount of room for growth. Despite India's rapidly growing biotechnology industry, critical gaps impede its development. Researches that are innovative would facilitate theses sectors to their full potential and to achieve long-term sustainable growth at a global level. Entrepreneurship in biotechnological interventions in integrated farming system have the potential to bloom and impact agricultural scenario as a whole.

Over the past year, scientists, researchers, and investors have benefited from biotechnology investments. Here are a few things to look forward to:

1. Government research results or internationally conducted research can be turned into services and/or products for commercial use. Tissue culture, for instance, can be used to multiply improved planting materials for farmers.

2. Creation of a biotechnology industry that exploits the unique features of plants and crops.
3. Environmental assessment, genetic testing, and quality control.
4. Other institutes and companies offering auxiliary services, such as biotechnology training and meeting facilities, books, venture capital firms, and consulting firms.

REFERENCES

Atlavinyte, O., and Vanagas, J. (1982). Effect of earthworms on the quality of barley and rye grain. *Pedobiologia*.

Bhardwaj, D., Ansari, M. W., Sahoo, R. K., and Tuteja, N. (2014). Biofertilizers function as key player in sustainable agriculture by improving soil fertility, plant tolerance and crop productivity. *Microbial cell factories, 13*(1), 1-10.

Dash, M. C., and Senapati, B. K. (1986). Vermitechnology, an option for organic waste management in India. *Verms and vermicomposting*, 157-172.

Eilenberg, J., Hajek, A., and Lomer, C. (2001). Suggestions for unifying the terminology in biological control. *BioControl, 46*(4), 387-400.

García-Fraile, P., Menéndez, E., and Rivas, R. (2015). Role of bacterial biofertilizers in agriculture and forestry. *AIMS Bioengineering, 2*(3), 183-205.

Gatehouse, J. A. (2013). Genetic engineering of crops for insect resistance. *Sustainable Food Production.(Eds.: P. Christou, R. Savin, BA Costa-Pierce, I. Misztal and CBA Whitelaw). Springer, New York.*

Gonsalves, D. (1998). Control of papaya ringspot virus in papaya: a case study. *Annual review of phytopathology, 36*(1), 415-437.

Guha, S., and Maheshwari, S. C. (1966). Cell division and differentiation of embryos in the pollen grains of Datura in vitro. *Nature, 212*(5057), 97-98.

Khor, E., and Loh, C. S. (2005). Artificial seeds. In *Applications of cell immobilisation biotechnology* (pp. 527-537). Springer, Dordrecht.

Kopittke, P. M., Blamey, F. P. C., Asher, C. J., and Menzies, N. W. (2010). Trace metal phytotoxicity in solution culture: a review. *Journal of experimental botany, 61*(4), 945-954.

Lavelle, P., and Martin, A. (1992). Small-scale and large-scale effects of endogeic earthworms on soil organic matter dynamics in soils of the humid tropics. *Soil Biology and Biochemistry, 24*(12), 1491-1498.

Liu, X., Yang, Z., Gao, G., Lin, Y., Zhu, X., Yu, J., and He, Y. (2010). Development of Bt rice by molecular marker-assisted selection and assays for insect-resistance. *Molecular Plant Breeding, 1*.

Miller, C. O., Skoog, F., Von Saltza, M. H., and Strong, F. M. (1955). Kinetin, a cell division factor from deoxyribonucleic acid1. *Journal of the American Chemical Society, 77*(5), 1392-1392.

Murashige, T, and Skoog, F, (1962). A revised medium lor the rapid growth and bioassay with tohacco tissue cultures. - Physiol. Planl. 15:73-497.

Pal, K. K., and Gardener, B. M. (2006). Biological control of plant pathogens. doi:10.1094/PHI-A-2006-1117-0223.

Ponomarera, S. T. (1962). Soil macro and micro-organisms and their roles in increasing fertility. *Vitoraya Zoologi Cheskaya Konferenciya Litovoskoi, SSR*, 97-98.

Senapati, B. K. (1993). Earthworm gut contents and its significance. *Earthworm resources and vermiculture.(Ed. AK Ghosh). Zoological Survey of India, Kolkatta*, 97-99.

Suh, J. P., Jeung, J. U., Noh, T. H., Cho, Y. C., Park, S. H., Park, H. S.,.. and Jena, K. K. (2013). Development of breeding lines with three pyramided resistance genes that confer broad-spectrum bacterial blight resistance and their molecular analysis in rice. *Rice, 6*(1), 1-11.

Thorpe, S. A. (2007). *An introduction to ocean turbulence* (Vol. 10). Cambridge: Cambridge University Press

Vasil, I. K. (2008). A history of plant biotechnology: from the cell theory of Schleiden and Schwann to biotech crops. *Plant cell reports, 27*(9), 1423-1440.

Wally, O., and Punja, Z. K. (2010). Genetic engineering for increasing fungal and bacterial disease resistance in crop plants. *GM crops, 1*(4), 199-206.

Wieczorek, A. (2003). Use of biotechnology in agriculture—benefits and risks.

Index

D

E

F

G

H

I

R

S

T

V

W

Printed at Repro India Ltd.